Principles of Public Utility Rates

Principles of Public Utility Rates

by JAMES C. BONBRIGHT

NEW YORK
COLUMBIA UNIVERSITY PRESS

SBN 231-02441-X

Library of Congress Catalog Card Number: 61-6569

Printed in the United States of America
10 9 8 7

To Martha

PREFACE

"I must admit that I possess no instinct by which to know the 'reasonable' from the 'unreasonable' in prices and must seek some conscious design for decision." Justice Jackson, dissenting, *Federal Power Commission v. Hope Natural Gas Co.*, 320 U.S. 591 at 645 (1944).

Public utilities, including railroads, conduct their business under a legal duty to supply satisfactory service at rates that are "reasonable" and not "unduly discriminatory." But what attributes must their rates possess in order to meet this twofold standard of validity? This is a question with which I have long been concerned from the standpoint of an economist, alike in the university classroom, as a public official, and as a participant in disputed rate cases. Having written from time to time on particular aspects of this many-sided question, I am now surveying the subject in a more comprehensive manner. But the survey is not a descriptive treatise on public utility rates or rate regulation. Instead, it is designed to complement the existing treatises by centering attention on the basic criteria of reasonable rates rather than on the many problems of application and administration.

In writing this book it has been my hope to make some contribution toward bridging the wide gap which unfortunately exists in this country between the thinking of the academic economists and that of the actual rate makers and rate regulators with respect to the goals of rate-making policy. As an economist, my own views are naturally colored by those of my profession. But I have tried to present these views in a form understandable to readers outside this profession, and with an awareness that rate-making practice must be governed partly by considerations foreign to the more abstract types of utility rate theory.

The major theme of the entire book, first set forth in Chapter II and developed in the following chapter, can here be stated briefly. Reasonable public utility rates, like reasonable prices in general, are rates designed to perform with reasonable effectiveness multiple functions as instruments of social control. But a system of rates that would be best designed to perform any one of these functions is unlikely also to be the best that could be designed to perform any of the others. Hence, to a substantial extent, sound rate-making policy is a policy of reasonable compromise among partly conflicting objectives.

A word is in order as to the division of the twenty chapters into three major parts. The textbooks on public utility economics, as well as the actual rate cases before courts and commissions, draw a primary distinction between the criteria of reasonable rate *levels* and the criteria of reasonable rate *structures* or rate differentials. This is an important distinction, and I have followed it in devoting Part Two to the former subject and Part Three to the latter. But cutting right across this distinction, the chapters of Part One (except for Chapter IX) will first review possible alternative or complementary standards and measures of reasonable or optimum rates that are germane to the whole field of rate making. The resulting threefold division of subject matter leads unavoidably to some repetition of earlier material in later chapters. But I trust that the insight to be gained from the cross-classification is worth its necessary cost.

In illustrating the basic principles of public utility rate making, I have cited examples from different utility industries including, on occasion, the railroads. For the most part, however, I have been concerned with the more nearly monopolized types of utility enterprise, and especially with the electric utilities, with whose problems I have had the most experience. It was my early intention to include a special chapter on that most frustrating of all current public utility problems—the regulation of natural-gas rates. But a tentative draft of such a chapter left me so dissatisfied that I have hopefully postponed publication pending further study.

I am indebted to so many public utility specialists for conferences and correspondence which have influenced the writing of these chapters that any attempt to express this debt here would be almost hopeless. But it gives me pleasure to thank particularly the following friends who have kindly read chapters in manuscript and who have given me the benefit of their suggestions and criticisms: Mr. John Alden Bliss, Mr. Gordon R. Corey, Professor James L. Dohr, Professor Robert Eisner, Mr. Franklin J. Leerburger, Mr. O. Townsend MacMillan, Mr. Hubert Nexon, Mr. Maurice R. Scharff, Professor William S. Vickrey, and Mr. Joseph L. Weiner. To Mr. Raymond J. Dixon, Editor of the Columbia University Press, I also express my keen appreciation for his expert and painstaking editorial work in the preparation of the manuscript for publication.

Finally, I owe a debt of gratitude to the Ford Foundation and to the Columbia University Council for Research in the Social Sciences for grants-in-aid that have permitted me to secure leaves from teaching duties—leaves without which the completion of this study would have been postponed indefinitely.

JAMES C. BONBRIGHT

Columbia University
October 10, 1960

CONTENTS

PART ONE

Basic Standards of Reasonable Rates

I

THE PUBLIC UTILITY CONCEPT

This is a study of the standards of reasonable or optimum prices applied, or proposed for application, to that limited but vitally important class of business enterprises called "public utilities." Since the relevant standards must depend in part on the special character of the enterprises under review, a foreword on this character is in order. But the foreword will be brief, since it is designed merely to supplement the more extensive discussions in the general treatises on public utility law and economics.

The term "public utility" is one of popular usage rather than of precise definition, and writers are not uniform in extending its scope to newer types of regulated enterprise, such as radio and television broadcasting.[1] For present purposes, however, the extension of the public utility concept need not concern us, since the basic principles of reasonable rates can best be developed by reference to the traditional public utilities—to those enterprises that have long been subject either to outright public ownership or else to government regulation of prices and of services. In the United States the control has usually taken the form of regulated private ownership, and this is the form that will be assumed throughout most of our discussion of price policy. It should not

[1] The American judicial opinions on the constitutionality of price-fixing statutes have used the phrase "business affected with a public interest" more frequently than the term "public utility" as a designation of a type of enterprise subject to special regulation. There has been a tendency to use the former phrase in a broader sense, restricting "public utility" to an enterprise enjoying special grants of authority and operating under an obligation to serve all applicants without "undue discrimination." Statutes sometimes apply special-purpose definitions, as does the Federal Public Utility Holding Company Act of 1935, which restricts the term "public utility company" to a company doing a gas or electric business. In Wall Street parlance, though not in law, the railroads are not classed as "public utilities."

be inferred, however, that an economic theory of public utility rates which is valid under private ownership becomes invalid under public ownership, and vice versa. On the contrary, the essential principles, as developed in these chapters, apply with modification under both forms of organization.

For the purpose of this study an enterprise is not regarded as a public utility unless the regulation to which it is subject includes direct control of its rates of charge for services. But governmental price control alone is not enough to confer public utility status upon an enterprise or an industry. By a generally, though not universally, accepted linguistic convention, there is the further requirement that the primary purpose of the regulation must be, ostensibly at least, the protection of the public in the role of consumers rather than in the role either of producers or of taxpayers.[2] For this reason if for no other, milk production was not converted into a public utility industry by the passage of state and Federal milk-price control laws; and the same statement applies to coal production under the former Guffey Act or to agriculture under the Federal farm-product price supports. To be sure, defenders of these price-support laws have often contended that they are really in the long-run interests of the consumers. But the defense comes from spokesmen for the producers.

Most public utilities can be divided conveniently into two major classes: (1) those enterprises which supply, directly or indirectly, continuous or repeated services through more or less permanent physical connections between the plant of the supplier and the premises of the consumer; and (2) the public transportation agencies. The most important members of the first class are the enterprises supplying electricity, gas, water, and telephone communication. The transportation agencies are sometimes divided into (1) the "steam" railroads along with competing forms of intercity public transportation, and (2) the local transit systems. Transportation, however, presents problems of unusual complexity and is therefore often reserved for special treatises that pay detailed attention, among other things, to the highly competitive nature of modern transport.[3] This book will cite illustrations of

[2] "Protection of consumers against exploitation at the hands of Natural-Gas companies was the primary aim of the Natural Gas Act." Phillips Petroleum Co. v. Wisconsin, 347 U.S. 672, 685 (1954).

[3] One reason why transportation is so frequently treated as a separate subject

rate-making problems in the railroad or local transit fields. But the systematic development of principles will have primary reference to the nontransport utilities, and especially to the electric companies. The reason for this narrowed emphasis lies in the closer approach to monopoly enjoyed by these latter companies along with the telephone systems.

Despite the distinction just drawn between the transportation agencies and the nontransport utilities, even most of the latter utilities do a transportation business if we use "transportation" in a broad sense to include what are more frequently called "transmission" and (in gas and electricity parlance) "distribution." True, a local utility company may have a production or manufacturing department, as does an electric company which generates its own power or a gas company which manufactures its own gas. But the transmission-distribution phase of the business is a vital part of most public utility systems and may constitute the major component of the total cost of service. Moreover, even though the entire utility system is usually subject to regulation, it is likely to have derived its recognized utility status from the department of the operations concerned with the transfer of the gas, or the electricity, or the telephone messages from one location to another.[4] The economic significance of this fact will be noted in a later paragraph in this chapter.

"PRIVATE" BUSINESS VERSUS BUSINESS "AFFECTED WITH A PUBLIC INTEREST"

We have already in effect defined a public utility as any enterprise subject to regulation, including price regulation, of a type designed primarily to protect consumers. And in order to come still closer to traditional usage, we may amend the definition so

---

is that not all forms of transport fall within the public utility category. This statement applies obviously to the unregulated, private trucking business. But even the regulated carriers by road or water enjoy the use of public highways and waterways, the services of which are not sold on a compensatory basis.

[4] The more clearly entrenched public utility status of the transmission aspect of a regulated industry is illustrated by the pending controversy as to the extent and nature of jurisdiction by the Federal Power Commission over natural-gas production as distinct from pipeline transmission. In the field of electric power supply, companies producing power merely for sale at the bus bar have sometimes contended—I think, on occasion and in some jurisdictions, successfully—that their business was not subject to regulation by state public service commissions. In Great Britain, the first step toward the complete nationalization of the electric power industry was the nationalization of the main transmission system, the "grid."

as to make it apply only to those enterprises subject to regulation as a matter of long-run policy rather than as a temporary expedient in wartime or in some other emergency. But what are the special attributes of an enterprise, or of an industry of which the enterprise is a member, that give it "utility status" even in a country that has gone as far as has the United States in its reliance on the "automatic forces" of market competition for protection of consumer interests?

Down to the decade of the 1930s, the question just raised was often discussed as a legal problem—specifically, as a problem in constitutional law. Except in times of emergency, state and Federal legislatures were held by the Supreme Court of the United States to have no power to impose price restrictions on ordinary business enterprises. Statutes imposing such restrictions were held void as violations of constitutional guaranties of property rights, including the guaranties of the Fifth and Fourteenth Amendments. But exception was made of certain types of business said to have been "dedicated to a public use" or "affected with a public interest," and these types included the railroads and the familiar municipal utility companies. A layman might suppose that a list of all businesses affected with a public interest would be very long and that it would exclude only the producers of frivolous or luxury goods, which the community could very well do without. In fact, however, the early Supreme Court rulings were much more restrictive and did not go very far beyond the traditional public utility field in recognizing legislative power to fix prices or to impose upon private businesses restrictions not merely designed to protect "health, safety, and morals." [5]

Perhaps the most plausible way to rationalize these early legal cases, which seem to deny any public interest in the production of vitally important goods and services, is to infer that what the courts were denying was the public importance of any single producing firm or enterprise rather than the public importance of an entire industry. But this rationalization would not fit *all*

[5] For a review of the earlier judicial pronouncements on "business affected with a public interest," see "Rate Regulation," by Felix Frankfurter and Henry M. Hart, Jr., *Encyclopaedia of the Social Sciences*, Vol. XIII (New York, 1934). The article cites other literature on the subject, including Walton H. Hamilton's distinguished paper on "Affectation with Public Interest," 39 *Yale Law Journal* 1089–1112 (1929–30).

the cases; and it would be cogent only under the assumption of competition among many producers, no one of which has a sufficiently large share of the market to make its output or price-fixing policies a matter of general concern.

Today, however, any attempt to explain the early judicial distinctions between a public and private business has little more than historical interest, since the Supreme Court has now changed its own position, as indicated by the famous *Nebbia* case of 1934.[6] Legislative proposals to place a given industry under price regulation may now be considered on their merits from the standpoint of economic and social policy, and without serious danger of upset by reason of conflict with the older, traditional legal doctrines. But this does not solve the problem; it merely shifts the emphasis from considerations that have seemed of special importance to lawyers and judges to considerations that seem valid to persons unindoctrinated in legal lore.

ESSENTIAL NATURE OF THE SERVICE AND PUBLIC UTILITY STATUS

The preceding paragraphs have followed custom in defining a public utility as any enterprise actually subject to regulation as a public utility. But this definition begs a question that must now receive attention: why certain types of enterprise are, or should be, singled out for this treatment whereas others are free from direct price control and from related types of regulation except, perhaps, in a period of emergency such as a war. Modern writers generally agree that no simple or single answer will suffice. The economic and social forces that have imposed regulation, say, on the electric power companies are multiple and complex.

[6] Nebbia v. New York, 291 U.S. 502 (1934), upholding the constitutionality of a New York State statute creating a milk control board with power to set minimum and maximum retail prices. Speaking for the Court, Justice Roberts conceded that "the dairy industry is not, in the accepted sense of the phrase, a public utility." But he declared that the legislative power of price regulation is not limited to a public utility, and he also stated: "Many other decisions show that the private character of a business does not necessarily remove it from the realm of regulation of charges or prices. . . . The guaranty of due process demands only that the law shall not be unreasonable, arbitrary, or capricious, and that the means selected shall have a real and substantial relation to the object sought to be attained." Justice McReynolds wrote a vigorous dissent, concurred in by Justices Van Devanter, Sutherland, and Butler, in which he adhered to the older philosophy.

Moreover, they are not precisely the same as the forces that have imposed somewhat similar regulation, say, upon the interstate railroads.

Nevertheless, two attributes of a public utility business have received emphasis in the textbooks, and they will be discussed in turn. The first is the special public importance or necessity of the types of service supplied by utility enterprises; the second is the possession by utility plants of technical characteristics leading almost inevitably to monopoly or at least to ineffective forms of competition. As Clemens neatly puts it: "Necessity and monopoly are almost prerequisites of public utility status."[7]

As to the character of public utility services considered as a group, few persons would deny that they are essentials of modern living rather than mere luxuries or conveniences. A well-functioning transportation system, for example, is a matter of life-and-death importance to the nation. Especially in a large city, even a temporary stoppage of electric power service is serious, and a prolonged cessation would be disastrous. This recognized public importance of adequate utility service, available without delay at reasonable rates and without unjust discrimination, certainly helped to account for the public demand for regulation even in a period of American history which was notably unfriendly toward "government interference with business."

But what the recognized importance of public utility service fails to account for is the restriction of regulation to services which, however essential they may be to the life of a community or the whole nation, are no more so than are the supplies of many commodities and services produced and distributed by unregulated business. Granted that electric power and telephone service are necessities of modern living rather than mere luxuries,

[7] Eli W. Clemens, *Economics and Public Utilities* (New York, 1950), p. 25. Earlier writers sometimes stressed the special privileges usually accorded to public utilities as justifying special regulations—privileges including the power to take private property under the law of eminent domain and the right to use the public streets. But current writers, while conceding a relationship between possession of privilege and subjection to special duties, no longer view this relationship as one of cause and effect. Indeed, even in early years, the courts did not limit their list of "businesses affected with a public interest" to those businesses enjoying legal privileges denied to ordinary businessmen. See, e.g., Chief Justice Taft's opinion for the Court in Wolff Packing Co. v. Court of Industrial Relations of Kansas, 262 U.S. 522 (1923), distinguishing three classes of business "clothed with a public interest" justifying some public regulation.

so also are food, clothing, and housing. Yet the prices of these essential products are not subject to peacetime control of a public utility nature, with the partial exception of housing rentals in limited areas and under assumed conditions of abnormal scarcity. Indeed, if the supply of electric power is a necessity, so also must be the supply of the turbines, the generators, and the boilers needed for the production of this power. Yet the electric equipment companies are not treated as public utilities and are left as free to determine their own price policies as are other "private" industrial companies.

What must justify public utility regulation, then, is the necessity of the regulation and not merely the necessity of the product. Indeed, one may go further and note that modern public policy is far from satisfied with regulation limited to the protection of consumers in securing *essential* types and amounts of service at fair prices. Instead, it extends to the encouragement of abundant use of service, especially of electric service, stimulated by promotional rates that are nevertheless high enough to discourage wasteful consumption. The theory of public utility rate making would be a very limited and rather dull subject if it were merely a theory of the proper pricing of economic *necessities*.

What has just been said may seem so obvious as to be hardly worth mentioning. Indeed, I would not have stressed the point but for its bearing on the criteria of "reasonable" utility rates. One sometimes reads the contention that charges for public utility services should not be based on the ordinarily accepted standards of cost pricing, since these services are essentials of living and hence should be made freely available even to persons who cannot afford to pay the costs of production. What this contention ignores is the very weak correlation between necessities versus luxuries on the one hand and utility services versus non-utility products on the other. If the prices chargeable for all necessities were to be based on standards of ability to pay rather than on standards of cost pricing, then a reorganization of the country's entire price system would be in order.

The above-noted reference to essentiality of service as an earmark of a public utility should be distinguished from a related point of a more subtle character to be found in some of the literature, namely, the assertion that public utility services have a peculiarly *social* or *community* value, not reflected by the prices that individual consumers are willing to pay for them. This "social-value" theory will be discussed briefly at the end of this chapter and treated in more detail in Chapter VII. But whatever its merits, I feel sure that it has been of only secondary influence in the American and British development of public utility regulation.

NATURAL MONOPOLY AND PUBLIC UTILITY STATUS

It is the general consensus of economists that the primary, even though not the sole, distinguishing feature of a public utility enterprise is to be found in a technology of production and transmission which almost inevitably leads to a complete or partial monopoly of the market for the service. Public utility regulation, if chosen in preference to outright public ownership, is therefore said to be a substitute for competition. Whether or not it should be a closely imitative substitute is quite another question, and one reserved for discussion in Chapter VI.

This "natural-monopoly" theory of public utility regulation reflects an old and orthodox point of view. Properly qualified, I believe it to be sound. But it must be expressed today with more caution than would have been deemed necessary in earlier years. For, as modern economists have shown, the differences between a competitive industry in a realistic sense of competition and a monopoly, natural or unnatural, are far less sharp and less simple than was once assumed. Close approximations to "pure" or "perfect" competition are thought to be rarely if ever found in manufacturing or trading industries. On the other hand, even public utilities may face severe competition, typically of a substitute-product type,[8] with respect to a large fraction of their services—sometimes with respect to the major fraction.

It follows that the traditional distinction between monopolistic public utilities and competitive private enterprises is an oversimplification, since the true distinctions are those of degree rather than of kind. In attempting to draw such a distinction, moreover, one should not overlook the possibility that some im-

[8] The term "monopolistic competition" has been applied to this type of competition: "monopolistic" because each firm has a monopoly of a significantly distinct type of commodity or service. See Edward Chamberlin, *The Theory of Monopolistic Competition*, 6th ed. (Cambridge, Mass., 1950).

portant industrial companies have avoided regulation, not because they have been more competitive than most utilities but rather because of the recognized difficulties of an extension of effective regulation to the manufacture of commodities. Efficient regulation of the American steel industry, or even of the aluminum industry, would present far more serious problems than has regulation of the electric power utilities or of the telephone companies.[9] Reliance on a certain degree of competition, fortified by antitrust laws, may therefore be deemed the lesser evil.

ECONOMIES OF LARGE-SCALE PRODUCTION AND NATURAL MONOPOLY

The familiar statement that a public utility is a "natural monopoly" is meant to indicate that this type of business, by virtue of its inherent technical characteristics rather than by virtue of any legal restrictions or financial power, cannot be operated with efficiency and economy unless it enjoys a monopoly of its market. So great are the diseconomies of direct competition that, even if it gets an effective start, the competition will probably not long persist if only because it will lead to the bankruptcy of the rivals. But even if the competition is long lived, as has occasionally happened when the rivalry has taken a restrained form, it is wasteful of resources because it involves unnecessary duplication of tracks, of cables, of substations, etc.[10]

[9] Professor Wilcox makes the further point that public utilities deal directly with small consumers and hence are not faced so generally with the "countervailing power" that even large manufacturing companies face. Clair Wilcox, *Public Policies toward Business*, rev. ed. (Homewood, Ill., 1960), p. 541.

Note that in Great Britain, the Conservative Government, in 1954, denationalized the steel industry, which had been nationalized along with other activities by the previous, postwar Labor Government. But it has not expressed any intent to denationalize electricity, gas, or railway transport. Arthur R. Burns, *Comparative Economic Organization* (New York, 1955), p. 245.

Few American economists, even among those who have insisted upon a serious decline of competition among the country's great industries, have looked for a solution to an "extension of the public utility concept." Instead, most of them have sought means of securing more effective, "workable competition." The dismal experiments with the N.R.A. Codes in the 1930s, which embodied some fair-price ideas akin to those of utility regulation, would be enough to give pause to further experiments in the same direction.

[10] With the electric power business in the United States and Canada, the most usual form of duplication within recent years has been that between publicly owned and privately owned electric plants. In some of the cities formerly receiving this duplicate service, such as Seattle, Wash., and Ottawa, Ont., rates were unusually low—a fact which might be cited as justifying a community in abandoning reliance

What then are the special characteristics of a public utility enterprise or plant which give it a natural monopoly character not conceded to other industries? An answer frequently given is that public utilities operate under conditions, or "under the law," of decreasing costs, whereas competitive enterprises operate under conditions either of constant cost or else of increasing cost. This means that the larger the output of a utility plant per day or per month or per year, the lower will be the cost of production and distribution per kilowatt hour, or per thousand cubic feet of gas, or per passenger mile, etc. Consequently, only a company enjoying a monopoly of the supply of service in a given area can operate at maximum economy.

This rationale of the natural monopoly status of the public utility industries was given currency, years ago, by the late Professor Henry C. Adams,[11] and has been repeated, with many variations, down to this day. Properly qualified, it remains valid. But a restatement is required. For, taken alone, the well-known economies of large-scale enterprise are by no means peculiar to the utility business. Instead, they are enjoyed by utilities in common with many unregulated types of enterprise, including steel companies, automobile companies, and chemical companies. Compared to some of the giant manufacturing companies and manufacturing plants, most utility systems are of a small-scale nature if measured by any of the conventional units of size.

What favors a monopoly status for a public utility is not the mere fact that, up to a certain point of size, it operates under conditions of decreasing unit cost—an attribute of every business, including a farm or a hand laundry. Nor is it even due to any indefinite extension of the declining-cost portion of a curve relating unit costs of production to scale of output. It is due, rather, to the severely localized and hence restricted markets for utility services—markets limited because of the necessarily close connection between the utility plant on the one hand and the con-

---

on rate regulation and in returning to public competition as a superior alternative. In view of the fact, however, that the public systems in Seattle, in Ottawa, and elsewhere have subsequently taken over the private systems and seem to have benefited by the resulting economies of a monopoly, these examples cannot be cited in support of direct competition.

[11] See Joseph Dorfman, ed., *Relations of the State to Industrial Action* (New York, 1954), which reprints two articles by Professor Adams. On p. 110 he is quoted as declaring: "The control of the State over industries should be co-extensive with the application of the law of increasing returns."

sumers' premises on the other. As a deterrent to successful competition, this market restriction is far more serious than is the case with manufacturing companies which can ship their products throughout a wide region or even throughout the nation. An automobile plant may be large enough to enjoy the full economies of large scale without requiring more than a fraction of the American car market to sustain full-capacity output. But a gas distribution system in Yonkers, N.Y., or an electric distribution system in Evanston, Ill., has a market limited to the load of one city. Even if permitted to supply the entire load, the local distribution system will still be engaged in fairly small-scale business. Were it compelled to share its limited market with two or more rival plants owning duplicate distribution networks, the total cost of serving the city would be materially higher.

The reader may note that Yonkers is served by the gigantic Consolidated Edison Company, which also supplies gas and electricity to much of the area of New York City, and that Evanston is one of many cities served by the Commonwealth Edison Company, which also carries the entire Chicago electrical load. But growth in size through an extension of territory, while it may result in substantial economies of scale, is no adequate substitute for the opportunity of a utility to cultivate intensively whatever area it does serve.

What has just been said about the interplay of the factors of economies of scale and of localized markets points to the significance of the fact, already noted earlier in this chapter, that public utility companies are essentially transportation or transmission agencies. The technology of electric, gas, or telephonic transmission is such as to require a close connection between the plant on the one hand and the consumers' homes or factories on the other. This is even true, though less rigidly so, for a railroad plant. Not all forms of transportation are so strictly localized—not ocean shipping, for example, nor truck transport. But for this very reason, these two forms of transportation have a less well established utility status.[12]

[12] Compare the significance attached by a British economist to the fact that public utilities are suppliers of service, and not (or not simply) of commodities. J. F. Sleeman, *British Public Utilities* (London, 1953), Chap. 2. Along somewhat similar lines, H. S. Houthakker cites nontransferability and nonstorability of service as the two basic features of the market for electric supply which give it a special place from the point of view of price policy. "Electricity Tariffs in Theory and Practice," 61 *Economic Journal* 1–25 at 2 (1951).

ARE PUBLIC UTILITY SERVICES NECESSARILY PRODUCED AT DECREASING UNIT COSTS?

One further point about the monopoly attributes of a public utility should be emphasized in this chapter because of its important bearing on measures of desirable or "optimum" rates of charge. The point concerns the assumption, sometimes accepted by American writers as if it were axiomatic, that a public utility, by virtue of the very characteristics which put it into this category of enterprise, supplies service under conditions of decreasing unit cost with increasing rates of output. For example, a 50 per cent increase in the annual output of electric power (other things equal, including price levels, load factors, and number of customers) may result in an increase in total cost of only, say, 25 per cent. In consequence, the incremental cost of the service—the unit cost of supplying more of it—will be lower than the average cost. And this excess of average over incremental or marginal cost is thought to prevail, not merely as a short-run phenomenon when existing plant capacity is temporarily redundant, but even as a long-run or chronic phenomenon exemplifying the economies of large-scale output. The implications of this supposed cost differential for policies of rate making, which will be discussed at length in Part Three, will be apparent to anyone familiar with modern price theory.

Stated as a general tendency or typical situation applicable to most types of utilities in the United States in the present stage of their expansion, this assumption that a utility operates on the declining-cost portion of a long-run unit-cost curve may be valid, although the available data in support of any such generalization are rather sketchy because of the currently inadequate state of public utility cost analysis. But the point to be stressed here does not concern the question of general tendency or typical situation. It concerns rather what seems to be a widespread assumption that a public utility *must* be producing on the declining-cost segment of a unit-cost curve in order to justify its claim to acceptance as a natural monopoly. This assumption is quite unwarranted. It ignores the point that, even if the unit cost of supplying a given area with a given type of public utility service must increase with an enhanced rate of output, *any specified* required rate of output can be supplied most economically by a single plant or single system.

By way of illustration, let us assume a region served by one electric utility which generates its own power in local plants. Three fourths of the generation is from economical hydroelectric plants at production costs (including capital charges on investment) of, say, ¼¢ per kilowatt hour. But since no additional water power is available within economical transmission distance, the remaining one fourth of the generation is by steam plants at a cost of, say, ¾¢ per kilowatt hour in an area of high fuel prices. Under these assumptions, the marginal or incremental cost of power generation will be ¾¢ per kilowatt hour, whereas the average cost is only 3¾ mills. In short, power is being produced under conditions of increasing unit cost. The increasing-cost behavior will be further emphasized if we also assume that, as more steam power is produced to meet the growing requirements of the area, even this power becomes more and more expensive to generate and transmit because of the absence of adequate condensing water or of good sites near the distribution network. To be sure, the increasing-cost tendency of the company's production department *may* be offset, or more than offset, by the declining-cost behavior of its distribution system. But there is no a priori ground for the assumption that the offset will be complete. Conceivably, the total kilowatt-hour costs of electric power in the area under review may rise as the rate of output increases, even assuming, as we do, no change in general price levels.

This hypothetical example of an electric utility company supplying power under conditions of increasing unit cost, irrespective of any change in price levels, may seem farfetched as an example applicable to this country today, although it may now apply to the Tennessee Valley Authority with its increasing reliance on steam power. But even when applicable, it does not belie the assumption that any given area can be supplied more economically by a public utility monopoly than by two or more companies operating in direct competition. For, on the one hand, the single company can secure the maximum advantages of economies of scale and of density, while on the other hand it is no more subject to the diseconomies of enhanced output resulting from scarcity of water power and of other natural resources than would two or more companies if called upon to supply the region with the same total output.[13]

The current literature on electric power rates in Great Britain and France reveals no such general tendency as one finds in the American literature to take for granted long-run trends of decreasing costs with increasing rates of output. In France available water-power resources have been pretty well exploited, and additional output must come largely from thermal plants. In England, where water power has always been of minor importance, emphasis has been placed on the increasing-cost character of coal supply and also on the difficulties of securing desirable sites and good condensing water for new power plants. (Hence the vigor with which the British government is now pushing the development of atomic-power plants.)

What the parallel situation may be in the future with electric power in this country is a question on which I venture no opinion, although the limited opportunities for further water-power development and even the scarcity of good condensing water (in some areas) are well recognized. It may be that, with expected further increases in demand for power, economies of scale can still more than offset limitations of sites and of raw materials. Moreover, the coming of atomic power may change the situation. But with natural gas, the increasing-cost tendencies of the production end of this wasting-asset industry threaten to outpace the economies of large-scale transmission and distribution. Whether or not the American railways, as a whole, reflect a type of industry still subject to long-run decreasing costs is a controversial question, to which a confident answer is precluded by lack of adequate data.[14] The telephone utilities, at least by their own contention

[13] In the economic textbooks, which usually discuss economies and diseconomies of scale under assumptions of competing firms, no one of which produces more than a small portion of the output of an entire industry, the above-noted distinction between simultaneously operative economies and diseconomies of scale can be expressed with less confusion. The diseconomies of the type illustrated by the scarce water-power resources are referred to as "external diseconomies"—i.e., as diseconomies external to any one firm in the industry; whereas the economies enjoyed by a monopolistic utility company through its ability to make use of larger generating equipment and of a more capacious distribution network are referred to as "internal economies"—economies internal to a given firm or company. But when a single firm enjoys a monopoly, the distinction between the firm and the whole industry disappears or, at least, becomes blurred.

[14] One must distinguish here between the question whether the railroads today, with their existing plant and equipment, are capable of handling additional

and under the usual assumptions of the textbooks, are subject to increasing unit costs if the telephone subscriber or station rather than the telephone call is taken as the unit of measurement. But, so far as I am aware, the telephone industry, despite its vast facilities for statistical and economic research, has never seen fit to publish elaborate studies of its cost functions; and until these studies have been made, a degree of skepticism is justified.[15] In any event, a telephone company does not present a standard example either of a firm or of an industry subject to "the law of increasing costs," since the character of the service rendered to any one subscriber changes significantly with a change in the number of other subscribers.[16]

COMPETITION OF SUBSTITUTE SERVICES AS AN ALTERNATIVE TO RATE REGULATION

This brief discussion of the reasons that account for rate regulation of public utility companies in nearly all countries which have adhered to private ownership should not ignore a minority position expressed in this country, some years ago, by a group of economists with a strong antipathy toward "government interference with business." While their views have won but little favor either among other economists or with the general public,

---

traffic at relatively low incremental costs, and the very different question whether railroads could enjoy further economies of scale even if they were not redundant with respect to the present, peacetime traffic. As to the first question, an affirmative answer is clearly indicated for American railroads as a whole. As to the latter question, my colleague Professor Vickrey has made studies on the basis of which he estimates that long-run marginal costs for freight transport on American railroads are materially below average costs—perhaps only 80 per cent of such costs. William S. Vickrey, "Some Implications of Marginal Cost Pricing for Public Utilities," 45 *American Economic Review, Proceedings* 605–620 at 614 (May, 1955). But he has necessarily relied on very inconclusive data.

[15] The telephone companies refer, quite persuasively, to the increasing-cost factor of their switching apparatus as the number of subscribers in each area increases. But there is a possible decreasing-cost offset in the cable and wire systems outside the telephone exchanges. There is the further question how the costs per station will respond to an increase in the number of stations in any given area, as distinct from an increase in the number of stations accompanying an extension of this area.

[16] Telephone company spokesmen insist that what they call "the value of the service" increases with an increase in the number of potential connections with other customers. This contention is more plausible for business users than for residential subscribers. I have heard several of these latter subscribers assert vigorously their wish that they could avoid being bothered with so many potential connections!

they are nevertheless entitled to careful attention as pointing to serious limitations of orthodox rate regulation.[17]

The contention of these economists has been that the assumed monopoly status of a public utility is an illusion, or at least a gross exaggeration. True, there seldom exists that primary form of competition illustrated by two electric companies vying with each other for the patronage of the same customers. But what does exist is the competition of substitute services or products. For many purposes, the use of electricity is alternative to the use of gas, oil, or coal. Moreover, the large industrial and commercial customer has a feasible option to produce his own electricity if the power company will not quote him a favorable rate. Similarly, communication by telephone must compete with possible communication by telegram, by post, or by direct contact.

Those writers who have stressed this point of view would concede that there are limited uses and amounts of a utility service for which the consumer may have no feasible substitute. This is notably true, for example, of electricity for lighting, where gas, even if available and even if provided by a rival utility company, must be dismissed as obsolete. But the contention is that the high potential profits from a modern utility business do not lie in a policy of high prices designed to exploit the most urgent uses of service. On the contrary, they lie in a policy of low "promotional" prices, of the type that will maximize profits by a

[17] The late Professor Philip Cabot of the Harvard Business School took the lead in presenting this point of view during the late 1920s and early 1930s. My incomplete references include: "Public Utility Rate Regulation," 7 *Harvard Business Review* 257–266, 413–422 (1929); "Ethics and Politics," *Atlantic Monthly*, Nov., 1929, pp. 686–694; "Four Fallacious Dogmas of Utility Regulation," 7 *Public Utilities Fortnightly* 719–729 (1931); "The Dangers of Rigid Rate Structures," 12 *Public Utilities Fortnightly* 183–191 (1933); "Rate Making and Rate Regulation," *American Bar Association Journal*, Oct. 11, 1932. For a similar point of view, see the testimony of the late President A. T. Hadley of Yale before the New York State Commission on the Revision of the Public Service Commissions Law, *Hearings*, Vol. II (1930), pp. 722–753.

The views of these economists, to the effect that utility rate regulation is both unnecessary and undesirable, bear a partial resemblance to the views expressed by those later writers who declare that rate regulation by public service commissions is ineffective because of its acceptance of a "static" cost-price standard of fair rates rather than a dynamic, competitive standard under which promotional rate reductions take the lead and under which unit-cost reductions result from increased rates of output. See, e.g., an address by Leland Olds, former Chairman of the Federal Power Commission: "The Economic Planning Function under Public Regulation," 48 *American Economic Review, Proceedings* 553–561 (May, 1958). But these latter writers insist on the need for "public yardstick" plants as a means of impelling private company managements to seek adequate profits by a policy of low pricing and of volume output.

heavy volume of sales. Thus, even the small user of electricity for lighting and for minor appliances will be protected, without government regulation, by the self-interest of a company in setting rates low enough to encourage bountiful consumption for uses less urgent or subject to feasible substitutes.

The writers who thus insist upon the competitive character of a modern utility business do not limit their objection to regulation to the claim that it is unnecessary. Instead, they have urged that it is positively harmful since it must result, at least in the long run, in utility rates higher than would prevail with unregulated private ownership. Regulation of the traditional type purports to restrict companies to a standard "fair rate of return" on invested capital or on property "values." Hence, it is believed to stifle the initiative and the risk taking inherent in an effort by management to maximize profits through cost reductions and through the promotion of heavy volumes of sales. If free to set its own rates, an intelligently managed, profit-seeking company will be led to make experimental rate reductions in the hope that, perhaps after a delay of one or more years, the resulting increase in the demand for the service may yield higher profits than ever. But if subject to regulation, the same company will hesitate to make a rate cut, since the anticipated increase in demand is far from a certainty, and since the company will not long be permitted to enjoy the benefit of the increase even if it should be realized. Moreover, under regulation in actual practice, a voluntary decrease in rates is not easily and quickly reversed even if the resulting rate of return should prove disappointing.

These are forcible arguments, if not against all feasible forms of regulation then against the more orthodox forms. They will be discussed again in Chapter XV, "The Fair Rate of Return." But the question now to be raised is why these arguments have failed to carry widespread conviction. A really adequate answer will not be attempted here, since it would involve a detailed review of the actual history of corporate rate-making policy in this country under the influence of different forms and degrees of regulation including, in some areas, the almost complete absence of regulation.[18] But two partial answers will be suggested.

[18] The extent to which regulation has actually restrained utility profits is largely a matter of surmise. A recent report by the Federal Power Commission enumerates six states in which there is still no commission regulation of electric rates, or only a limited form of regulation. *State Commission Jurisdiction and Regulation of*

The first answer is that, in actual experience, even those public utility managements that have been completely or relatively free from rate regulation of the "fair profit" type have not generally espoused a philosophy of low pricing designed to maximize long-run profits by the encouragement of widespread use of their services.[19] Instead, they have acted in much the same way as have the managements of the more rigidly regulated companies, by revealing a marked degree of skepticism as to the alleged price elasticity of the demand for their products. On occasion, to be sure, laxly regulated private companies have made drastic rate reductions on their own initiative rather than on order from a public service commission. But these reductions may be attributed to the example of adjacent publicly owned plants or to the fear of public-plant competition. The famous and oft-cited example of the late Henry Ford's dramatic cut in automobile prices, followed, rather than preceded, by a cut in production costs, has not found many imitators in the public utility field. One may perhaps add that it has not been widely imitated in other fields nor even often repeated in automobile production.

Of more interest for the theory of public utility rate making is a second reason why any proposal to rely on competition has failed to win more converts. The point is that public utility companies, if free from rate regulation, would seldom be under competitive necessity to make *general* rate reductions in harmony with opportunities of cost reduction. Instead, they would be more likely to follow the policy of rate discrimination, otherwise known as "charging what the traffic will bear," thus enjoying at one and the same time the commercial advantages of a high price policy with respect to services for which the demand is inelastic, and of a low price policy with respect to services for which the availability of substitutes makes the demand highly elastic.

For this practice of price discrimination the mechanism of

Electric and Gas Companies (Washington, D.C., 1954). But in some of these states city regulation is significant; while in some states with commissions having nominal power to regulate, actual regulation has been notoriously ineffective.

[19] This point of view was expressed by Commissioner Lewis Goldberg of the Massachusetts Department of Public Utilities in taking issue with President Hadley's denial of the need for strict rate regulation. New York State Commission on the Revision of the Public Service Commissions Law, Hearings, Vol. II (1930), pp. 815–817. The Commissioner stressed his belief that, in actual practice, public utility companies have based their rate-making policies on *short-run* profit objectives.

modern utility rate structures is admirably adapted. Through block rates, sometimes applied both to the "demand" charges and to the "energy" or "commodity" charges, favorable terms can be offered to large consumers who have a feasible option to generate their own electric power, while stiffer terms can be offered to small consumers. Still other devices of differentiation are available, such as special, low rates for services designed for special use. Not all of these differentials, to be sure, need be discriminatory in character. Indeed, all of them can be so administered as to conform to cost-of-service standards rather than to value-of-service standards. But an unregulated public utility is under no such limitation in the use of these instruments of monopoly power.

Let it not be supposed from the foregoing remarks that all forms and practices of rate discrimination are to be condemned as against the public interest. On the contrary, wise resort to discrimination of a value-of-the-service character is an essential tool of rate-making policy with respect to unsubsidized utilities that supply services at incremental costs lower than total average costs. But this very fact enhances the need for regulation. For it precludes the adoption of the proposal, sometimes made by persons who have not considered its logical consequences, to leave public utility companies the complete freedom to set their own rates subject only to a statutory mandate to avoid discrimination. Literally construed, the imposition of such a mandate would be fatal to private ownership. To be sure, even the traditional public utility laws are sometimes carelessly said to forbid discrimination. In fact, however, what they forbid is merely "undue" or "unjust" discrimination, thus placing upon commissions the major burden of interpreting and policing these ambiguous terms.

As already noted, the arguments against rate regulation have never won widespread support in their application to the so-called "municipal utilities" such as the electric power companies and the telephone companies. But arguments for much less restrictive standards of regulation have been receiving increasing attention and support with respect to the railroads. Largely because of the competition from the newer rival forms of transport, the railroads as a whole have been unable to earn rates of return that are adequate when measured by any of the conventional tests of adequacy. Even if completely free from rate control, only a few of them could expect to enjoy excessive rates of profit. Under these circumstances, and for the purpose of enabling them to compete more effectively with road, water, and air carriers, it has been proposed to amend the Interstate Commerce Act so as to permit them to set their own rates within a presumably wide band of minimum and maximum tolerable rates set by the Interstate Commerce Commission. Even within these limits, however, the Commission would still have the duty to enforce the rule against "unjust discrimination." [20]

The merits of this proposal will not be discussed in this monograph on public utility rates, since only a transportation specialist, with an intimate knowledge of the technicalities of the different forms of transport, including a knowledge of their cost functions, is competent to have an expert opinion on its merits. In its favor is the argument that substitute-service competition is more nearly all-pervasive for intercity transportation than it is for the local utilities. But if the proposal should be adopted by an act of Congress, one may guess that part of the trouble associated with the enforcement of reasonable rates under the current standards of the Interstate Commerce Commission would be transferred to the problem of redefining and reapplying the rules against "unjust discrimination." In any event, the need for some form of government regulation will remain.

PUBLIC UTILITY SERVICES VERSUS "SOCIALIZED" SERVICES

The emphasis placed by the legal and economic literature on the distinction between a public utility and an unregulated "private" business has unfortunately tended to obscure another aspect of "the public utility concept" that is quite as important from the standpoint of rate theory. The fact is that a public utility enterprise is nonetheless a "business" even though subject to

[20] This was a proposal of the controversial "Weeks Report": *Report of the Presidential Advisory Committee on Transport Policy and Organization*, April, 1955. Its recommendation on maximum-minimum rate control was designed to "limit regulatory authority of the Interstate Commerce Commission to determination of reasonable minimum or maximum rates with no change in existing provisions making undue discriminations and preferences unlawful." In Great Britain under railway nationalization, the freedom of the public corporation, called the British Transport Commission, in the fixation of railroad rates, has been made much more extensive, since the amended Transport Act seems to have gone far toward eliminating the old rule against "unjust discrimination." See Otto Kahn-Freund, *The Law of Carriage by Inland Transport*, 3d ed. (London, 1956).

regulation and even when directly owned and operated by the government. This statement, at least, is in accord with the spirit of the concept both in British and in American usage. Stated more concretely and without the many qualifications required for strict accuracy, public utility services are designed to be sold at cost, or at cost plus a fair profit.[21]

The thrust of this statement will be apparent if we compare the services supplied, say, by a municipally owned electric plant with other services supplied by the same city—for example, the construction and maintenance of streets and sidewalks. In all probability, the electricity will be sold at rates yielding a total revenue at least equal to cost including debt-service charges. Moreover, while the specific rate structure will not bear any very *close* relationship to the relative costs of different classes and amounts of service, a cost relationship will be present to an important degree. In short, the rates will be designed in accord with the old legal maxim, "Let the beneficiary bear the burden."

Not so with the city streets, nor with its parks, its zoological garden, its public schools, its health services, or its police system. These services are also costly to render, but the costs will be met directly or indirectly by taxation; and even the taxes will not be designed with any serious attempt at apportionment in relation to benefits received. The public schools, for example, are not only tuition free but are subject to compulsory attendance, and the payment of school taxes is not lightened for persons without children or for parents who choose to send their children to private schools. In short, the schools are not operated as a business, not even as a public utility business. They are "socialized" in a sense quite different from that which we have in mind if we refer to the "socialization" of an electric utility plant through municipal or state ownership. Only if a city, having taken over a private electric plant, were to abandon the effort to make it financially self-supporting in favor of free service, or of service

[21] "Within a more comprehensive legal framework, the public utility classification is designed on the one hand to distinguish industries best conducted as monopolies and on the other hand to assure that these industries will not, under normal conditions, look to government funds for support, but that they will be required to *sell* their services and look for support only to those who are the actual users." Martin G. Glaeser, *Public Utilities in American Capitalism* (New York, 1957), p. 8. Glaeser adds that city water supply presents a marginal case in which "the public utility classification limits the spread of a legal framework of collectivist economics."

supplied on an ability-to-pay basis, would it have socialized the electric service as it has socialized schools and public hospitals. In that event the city's electric department would have ceased to be a public utility.

Having drawn this sharp distinction between a public utility service and a fully "socialized" service, we must of course hedge the distinction in a manner required of nearly all attempts to classify social institutions. On the one hand, even private utility companies and railroads are sometimes required or induced to violate "business" or "economic" principles by supplying free service to indigent people, or by operating trains that fail to cover even their out-of-pocket costs, or by adherence to rate schedules designed to deal gently with low-income consumers. On the other hand, services supplied directly by government run the whole gamut from those rendered completely without charge (public vaccination) to services rendered at high profits (a European tobacco monopoly). Moreover, linguistic convention applies the term "public utility" to a type of enterprise *traditionally* supplying service at cost or at cost plus a profit, even though the cost principle is violated in a given instance; for example, the New York City subway system, which has not even been covering its operating expenses to say nothing of interest costs on the invested capital.

Despite these qualifications, the distinction just drawn between a public utility service and a completely socialized service is basic to the prevailing theories of reasonable public utility rates. For, with important reservations, these theories are variations of the major theme that the consumers of public utility services (1) should be free to take whatever types and amounts of service they are ready to pay for but (2) in return therefor should be required to pay rates not seriously out of line with costs of rendition.

This book would be a very different work both in its development of rate theory and in its conclusions on wise public policy if the author did not accept, in the main and with important qualifications, the orthodox view that those services presently known as public utility services should be sold on a cost principle. But the reader should be warned that this view is not universally accepted. Among those who have apparently rejected it is Professor Horace M. Gray, whose article on "The Passing of the

Public Utility Concept" should be familiar to every student of public utilities.[22] Gray believed that the concept had become outmoded partly because of the failure of regulation to accomplish the very objectives which it purported to accomplish, namely, to serve as an effective substitute for competition. But he also insisted that a cost-price standard of rate making is inappropriate for application to electric power supply or to the other vital services now largely supplied by private enterprise on "business principles." [23] Social and national objectives, he believed, should supersede the limited objectives of a fair-profit or of a cost-price philosophy of rate making. At least, so I interpret his readiness to move away not only from private corporate ownership in its present form but also from the public utility concept as applied to a plant owned and operated directly by an agency of the government.[24] The merits of this "social theory" of price fixing, to which I would dissent except in special situations, will receive a brief discussion in Chapter VII.

[22] 16 *Journal of Land and Public Utility Economics* 8–20 (1940). Reprinted in *Readings in the Social Control of Industry* (Philadelphia, 1942), pp. 280–317.

[23] The reader should note carefully this twofold nature of Gray's devastating attack on the public utility concept in action. Gray himself fails at times to maintain a sharp distinction, since his article merges criticism number one (that regulation has failed to accomplish what it pretends to accomplish) with criticism number two (that even its pretense is shoddy). As to the first point, it was written with the lessons of the experience of the 1920s in mind, with their disgraceful episodes of holding-company finance and with a type of rate regulation seriously crippled by the Supreme Court's pronouncements on the "fair value" rule of rate making. But in 1935 came the Public Utility Holding Company Act and in 1944 came the Hope Natural Gas case, 320 U.S. 591, in which the Supreme Court accepted a very different, and less restrictive, philosophy of rate regulation. These and other developments associated with the "New Deal" helped to give a new lease of life to private ownership! But a recent book under Professor Gray's co-authorship indicates his continued hostility to regulated private monopoly and his support of limited degrees of competition even among utilities. Walter Adams and Horace M. Gray, *Monopoly in America* (New York, 1955), Chap. 3.

[24] How far Professor Gray would carry the departure from this concept under public ownership I am not able to say. Under Congressional mandate, the Tennessee Valley Authority, e.g., still adheres in the main to public utility principles as to its sale of power, though not as to its supply of navigation or flood control services. As to the nationalized British public utilities, Sleeman writes: "In general the commercial view of public utility undertakings still prevails strongly, and the 'social service' elements considered above are merely modifications on a predominantly commercial pattern." *British Public Utilities*, p. 278. To the same effect, see the *Report of the Committee of Inquiry into the Electricity Supply Industry Presented by the Minister of Fuel and Power to Parliament by Command of Her Majesty*, Jan., 1956, Cmd. 9672, p. 139: "We attach great importance therefore to the industry being run on business lines. It should have one duty and one duty alone: to supply electricity to those who will meet the costs of it and to do so at the lowest possible expenditure of resources consistent with the maintenance of employment standards at the level of the best private firms."

II

THE PUBLIC INTEREST AS THE ASSUMED GOAL OF RATE MAKING

The introductory chapter, on the public utility concept, has stressed the point that this concept itself carries with it important implications for rate-making policy. Thus, an assertion that the supply of telephone service or of electric power should be treated as a public utility implies that the service should be offered for sale instead of being given away and that the sale prices should bear a fairly definite relationship to cost, or to cost plus a fair return typically well below the point of monopoly profits. In other words, the so-called "theory of public utility rates" already starts with certain presumptions about the relevant principles of price determination.

We now turn directly to a study of these principles. But the principles cannot be derived from the public utility concept as a corollary is derived from a proposition in geometry. For in the first place, the concept is too indefinite; in the second place, it is sufficiently flexible to bring within its purview many divergent standards of reasonable rates; and in the third place, it admits of exceptions or deviations based on so-called "social considerations" of a type discussed in Chapter VII, "Social Principles of Rate Making." These possible exceptions are properly considered on their merits instead of being ruled out on the "jurisdictional" ground that their acceptance would violate the logic of the public utility philosophy.[1] Only if the "social considerations" are deemed

[1] The literature on public utilities, which includes its full share of efforts by class interests to practice the art of influencing people, has many examples of attempts to destroy the force of unwelcome proposals by an appeal to the inexorable logic of some basic social philosophy which the opponent may disavow

so pervasive that they cease to be thought of as exceptions or deviations does the public utility concept become a handicap rather than a useful tool of economic thought, as Professor Horace M. Gray seems to have considered it in an article noted in the preceding chapter.

PUBLIC-INTEREST OR SOCIAL-WELFARE CRITERIA OF REASONABLE RATES

As used in this book and in most of the treatises on public utility economics, "the theory of rates" is a normative, not a positive, study. Its task is the systematic development of principles of rate-making policy, the complete or qualified observance of which would subserve "the public interest" or "the social welfare." In its acceptance of a norm or goal by which to appraise the relative merits of alternative rates, it is in the same class with other purposive theories, such as a theory which defines optimum rates as whatever rates will maximize corporate profits, or such as would be a theory of electric rates designed to minimize the charges for service supplied to residential customers. But it is almost unique in the extreme vagueness of its ultimate verbal norm and in the highly indirect and unprovable relationship between changes in utility rate-making policies and effects on social welfare, however defined.[2] Indeed, one is tempted to say that the so called standard of public interest is not a real standard at all; that, instead, it is a mere form of words of highly emotional

at his peril. Thus, a project of public ownership, such as that of the Tennessee Valley Authority, is attacked as "creeping socialism." And thus the New York Natural Gas and Oil Resources Committee objects to Federal control of the prices charged by natural gas producers—because the control threatens the producers' opportunities to make money? Oh, no! Because "the proposed controls are against the public interests and *can benefit no one*" (*original italics*). "The issue," continues the committee, "goes far beyond gas. It goes to the roots of America's greatness." Advertisement in the New York *Times*, March 16, 1955, p. 36.

[2] But even a maximum-profit standard of rate-making policy is by no means free from ambiguities. It can be cleared of these ambiguities only by factitious definitions of "profits" and of applicable maxima that would convert it into a standard which no corporate management would be interested in following. The ambiguity applies notably, though by no means solely, to the assumed standard of maximum "long-run" profits. In the words of Professor Joel Dean, an expert on the pricing policies of unregulated, industrial enterprises, the modern economist's conception of profit maximization as a practical objective of private enterprise "has become so general and so hazy that it seems to encompass most of man's aims in life." *Managerial Economics* (New York, 1951), p. 28.

content, invoked as an instrument of persuasion by people who have at heart much more immediate interests in public utility tariffs—interests often, but not always, of a self-seeking nature.

At least as applied to prevailing conceptions of public interest or social welfare in this country at the present time, such a belittling statement would go too far.[3] But it has sufficient validity to suggest the formidable problem faced by any economist or rate specialist who undertakes to appraise alternative proposed measures of reasonable rates from a public-interest standpoint—the problem of developing less indefinite, less controversial, and more nearly objective goals of rate making that nevertheless make sense when viewed as instruments of sound public policy.

This serious handicap imposed upon any social-welfare theory of public utility rates by the indefinite nature of its assignment might be supposed to doom it at the outset to complete failure. Indeed, some such conclusion may be implied by those rate experts who reject *any* "theory" on the ground that "rate making is an art, not a science." But the situation is not quite as frustrating as it might seem. Indeed, it is not as serious as that presented by other and grander issues of wise economic policy, such as issues of socialism versus capitalism. The reasons for this hope-

[3] Ostensibly, at least, most of the disputes about public utility rate-making policies that come before legislatures, courts, and regulating commissions in this country involve no basic disagreements about "ultimate human values" or even about the proper economic organization of society. Instead, they appear to involve, for the most part, controversies about means to the attainment of ends that all disputants would concede to be desirable. To this extent, they would support the views of the late Professor Henry C. Simons and of Professor Milton Friedman, to the effect that, in America today, there is general agreement on the larger objectives of economic policy and that the real issues have to do with means, not with ends. Henry C. Simons, *Economic Policy for a Free Society* (Chicago, 1948), p. 40; Milton Friedman, *Essays in Positive Economics* (Chicago, 1953), pp. 4–7.

But in its application to public utility problems, the statement that controversies are confined mainly to means rather than to ends requires two important qualifications. The first is the obvious one that most of these controversies involve a clash of individual or class interests such as that between corporate stockholders who seek higher dividends and residential consumers who seek lower domestic rates. Each side invokes public interest issues as weapons to throw at the enemy. The real objectives of the disputants are, in a sense, directly opposed to each other, although in another sense they are the same (or are "symmetrical") in that each side is out for itself, as in a football game. The second qualification, suggested by the social philosophy of John Dewey, is that we live in a means-ends continuum, not in a world which neatly divides goals of action into instrumental goals and ultimate ends. This point of view, unfortunately too seldom recognized in standard economic theory, is implicit throughout the following pages.

ful conclusion are fairly obvious, but they may nevertheless be worth a brief summary.

In the first place, public utility economics may usefully accept as "given" those basic conceptions of social welfare that prevail in the country and in the time period under review. At least in Western Europe and in the United States, this would mean, among other things, the identification of the public interest with the welfare of the people in the community or nation, the state being regarded merely as an instrument for the attainment of this welfare.

In the second place, and related to the first point, the public utility economist is justified in going a long way toward the acceptance, as final for his restricted assignment, of widely held goals of economic policy that a social scientist or social philosopher might properly regard as subjects for intensive and critical analysis. A vitally important example of such a goal is that of "consumer sovereignty," under which the allocation of the community's scarce resources is made to depend on consumer choices or preferences rather than on governmentally determined decisions as to relative needs or national interests. This goal is basic to the whole modern theory of public utility prices and forms the underlying rationale of a cost-price standard of utility rates. But it is no more fundamental to the theory of utility rates than to the theory of economic organization in general. Hence the author even of a large treatise on public utility economics is under no obligation to present either an elaborate defense or an elaborate critique of the standard. Instead, he is justified in limiting his inquiry to the question whether and to what extent the *special character* of public utility services calls for departures from the basic standard of consumer sovereignty in favor of rates designed to attain social objectives (say, national defense) that may not be attained if the service is offered for sale at cost of production.

In the third place, public utility rate theory has a great advantage over broader theories of sound economic policy because its interest is limited to those highly restricted goals effectively attained by programs of rate control. By and large, the task of rate making or rate regulation is that of adapting utility rates to a larger economic environment, including a universe of non-utility prices and wages, on which these rates have only a limited repercussion. This means that the role of public utility rates is severely limited, though by no means completely predetermined, by the country's general price and wage system. To be sure, any grandiose scheme for the economic reorganization of society would include changes in the pricing of services now classified as public utilities. Indeed, a socialist, concerned to carry the country over gradually to the wholesale nationalization of industry, might well begin with a program for subsidized, publicly owned, utility systems. But the rate-making policies involved in such a program would be policies of socialist strategy; they would not be principles of public utility rate theory in the accepted sense of that term.

The opportunity, enjoyed by public utility theory, to cut down social-welfare issues to more nearly manageable size by keeping within the restrictive bounds of its subject matter will be discussed at length in the next chapter, on the role of utility rates, as well as in Chapter VII, on social theories of rate making. But one possible example of such an opportunity may serve here as an illustration. It concerns the question whether or not public utility rates, like income taxation, should be based on the relative abilities of rich and poor consumers to pay for the service, thereby serving partly to offset inequalities in personal cash incomes. If, in answering this question, the public utility specialist were under obligation to pass judgment on the whole public policy of programs of social control looking toward reduction of personal-income differentials, he would be carried hopelessly out of his field into a controversial area in which he has no special competence and about which he could probably say nothing not already said more ably and succinctly by other writers. Yet, in appraising the merits of "ability-to-pay" criteria of reasonable utility rates, he is not completely silenced by a lack of professional competence in income-distributive philosophy in general. A significant answer to the question just raised—admittedly not conclusive in all situations, yet persuasive for general rate-making policy—is that public utility rates are ineffective instruments by which to minimize inequalities in income distribution and that alternative instruments (including public education, social security laws, progressive taxation, and possibly even some form of socialized medicine) are better designed to accomplish this objective even on the assumption that the objective itself is desirable. Reasons for this conclusion are suggested in Chapter VII.

ASSERTED RESTRICTION OF PUBLIC UTILITY RATE THEORY TO "ECONOMIC" PRINCIPLES

Writers on general principles of public utility rates naturally welcome feasible opportunities to simplify their assignment by limiting attention to those objectives of rate-making policy, the attainment of which can be aided by fairly definite standards of optimum or reasonable rates. Thus, a study of rate theory may ignore, or dismiss with brief comments, political considerations, special statutory provisions, and important technical details or special situations that call for close consideration by persons engaged in the actual practice of rate making or rate regulation. These "practical" or legal issues do not lend themselves to useful generalizations. Moreover, they can be discussed more intelligently by actual practitioners than by those professional economists on whom has fallen the major responsibility for the development of general rate theory.

But what must now be noted is the assertion, frequently found in the literature of rate theory, that this theory is concerned solely with *economic* principles of rate making, or solely with considerations of *economic welfare*. The significance of such statements requires notice in this chapter, since they raise the puzzling question how *any* public utility rates or rate-making policies could have an effect on individual or social welfare other than an *economic* effect.

In the current publications on rate theory by academic economists, the most frequent use made of this self-imposed restriction to "economic" principles is to absolve the economist from any professional concern for considerations of *fairness* or *equity* as between investors and consumers, or as among different classes of consumers. Instead, the merits of alternative rules of rate making are to be judged solely by reference to their functional efficiency in getting the work of the world accomplished—in attracting capital to public utility enterprises, in supplying incentives to high-grade management, in controlling the demand for the service, etc. Thus, a recent monograph by an academic economist on public utility rate discrimination, in its discussion of the controversial problem of capacity-cost allocation, declares that it will analyze the problem solely "from an economic point of view. Whether or not the various methods proposed could give results that are 'just' or 'equitable' will not be discussed." They are omitted, presumably, because they raise ethical questions, as to which an economist has no professional competence.[4]

A different distinction between "economic" and "noneconomic" considerations of rate-making policy has been expressed or implied by other writers including, especially, those from the engineering professions. Here, "economic" principles of rate making are contrasted with "social" principles or social objectives. For example, rates of charge for electric service that fail to cover the out-of-pocket cost of supplying small consumers may be called "economically unsound" even by a writer who concedes that they may nevertheless be "socially desirable." Or again, a policy of transportation rate control that may result in driving a high-cost transportation agency out of business (perhaps the road hauler, or perhaps a branch railroad), while declared sound if judged solely by "economic" criteria, may nevertheless be opposed for ("noneconomic") reasons of national defense.

The above-noted distinctions between so-called "economic" and "noneconomic" standards of rate making will be discussed in later chapters, especially in Chapters VII and VIII. But one may question the accuracy, without denying the convenience, of the use of the adjective "economic" as the term by which to characterize the distinctions. For even a national-defense principle or an ability-to-pay principle of rate-making policy is an "economic principle" according to the most widely accepted modern definition of economics as "the study of the principles governing the allocation of scarce means among competing ends." [5]

I confess my own inability to find any consistency in the various distinctions that writers, including myself on occasion, have

[4] Ralph Kirby Davidson, *Price Discrimination in Selling Gas and Electricity* (Baltimore, 1955), p. 111.

[5] In thus defining economics, Professor George J. Stigler adds the proviso that the objective of the allocation of the scarce means must be "to maximize the attainment of the ends." *The Theory of Price*, 1st ed. (New York, 1946), p. 12. This proviso might possibly be held to exclude the objective of fairness as an "economic" objective. But it could hardly exclude national defense or other so-called social objectives such as income redistribution or decentralization of population.

drawn between "economic" and "noneconomic" principles of public utility rates. "Economic" is perhaps best regarded as a term of convenience, the import of which must be derived from the context, as in a statement contrasting "economic" principles of rate making with "legal" principles. But the closest approach to consistency seems to me to lie in the use of the term "economic" to denote whatever general principles of rate making have been developed or systematized by professional economists, and defended by them as valid under simplified but useful assumptions both as to the fact situation and as to the normal role of utility rates.[6]

"REASONABLE" RATES VERSUS "OPTIMUM" RATES

It is a general doctrine of American law, almost universal in its application to public utility companies operating under special franchises or "certificates of convenience and necessity," that these companies are under a duty to offer adequate service at "reasonable" (or "just and reasonable") rates. In addition, the governing state or Federal statutes require that, in its rates of charge as well as in its supply of services, a company must avoid "unjust" or "undue" discriminations or preferences among consumers. But the rule against undue discrimination is a mere extension of the mandate of reasonable pricing to reasonable price relationships, and it need not be distinguished for present purposes.[7]

[6] Compare I. M. D. Little's contention that the phrase "economic welfare" involves a misplacement of the adjective. The proper distinction, he insists, is between economic and noneconomic means of attaining human welfare. *A Critique of Welfare Economics* (Oxford, 1950), p. 6.

[7] See Chap. XIX. While some of the public utility statutes rest content with the requirement that rates be reasonable and not unjustly discriminatory, others go a certain distance toward prescribing or implying standards of reasonableness. The prescriptions may take the form of an enumeration of objectives of rate-control policy, as in the various amendments to the Interstate Commerce Act. Or they may take the form of a partial enunciation of measures or tests of reasonable rates which the regulating commission is instructed to follow or which it must "take into consideration" in reaching a rate decision. All of these statutory provisions leave much room for "interpretation" by a commission, subject to the rulings of the appellate courts.

In a brilliant article on "Commissions, Rates, and Policies," 53 *Harvard Law Review* 1103–1144 (1940), Professor Robert L. Hale comments on opinions of the Supreme Court restricting the standards of lawful rates under the Interstate Commerce Act to "transportation standards." United States v. Illinois Central R. R., 263 U.S. 515, 525 (1924); Texas & Pacific Ry. v. U.S., 289 U.S. 627, 637–638 (1933). In

In many of the pages throughout this book, we shall have occasion to compare the criteria of reasonable rates established by statutes, courts, and commissions with those rate-making principles that have had the support of economists. But what calls for present comment is the distinction between the traditional legal standard of *reasonable* rates or rate relationships, and the standard of *"optimum"* rates often set forth as the ideal of public utility rate theory. The law accepts results that are merely *satisfactory,* whereas economic theory seeks the conditions for the attainment of the ideal.

A full treatment of the import of the legal rule of "reasonableness" as applied to utility rates would go far beyond the scope of this study. But certain aspects of the subject are fairly elementary. In the first place, the law of public utility rates is, for the most part, a law of rate *regulation.* Instead of prescribing a complete set of principles or measures of rates, it leaves primary responsibility for rate-making policies to the management of the enterprise, private or public, so long as the management keeps within bounds set by public-interest or consumer-interest considerations. Only rarely will a commission feel called upon to take the initiative in dictating the precise rates that a company must charge. Its usual action is that of deciding whether or not existing or proposed rate schedules are reasonable or unreasonable.[8]

the latter case Justice Roberts (for the majority) declared: "The Act was passed for the protection of those who bear the rates. The standards it establishes are transportation standards, not criteria of general welfare." The term "transportation standards" was, and continues to be, ambiguous. But the statement that the Act was designed to serve limited purposes and not to accomplish unlimited objectives of public welfare is quite in keeping with the position of economists as to the limited functions properly assigned to public utility rates as instruments of social control.

[8] As a matter of practice rather than of legal authority, state public service commissions have tended to allow private companies much more freedom in determining rate structure or rate design than in determining the general level of their rates.

Under the Interstate Commerce Act and some of the state utility statutes, courts and commissions have thought in terms of a zone of reasonableness, within which zone existing rates may not be disturbed by commission fiat (barring a finding of unjust discrimination). But if the commission finds, "as a judicial fact," that existing rates lie outside this zone, it may, if it so chooses, set the precise new rates instead of directing the company to propose reasonable alternatives. See Robert L. Hale, "Commissions, Rates, and Policies," cited in the preceding footnote. For a general treatment of the concept of a zone of reasonableness under the Interstate Commerce Act, see I. Leo Sharfman, *The Interstate Commerce Commission,* Vol. III B (New York, 1936), pp. 417–421, 425–463, 652. Unlike this act,

But even if the scope of rate regulation were not so limited and even if the whole task of rate making were to fall upon regulating commissions, these commissions could not possibly hope to discover that particular rate structure, or even that particular complex of rate-making criteria, which is better than any other when judged by any plausible tests of goodness. *Satisfactory* results, not ideal or optimum results, are all that can be expected of the ablest group of rate makers.

Unlike actual practitioners of rate making, the rate theorist seldom has the task of putting his theory into practice. Any attempts on his part to set up principles of optimum rates are therefore mere attempts to state conditions, the attainment of which *would* result in the best rates *if* the fact situation and the objectives of rate making were those which he postulates. But even the economist, if he wishes to get beyond highly artificial, simplified assumptions as to the role of utility rates, so as to participate with the practical rate experts in developing *workable* standards of rate design, will be compelled to abandon the goal of optimum rates in favor of less lofty and less precise standards of adequate performance. Like the courts or the public service commissions, the economist must then rest content with principles of *reasonable* rates, although his standards of reasonableness may differ materially from those accepted by the law or by popular opinion.[9]

RATES AND EFFICIENT PERFORMANCE OF FUNCTIONS

The reasonableness of any public utility rates should be determined primarily by standards of efficient performance of their accepted functions. The reader may well regard this statement as a mere truism. Yet, truism or not, it emphasizes an approach to rate-making problems not always reflected in the legal cases or in popular controversies. It implies that any rule of rate making, such as a rule that rates should be based on average cost, or on marginal cost, or on "value of the service" instead of cost, must be adjudged good or bad, reasonable or unreasonable, primarily by reference to its effectiveness in securing rates that will per-

---

the Natural Gas Act of 1938 authorizes the Federal Power Commission to order a reduction in rates to "the lowest reasonable rates."

[9] For a discussion of important differences in emphasis, see Chap. VIII.

form the proper functions of a price system when applied to public utility services. What these proper functions may be with respect to any given type of utility enterprise is a controversial question, for reasons to be discussed in the next chapter. But the controversy merely adds to the difficulty of a functional approach to rate theory without offering any rational alternative.

A striking illustration of the difference between a functional and a nonfunctional appraisal of rate-making policy may be found in a comparison between the earlier opinions of the Supreme Court upholding the "fair value" rule of rate making as a mandate of constitutional law, and the later opinions of the Court in rate cases beginning with, or shortly before, the *Hope Natural Gas* decision in 1944.[10] True, even the old "fair value" rule lends itself to an interpretation under which a plausible, even if not convincing, case can be made in its favor on functional grounds—on grounds of the efficiency of the rule, when properly applied, in enabling a company to attract capital during a period of price inflation.[11] But it was not so interpreted and so defended by the Supreme Court in the days when it was held to reflect the constitutional "law of the land." On the contrary, it was rationalized as an application of the general guaranties of property rights established by the Fifth and Fourteenth Amendments to the Constitution of the United States. Whether or not it was feasible of administration, or efficient as a means of enabling a corporation to attract needed capital, or conducive to managerial efficiency, or otherwise effective as a means by which to harness the profit-making objectives of private investors to the job of supplying the nation or the community with railroad and public utility services, were questions which, if deemed relevant at all, were not recognized as the controlling issues in the ruling opinions.[12]

[10] Federal Power Commission v. Hope Natural Gas Co., 320 U.S. 591 (1944). See also Chief Justice Stone's ruling opinion in the preceding rate case, Federal Power Commission v. Natural Gas Pipeline Co., 315 U.S. 575 (1942).

[11] See Chap. XIV.

[12] Compare the following critical comment on the nonfunctional approach to rate regulation believed to have been current at least as late as 1938: "The entire machinery of present-day rate fixing bears little relation to the frequently-asserted purpose of regulation to provide maximum service at minimum costs. If, instead of attempting at great expenditure of time and money to ascertain what a utility is 'entitled to earn' (legalistically speaking), the accent were to be placed on the company's reasonable and essential needs, or 'costs,' including those of required

A very different point of view is reflected by the written opinions of the justices in the *Hope* case, including both the concurring and the dissenting opinions. Justice Douglas, speaking for the Court in sustaining the rate order of the Federal Power Commission, threw major emphasis on evidence tending to show that the prescribed rates were adequate to maintain sound corporate credit. In short, he stressed the capital-attraction function of utility rates. Justice Jackson wrote a dissent. But the dissent itself was a brilliant example of another functional approach to rate theory. It stressed the exhaustible character of natural gas as justifying special price policies designed to conserve a limited natural resource for its more important uses. In short, Justice Jackson emphasized a conservation function of utility rates—a special form of the so-called "resource-allocation function" to be discussed in the next chapter.

The following chapter, on the role of public utility rates, will develop further the thesis that the basic standards of reasonable rates should be primarily standards of functional efficiency. But just as the acceptability of a headache tablet must be determined in part by reference to its undesired side effects and not alone by reference to its effectiveness as an anodyne, so also must the reasonableness of any given rates or rate policy be determined in part by reference to its unintended consequences. The administration of *any* standard or system of rate making has consequences, some of which are costly or otherwise harmful; and these consequences may warrant the rejection of one system in favor of some other system admittedly less efficient in the performance of its recognized economic functions. Thus an elaborate structure of rates designed to make scientific allowance for the relative costs of different kinds of service may possibly be re-

capital, regulation would apparently be on its way toward a direct approach to this basic aim." *Final Report of the Telephone Rate and Research Department*, Federal Communications Commission, June 15, 1938, mimeographed, p. 23. Prepared under the direction of Carl I. Wheat.

The early failure of the Supreme Court to appraise the merits of its fair-value rule and its other rules of rate control by reference to their practical efficiency may be explained in part on the ground that these early rules were designed, not directly as tests of reasonable rates but rather as setting limits below which commission-imposed or statutory rates would be held to be so outrageously low as to amount to "confiscation" of private property. In actual practice, however, "reasonable" and barely "nonconfiscatory" rates became seriously blurred. See my *Valuation of Property* (New York, 1937), Chap. 30.

jected in favor of a simpler structure more readily understood by consumers and less expensive to administer. And thus a system of rate regulation that would come closest to assuring a company of its continued ability to earn a capital-attracting rate of return may be rejected in favor of an alternative system that runs less danger of removing incentives to managerial efficiency. The art of rate making is an art of wise compromise.

STANDARDS OF REASONABLE RATES INVOLVE REASONABLE RESOLUTIONS OF CONFLICTS OF INTEREST

In the preceding paragraph, a crude analogy was drawn between the problem of developing standards of reasonable public utility rates and the problem of developing standards of acceptable commodities, such as headache tablets. The rates, like the tablets, must be adjudged reasonable or unreasonable, satisfactory or unsatisfactory (in the light of available alternatives), not only by reference to their effectiveness in the performance of their intended functions but also by reference to the minimization of undesired side effects.

But the determination of reasonable rates introduces one complication not ordinarily involved in standards for satisfactory commodities, such as headache tablets or automobiles. By and large, standards for these commodities are those of satisfaction to the person who swallows the pill or who buys the car. But principles of reasonable rates are designed partly to resolve conflicts of interest among different parties, particularly between investors to whom a rate means a source of income and consumers to whom it means an item of expense.

This interest-conflict aspect of rate making is emphasized in the legal cases, which tend to associate, if not completely to identify, rates that are "reasonable" with rates that are *fair* or *equitable* as between buyers and sellers, and as among different classes of buyers. As will be noted in Chapter VIII, on fairness criteria (when distinguished from functional-efficiency criteria), it raises problems that defy scientific solution unless one accepts as "given" the standards of fairness prevailing in the community; and not even completely then because these standards are themselves in conflict with one another.

But while interest conflicts present serious difficulties, their rational solution is not as hopeless as one might assume at first thought. For in the first place, notions of fairness, such as those based on good-faith performance of earlier promises, are themselves partly utilitarian. In the second place, some of the most important principles of rate making, such as those designed to secure an optimum allocation of the country's scarce resources as between the production of utility services and the production of alternative goods or services, are related only indirectly to conflicts of interest among different individuals. And in the third place, under systems of private or public ownership that depend entirely on revenues rather than on taxes for financial support, there is an important degree of harmony between the interests of consumers and of investors. This partial harmony justifies a public service commission in going far toward the acceptance of the long-run interests of consumers as its sole responsibility. With an important qualification,[13] the legitimate interests of investors may be regarded as amply protected by the allowance of rates sufficiently high to maintain corporate credit and hence to assure the maintenance of adequate service.

THE PUBLIC INTEREST AND "WELFARE ECONOMICS"

No chapter on "the public interest" or on "maximum social welfare" as the ultimate objective of rate-making policy can afford completely to ignore a restricted conception of economic welfare developed by that modern branch of economics called "theoretical welfare economics" or (quite illegitimately) simply "welfare economics." [14] In line with Western European tradition, this school of thought identifies the welfare of any given community with the totality of the welfares of the individuals therein. Also in line with this tradition, it accepts the revealed choices or

[13] The "important qualification" lies in the possible obligation of commissions to protect the interests of investors who may have committed their funds in reliance on rules of rate making no longer accepted. See pp. 155–158.

[14] For an introduction to the significance and principles of modern theoretical welfare economics, see Melvin W. Reder, *Studies in the Theory of Welfare Economics* (New York, 1947); Kenneth Boulding, article on "Welfare Economics," in Bernard F. Haley, ed., *A Survey of Contemporary Economics*, Vol. II (Homewood, Ill., 1952); I. M. D. Little, *A Critique of Welfare Economics* (Oxford, 1950); J. de V. Graaff, *Theoretical Welfare Economics* (Cambridge, England, 1957). See also the references on marginal-cost pricing in Chap. XX. The great classic on welfare economics in a broader sense is A. C. Pigou, *The Economics of Welfare*, 4th ed. (London, 1932). But Pigou does not refrain from making interpersonal comparisons.

preferences of individuals as determining the relative satisfactions derivable by these individuals from alternative forms of action. This acceptance of the preferred position as the position more conducive to individual welfare is what gave specious support to the contention, now no longer advanced, that welfare economics can pass judgment on economic welfare without taking any position on ethical values.

Up to this point there is nothing esoteric in the welfare economist's conception of economic welfare. What is esoteric, however, is his denial of any scientific basis by which one may make interpersonal welfare comparisons—by which one may decide whether an economic change which adds to the welfare of some individuals while detracting from the welfare of others will enhance or diminish net social welfare. This self-denial limits the welfare economist (of the more rigorous persuasion) to attempts to pass judgment on the welfare implications of those proposed changes in economic policy which, while benefiting some members of the community, will not be adverse to any other members. Needless to say, such a limitation is a serious impediment to the resolution of controversies in the field of public utility rate making, since most of these controversies present a clash of interests among the parties to the dispute. And only to a minor extent is the impediment removed by the qualification, now generally accepted in welfare economics, that any proposed economic change will contribute to total economic welfare if the individual beneficiaries, after being actually made to indemnify all the individual losers, will still remain net beneficiaries.

For the reason just suggested, as well as because of the variety of oversimplified assumptions on which reliance must be placed to prove the validity of propositions as to what action will tend to enhance economic welfare, theoretical welfare economics has only a limited usefulness to persons concerned with practical problems of rate making or rate regulation. But "limited usefulness" by no means implies trivial usefulness. On the contrary, a study of the norms of "optimum pricing" set forth by modern welfare economists should help materially in the development of practical principles of rate making. This statement applies notably to the analysis of marginal-cost pricing in the American and

European literature of welfare economics—a subject to be discussed in Chapter XX.[15]

With this brief introduction to the problem of developing social-welfare or public-interest principles of utility rates, we now turn to the role that these rates are designed to play as instruments of economic control.

[15] But compare J. Wiseman, "The Theory of Public Utility Price—An Empty Box," 9 *N.S. Oxford Economic Papers* 56–74 (1957).

III

THE ROLE OF PUBLIC UTILITY RATES

The previous chapter, on the problem faced by rate theory in developing social-welfare criteria of reasonable rates, has emphasized the importance of a functional approach. That is to say, the merits of any established or proposed principles of rate making must be appraised primarily by reference to the estimated effectiveness of the resulting rate schedules in performing the functions properly assigned to utility prices as instruments of economic control. What, then, are these functions? Two possible functions (a capital-attraction function and a use-rationing function) were mentioned casually as illustrating the functional approach to the law of public utility rates taken by Justices Douglas and Jackson in their opinions in the *Hope Natural Gas* case.[1] But the subject now requires a systematic treatment—all the more so since it receives only incidental attention in the standard treatises.

A partial explanation of the tendency of the literature to deal only incidentally with the role of public utility rates may lie in a widespread assumption that this role is fairly obvious and that the only really tough problems lie in the development of policies of rate making and principles of rate control designed to secure effective performance.[2] In fact, however, no such assumption is

[1] P. 37, *supra*. For a further discussion of the distinction between a functional and a nonfunctional conception of reasonable rates, see Chap. VIII.

[2] A further partial explanation may lie in the tendency of the public utility treatises to expound the principles of rate control as they have been developed in the case law. But legal thinking has not been highly analytical in its use of economic concepts; and its criteria of reasonable or nondiscriminatory rates are based, not so much on notions of functional efficiency as on traditional notions of "equity" and of property rights. By way of example, note Professor I. L. Sharfman's valid criticism of the Interstate Commerce Commission for its failure to

warranted, since the question of precisely what tasks should be assigned to public utility rates is highly controversial.

THE ROLE OF THE PRICE SYSTEM

The purposes served by the imposition of any given rate of charge for a specific public utility service, no less than the purposes served by a market-fixed price for a particular commodity such as a given grade of wheat, are largely (though not completely) predetermined by the fact that this rate is merely a tiny part of a whole universe of prices, including prices for other commodities and services and prices for labor (wages) and for other factors of production. This same statement applies also to the entire schedule of rates charged for different classes and amounts of service by a great public utility enterprise like the Consolidated Edison Company of New York or the Tennessee Valley Authority. For both of these enterprises, the function of rates is one of adaptation to the outside world of prices. "Reasonable" rates may therefore be viewed as rates reasonably aligned with other prices, including the prices of the commodities and services supplied by the nonutility industries.

But the nature of this function of adaptation or alignment can be understood only by reference to the larger part played by the nation's entire price system in the control of economic activity. A survey of "the role of the price system" is therefore called for. But the survey will be brief and elementary, since a thorough review would cover a large part of the entire field of economics.

THE DEFINITE ROLE OF PRICES IN A PURE-MARKET ECONOMY

For reasons to be noted in the following section, it is impossible to make general but precise statements as to the functions performed by the many different kinds of prices in a "mixed economy" such as that of modern capitalism. This economy has at its command instruments of economic control, such as taxation, subsidy, and rationing, partly alternative and partly supplementary to the forces of the market place—instruments that leave

---

make articulate the economic rationale of its "primary policy of seeking to make rates conform to worth of service." *The Interstate Commerce Commission*, Vol. III B (New York, 1936), pp. 691–692.

considerable leeway for legislatures and administrators, influenced by their notions of public interest or by pressure groups, to choose what role they wish different kinds of prices to play. Here, the precise functions of specific groups of prices are decided upon as a matter of policy, not just discovered by reference to the inherent characteristics of the price system. Only under the fictitious assumption of a pure-market economy, operating without central direction, can one ascribe a definite role to the whole price system, symmetrical with respect to all component prices. Nevertheless, there is a decided similarity between this archetypal role of prices in a pure-market economy and the modified roles played by specific prices in real life.

In a pure-market economy, then, all goods are produced for sale rather than for direct use or for free distribution; and all incomes are realized through the sale of these goods or else through the sale of the labor or the other "factors of production" that go into the making and delivery of the goods. Here, the role of prices is coterminous with that of the entire economic organization of society. It is characterized as that of controlling the distribution of scarce resources among multiple and competing uses. If the resources were not scarce, there would be no need to "economize" their use by means of a price system. But since many resources are scarce, and since most of them have alternative uses (coal for the production of electricity versus coal for the production of gas or of steel products, etc.), there is need for a complex of prices by means of which shares of each such resource are apportioned among the different alternatives.

This resource-distribution role of the price system is composed of many subroles, which can be classified and subclassified in various ways according to the purposes and convenience of the analyst. The simplest classification will serve present purposes: a threefold division suggested by Professor Samuelson in his introductory textbook on economics.[8] Under this division, the price system determines (a) *what* things shall be produced, and in what amounts; (b) *how* they shall be produced, and by what producers; and (c) *to whom* they shall be distributed. Thus, in an unregu-

[8] Paul A. Samuelson, *Economics, an Introductory Analysis*, 3d ed. (New York, 1955), p. 15. For a more elaborate sixfold classification, expanded from an earlier classification by Frank H. Knight, see C. Lowell Harriss, *The American Economy* (Homewood, Ill., 1953), pp. 13–18.

lated, pure-market economy, the price system would determine, among millions of other things, whether or not 60-cycle alternating-current electric power shall be generated and supplied in a given area and at what rates of output; whether the generation in this area shall be entirely by steam plants or entirely or partly by water power; and how the supply of the power shall be apportioned as between industrial users and residential consumers, and as between rich consumers and poor consumers. In these determinations, prices act both as an attracting force motivating people to produce for sale, and as a repelling force leading prospective consumers to restrict their demands. These offsetting forces are illustrated in the textbooks by the familiar tables and diagrams of supply and demand.

In a pure-market economy, the prices that perform these vital functions of an economic organization are not themselves determined, or even influenced, by any central authority. Instead, they result from the interaction of buying, selling, producing, and consuming activities of individuals or groups of individuals, each seeking to maximize profits or gains on sale and to minimize costs on purchase. But as a result of these interactions, there emerges a genuine economic system instead of a condition of chaos.

The nature of these interactions, even under the simplified assumptions of pure competition, is complex, as the uninitiated reader would discover by studying even as skillfully clarified an exposition as that by Professor Stigler in his book on *The Theory of Price*.[4] But the resulting allocation of resources among alternative uses is supposed to reflect a condition called "consumer sovereignty"—a crude phrase since the implied analogy of the market place to the voting booth is very loose. The preferences of the consumers, expressed in their bid prices or in their responses to asked prices, determine the kinds and amounts of products to be produced and also their distribution among different persons or "spending units." But the voting power (or, more accurately, the drawing power) of these different persons is weighted by their relative money incomes—incomes which, in turn, are derived from the sale of products or of productive services at market-determined prices.

How well the price system of a pure-market economy would

[4] George J. Stigler, *The Theory of Price*, rev. ed. (New York, 1952).

subserve the public welfare if it could be, and actually were, operated under conditions of strict or perfect competition is a question of doubtful meaning, to which no confident answer can be given if only because it assumes a form of economic organization that never has existed and never can exist. Any answer, moreover, would need to assume a standard of social welfare that at best is only plausible, not axiomatic, and that would surely fail to win unqualified public acceptance. But having in mind a standard of "economic welfare" which identifies maximum welfare with maximum satisfaction of individual consumer preferences, many economists have declared that, with certain exceptions or qualifications, the prices that would result without regulation but under pure or perfect competition would be the "ideal" prices. The import of such a statement, however, might well mislead a layman, or even an economist himself. For no scientific proof (nor any common-sense proof, for that matter) can be adduced for the proposition that an unmodified competitive-price system will lead to an optimum distribution of income among individuals and among families.[5]

Despite its limitations, the familiar assertion that competitive prices are, in a significant sense, "ideal" prices is of much interest to a student of principles of public utility regulation. For it supplies the major argument for a contention that the primary object of rate regulation should be to secure, by deliberate price control, those charges for public utility services that *would* prevail in the absence of regulation if the services were rendered under conditions of competition. The merits of this contention will be discussed in Chapter VI, "Competitive Price as a Norm of Rate Regulation."

[5] The standard work on the relationship between competitive price conditions and optimum economic conditions was written by Professor A. C. Pigou of Cambridge University: *The Economics of Welfare*, 4th ed. (London, 1932). The original edition, itself expanded from an earlier work, was published in 1920. Pigou's techniques of analysis have been much modified by later writers in the field of "welfare economics." But I doubt whether the revisions of technique have led to conclusions on rate-making policies notably different from those suggested by Pigou's analysis as revised by him in his later editions. The more recent literature has laid great emphasis on marginal cost, rather than average total cost, as a measure of optimum rates. See Chap. XX. But the general philosophy of marginal-cost pricing is suggested in Pigou's treatise as well as in earlier literature on price or rate theory.

THE FLEXIBLE ROLE OF PRICES IN A MIXED CAPITALIST ECONOMY

In the theory of price determination under the assumption of a pure-market economy, the role of the price system is predetermined by the very definition of this economy. Here, price fixing or price regulation would not be available, say, as an instrument for the relief of drought-ridden farmers in Southwestern United States, or as part of a government program for the stimulation of industries deemed of critical importance for national defense, or for the purpose of discouraging the consumption of liquor through the imposition of heavy excise taxes. In short, specific prices or groups of prices are not determined by reference to specific price-making policies. Samuelson has this point in mind in warning the reader that the failure of a competitive market economy to distribute goods on the basis of relative needs rather than on the basis of money demand is no sign of its failure to perform efficiently the functions that it is designed to perform.[6] A charge of inefficiency on this score would be just as pointless as would be a complaint that a typewriter is inefficient because it fails to correct a typist's misspellings.

Even when we turn to a mixed economy like that in the United States of today, an exposition of the role of prices in a pure-market economy fits the situation fairly well if not pressed too far—much better, in fact, than would any alternative simple exposition. For the economy of modern capitalism is not *hopelessly* mixed, and the forces of the market place are still dominant in the determination of relative prices of commodities and services—even in the determination of the regulated prices charged by public utility companies. It is for this reason that we can still refer usefully to the role of the price system as an organic whole; and it is for this reason that we can go a long way toward a derivation of the functions of any one price or group of prices, such as the functions of the rates charged by the Consolidated Edison Company of New York or by the Pennsylvania Railroad, by assuming that these functions are not fundamentally different from those performed by other prices, including the prices of wheat, of electrical equipment, or of laundry service.

[6] *Economics*, p. 37.

The force of this statement will be brought out in the following pages, which distinguish four primary functions of public utility rates. For this fourfold classification of functions also reflects the major functions performed by the unregulated prices of any one commodity or group of commodities. But the limitations of this statement will also be apparent as the discussion proceeds. For a mixed economy has at its disposal a variety of economic controls ancillary or alternative to that of the price system, with the result that, in the performance of any given function, price may merely assume *some share* in getting the desired result. What this share shall be is a question of policy that cannot be decided in the same way for all types of prices, or even for all public utility prices. It follows that the writer on public utility rate theory, no less than the writer on any theory or program of reasonable or optimum prices in a mixed economy, cannot rest content to take as "given" for his purposes the precise tasks to be assigned to the prices under review, limiting himself to an attempt to secure prices best designed to perform these predetermined tasks. On the contrary, he must assume responsibility for an attempt to resolve controversies as to how programs of rate making should cooperate with other policies of public utility operation and regulation in securing basic objectives of public policy. This problem of deciding precisely what role should be assigned to public utility rates with respect to particular types of utility enterprises and under different schemes of regulation is one of the most difficult and most controversial issues of modern rate theory.

THE FOUR PRIMARY FUNCTIONS OF PUBLIC UTILITY RATES

Without here attempting to answer the question what functions *ought to be* assigned to public utility rates, let us now review the main functions that these rates, in combination with other instruments of social control, are actually called upon to perform for American railroads and public utilities, whether operated under direct public ownership or under regulated private ownership. These functions can be classified in different ways; indeed, a thorough analysis of the role of utility rates would call both for subclassification and for cross-classification. But a study of the many controversies about standards of reasonable rates to be found in the literature and in the litigated rate cases leads me to distinguish four primary functions. The significance of the distinctions, for purposes of rate theory, lies in the association of these different functions with different criteria and measures of reasonable or optimum utility rates. Indeed, the standards of efficient performance of these functions are in partial conflict with one another, with the result that sound rate-making policies are necessarily policies of wise compromise.

In default of more clearly established terms, the four functions to be distinguished may be called, respectively, (1) the producer-motivation or capital-attraction function; (2) the efficiency-incentive function; (3) the demand-control or consumer-rationing function; and (4) the income-distributive function. All have their counterparts in the role played by any unregulated competitive price or group of prices. For example, the prices charged for hats, viewed from the standpoint of the community or the nation, have the economic tasks, operating in participation with the surrounding universe of prices, (1) of stimulating or encouraging the production and marketing of hats, (2) of rewarding or penalizing hat makers and hat sellers for efficiency or inefficiency, (3) of restricting the effective demand for hats, thereby making overt hat rationing unnecessary, and (4) of transferring a compensatory amount of purchasing power from those who wear the hats to those who make and market them. But the relative importance of these four functions, as well as the accepted standards of good performance, may differ as between regulated utility services and commodities or services produced without regulation and in a highly competitive market.

1. THE PRODUCTION-MOTIVATION OR CAPITAL-ATTRACTION FUNCTION

One of the most obvious functions of prices in general is that of motivating and enabling persons to participate in the production and distribution of commodities or services for which they themselves may have no direct use. This is also one of the most prominent and most widely recognized functions of public utility rates. Public utility companies are permitted to impose charges for their services largely in order to induce and enable them to supply these services and to make provision for their continuation and for their required expansion. If denied the opportunity to levy compensatory charges, they could not long continue operation in the absence of tax-financed subsidies.

This production-motivation function of prices gives rise to the capital-attraction standard of reasonable public utility rates. By this standard, "reasonable rates" are rates adequate to yield revenues that will cover all legitimate operating expenses plus a return on investment sufficient to maintain sound corporate credit and to attract required amounts of new capital. Rates below this level are deemed deficient because, at least in the long run, they will not enable the company to live up to its obligations to serve the community.

In public utility cases in which the general *level* of rates (as distinct from the rate *structure*) is at issue, the capital-attraction standard of reasonable rates tends to be accepted by commissions as the primary basis for their decisions. Even the representatives of the public utility companies will usually base their requests for a rate increase or their opposition to a rate decrease on the ground of need for credit-sustaining revenue. True, company counsel will also assert their *legal* rights under constitution or statute and will support these claims in part by considerations of fairness to investors. But, especially in recent years, these collateral claims are seldom asserted except in defense of rate increases which the company also claims that it would need to secure solely in the long-run interests of its present and prospective consumers.

Impressed not just with the desirability but with the utter necessity of command over capital as a means of corporate survival and growth, one might be tempted to accept the capital-attraction function of rates, with its correlative capital-attraction standard of a reasonable return, as the only function that needs serious recognition as a criterion of reasonable rates. Any other functions, such as the three others already mentioned, might be deemed purely incidental, with the result that the whole theory and practice of rate regulation would be limited to the determination of rates adequate to maintain the enterprise in sound financial health in the light of its needs for expansion. Indeed, one sometimes reads in the literature of public service regulation statements that suggest this conclusion.[7]

But no such conclusion would be justified, since the capital-attraction function is only one of several important functions of utility rates. Indeed, it is not even the most nearly indispensable function, which is that of demand-control or consumer-rationing. But postponing discussion of this latter point, let us now note reasons why efficient performance of the capital-attraction or production-motivation function cannot be accepted as the sole, or even as the overriding, objective of rate-making policy.

In the first place, the nation or the community is under no compelling necessity to require its public utility enterprises to maintain and expand service on a self-financing basis. If desired, deficiencies in revenues from the sale of service can be made good by tax-financed subsidies. Subsidized public utility and railroad enterprises, both privately and publicly owned, are by no means unfamiliar in American history. To cite two current examples, one may mention the heavy, though now greatly reduced, Federal subsidies to the domestic airlines and the subsidized service provided by the New York subway system since its acquisition by the city government. In recent years, this subway system has not even covered operating expenses despite an increase in fares from 5¢ to 10¢ and then to 15¢.

To be sure, practical or theoretical reasons may justify a conclusion that, save in unusual situations, resort to subsidies is undesirable and that a public utility should be permitted and required to charge rates sufficient to maintain financial self-sufficiency. Indeed, with important qualifications, this is the view taken in the present book. But in any attempt to develop sound principles of rate making, the undesirability of subsidized services cannot properly be taken for granted as a starting point for the theory of

[7] Thus a recently published book on electric rates declares: "Rates for electric service are nothing more than price tags which the electric utility places on the service it renders. As such, their ultimate purpose is to provide for the utility, in the aggregate, sufficient revenue to cover all operating costs and earn a fair return on the fair value of the property devoted to public use." Russell E. Caywood, *Electric Utility Rate Economics* (New York, 1956), p. 22. But Caywood himself later recognizes other objectives of rate-making policy, not merely ancillary to the objective of adequate over-all revenues.

utility rates. Instead, the merits or demerits of a subsidy should be subject to careful analysis.[8]

In the second place, even under the assumption that a public utility enterprise should be financially independent, the capital-attraction role of rates can reach only a certain distance in determining standards of reasonable pricing. For, under conditions of regulated monopoly, various alternative schemes of rate making, if skillfully administered, may all suffice to attract the required capital. This statement applies even to the determination of adequate rate *levels* for reasons discussed in the chapters on rate-level determination. But it applies even more obviously to the determination of proper rate patterns or rate structures. The good or bad attributes of the rate relationships within a given public utility system—the relationship, say, between domestic and industrial electric rates, or between rates for on-peak and off-peak service, or between railroad coach fares and Pullman fares—are by no means limited to those attributes bearing on the company's ability to earn an adequate over-all rate of return.

Finally, unless supported by standards of reasonable or optimum rates derived from other functions, the capital-attraction function offers only a standard of *minimum* rates. True, as invoked in rate cases and in the literature, this function is usually construed to imply that reasonable rates are the *lowest* rates needed to maintain corporate credit and to permit the attraction of necessary new capital. And this is a very sensible construction in the normal situation. But one cannot derive this construction as a mere corollary of the proposition that public utility rates must be sufficiently high to attract capital. One must look to other functions of rates, notably to the consumer-rationing function and to the income-distribution function, for a reason in favor of the *lowest* capital-attracting rates.[9]

[8] This view may be contrasted with that expressed by Admiral Ben Moreell, then chairman of the board of the Jones & Laughlin Steel Corporation, when he declared: "I believe that all public power which is subsidized by tax funds, whether on the federal, state, or local level, is economically and morally wrong." Quoted in *Electrical World*, May 12, 1958, p. 55.

[9] In one important respect, the capital-attraction role of utility rates differs from the similar role played by unregulated, competitive prices. A private, non-utility producer is under no legal obligation to expand output beyond the point that he deems desirable on grounds of profit-maximization. But most public utilities are under a legal duty to supply adequate service within their franchise territories. Hence, they lack the freedom enjoyed by private businesses to base their expansion program on an estimate of profitability.

2. THE EFFICIENCY-INCENTIVE FUNCTION

Under unregulated competition, the price system is supposed to function in two ways with respect to the relationship between the price of the product and the cost of production. In the first place, the rate of output of any commodity will so adjust itself to the demand that the market price will tend to come into accord with production costs. But in the second place, competition will impel rival producers to strive to reduce their own production costs in order to maximize profits and even in order to survive in the struggle for markets. This latter, dynamic effect of competition has been regarded by modern economists as far more important and far more beneficent than any tendency of "atomistic" forms of competition to bring costs and prices into close alignment at any given point of time.

In the regulation of public utility monopolies, the principle that rates should be set at levels designed to yield revenues covering cost including or plus a "fair rate of return" may be regarded as a substitute, though not a close substitute, for the tendency of prices and costs to come into accord under the forces of market competition. But where is the efficiency-incentive counterpart?

Under prevailing methods of rate regulation, such incentives are, indeed, provided to a limited degree. First, private companies receive no guaranty of their ability to enjoy a "fair rate of return," with the result that they may be under more or less severe pressure to practice operating economies and to stimulate growth of demand for service in order to earn the officially sanctioned rate.[10] Second, the standards of a commission-fixed "fair rate of return" are themselves somewhat flexible, and some commissions, in setting these rates, try to make allowance for supposed relative efficiency or inefficiency of operation and of financial planning. And third, there is the so-called "regulatory lag"—the quite usual delay between the time when reported rates of profit are above or below standard and the time when an offsetting rate decrease or rate increase may be put into effect by commission order or otherwise.

[10] The significance of this no-guaranty situation is enhanced by the general refusal of commissions and courts to recognize past deficiencies or past excesses in corporate earnings as grounds for offsetting allowances in later rate cases.

But these incentive-encouragement features of orthodox rate regulation are extremely crude, and one may suspect that they are very ineffective in comparison with the stimulation of direct and active competition. Whether or not the situation lends itself to material improvement, through the adoption of systematic differential rates of return based on estimates of relative efficiency, is a controversial question, about which something will be said in Chapter XV, "The Fair Rate of Return."

3. THE DEMAND-CONTROL OR CONSUMER-RATIONING FUNCTION [11]

Although the capital-attraction function of public utility rates is the one that has had the most influence in the determination of American principles of rate regulation, it is not the function given first place by those modern economists who have written on rate-making theory. Instead, the function which they have emphasized is that of "demand control" or "consumer rationing." [12] Here, the price is designed, not to induce production but

[11] This function of price is often referred to in modern economics as a "resource-allocation function"—the function of allocating scarce resources among competing uses. But the latter term seems to me to be unhappily chosen for the purpose of designating the transactions that it is designed to cover. Logically, the effect of a price on the distribution of income among different persons is a resource-allocation effect. Yet, in practice, economists have distinguished between income-distributive effects and what they have called resource-allocation effects. See, e.g., Abba P. Lerner, *The Economics of Control* (New York, 1947), who distinguishes "The Optimum Distribution of Goods," *given* the individual-income distribution (Chap. 2), from "The Optimum Division of Income" (Chap. 3).

[12] The difference in relative emphasis placed upon this function of price fixing marks perhaps the major distinction between the outlook typical of the professional economist and that typical of popular and legal thinking. As recent writers have noted, neither the American public utility statutes nor the decisions of the public service commissions give primary attention to consumer-rationing or resource-allocative objectives of rate-making policy. Instead, the emphasis has been on a variety of other objectives such as those of capital attraction, of income distribution, and of "fairness" considerations. In commenting on a paper in which I had referred to the consumer-rationing standard of utility rates, Professor Emery Troxel declared that, under the rate-making policies of the Tennessee Valley Authority or the Bonneville Administration, "the rationing standard of Professor Bonbright's paper does not operate." With these Federal projects, he said, so much emphasis has been placed on the long-run maintenance of rate stability that "the public pricing policy is not directed either to a rationing purpose in short periods or to control of efficient allocative (investment) limits in longer organizational views." 47 *American Economic Review, Proceedings* 404 (May, 1957).

The same popular tendency to belittle or ignore resource-allocation objectives of economic policy has been found to exist outside the realm of utility economics. Thus Professor Melvin W. Reder, referring to the modern economist's objection to business monopolies on the ground that they tend to create a misallocation of productive resources (as distinct from the more familiar ground that they lead

rather to restrict or influence demand. When this function is performed efficiently, as is by no means always feasible in the regulation of utility rates, every potential consumer has the opportunity to receive forthwith, or after only a minimum delay, whatever type of service the utility enterprise undertakes to supply in his area, and in whatever amounts he is willing to buy at the scheduled rates of charge. Thus public utility rates are designed to avoid the necessity for overt rationing by making the consumer, in effect, ration himself.

In order to illustrate the high importance of this rate-making function, let us assume that the council of a large city which directly owns and operates its electric power system is considering a proposal to supply free electric service to all residents, and in whatever amounts each consumer requests. Let us also concede the financial feasibility of the proposal in view of the city's powers to pay operating expenses and fixed charges out of the proceeds of various taxes. What objections can be raised against this pleasant arrangement?

There are several possible objections, including the criticism that the plan would be unfair to those residents who are subject to high taxes but who have little need for electric service—a criticism relevant to the income-distributive function of utility rates. But another objection, even more serious, is that the offer of free

---

to a bad income distribution) says: "But this complaint is almost exclusively a complaint of the professional economist; indeed, the lay public is scarcely aware of the existence of a resource-allocation problem, let alone the role of the pricing system in its solution, and the implications of 'Monopoly' in connection with it." *Studies in the Theory of Welfare Economics* (New York, 1947), p. 47.

The recent writings by professional economists on urban transit problems have laid stress on the failure of public authorities to base their fare-fixing policies on demand-control objectives. Thus Wilfred Owen writes: "Along with this failure of the urban transportation system to meet the needs of modern communities, the possibilities of influencing transportation demand to bring it more nearly into balance with the supply of facilities have been almost completely overlooked." *The Metropolitan Transportation System* (Washington, D.C., 1956), p. 249.

But some recent economists have insisted that the majority of their profession have overemphasized consumer-rationing or resource-allocative objectives of rate making and price fixing to the neglect of other, assertedly even more important, objectives such as those of income distribution or income stability. With respect to farm-price controls, this criticism of the academic economists has been voiced sharply by Professor J. K. Galbraith: "Economic Preconceptions and the Farm Policy," 44 *American Economic Review* 40–52 (1954). It is also implied by writers who defend the application of a cost-price standard to the regulation of well-head prices of natural gas. See, e.g., Joel B. Dirlam, "Natural Gas: Cost, Conservation, and Pricing," 48 *American Economic Review, Proceedings* 491–501 (May, 1958).

and unlimited service would surely result in wasteful use of electric power. The waste would be most obvious with respect to the demand for industrial power, say, for the production of aluminum or for electrochemical industries. But in these days of modern electric appliances, it would be serious even with respect to domestic consumption, as the British have occasion to know in view of the prevailing below-cost charges for the peak-time use of electricity for house heating.[13]

Faced with this prospect of an excessive demand for electric service in the absence of any charge therefor, our hypothetical city would be compelled either to abandon the proposal for free electricity or else to introduce complex restrictions as to the amounts, the times, and the types of use. But the imposition of such restrictions would be a very clumsy and ineffective instrument of economic control, and the case against it would be overwhelming save, perhaps, in a period of emergency. Under normal conditions, a far better practice is a resort to the "automatic rationing" of a price system.

It remains to note an ambiguity in the terms "demand-control" or "consumer-rationing" when used to characterize a specific function of rates or prices. In its broadest sense, this function refers to any use made of prices as devices by which to affect the demand for the services to which the prices are attached or the demand for alternative or complementary services. Thus, if the city council in our imaginary city, on receipt of reports from medical experts on the amount of light needed to minimize eyestrain, were to decide that every family "needs" to consume electricity for lighting at the rate of 50 kilowatt hours per month, it would be resorting to a consumer-rationing standard of rate making if it should attempt to fix rates so as to encourage this amount of lighting use. A more familiar example of the same interpretation of a consumer-rationing standard of price is the supply of city water to dwelling houses at less than cost when justified on the ground that a liberal use of water is in the interest of community health.[14] And at the other extreme, although out-

[13] See *Report of the Committee on National Policy for the Use of Fuel and Power Resources* (The "Ridley Committee" Report), Cmd. 8647, London, Sept., 1952; also I. M. D. Little, *The Price of Fuel* (Oxford, 1953).

[14] The desirability of encouraging free household use of water for sanitary purposes has often been given as a justification for nonmetered water service, especially for the smaller users. In a defense of the use of meters, despite political

side the field of the public utilities, is the imposition of high excise taxes on liquor for sumptuary purposes rather than merely for revenue-raising purposes.

But as the term is usually construed in price or rate theory, the "consumer-rationing" function does not refer to the use of prices by the government as a means of allocating commodities and services to different individuals on the basis of centrally determined needs. Instead, the prices are designed to let the consumer be the judge of his own needs or wants, subject, however, to the requirement of compensatory payment. This is the principle of consumer sovereignty, the most fundamental single principle of modern public utility rate theory.

Under conditions of unregulated competition, prices are supposed to adjust themselves "automatically" in accord with this principle of consumer sovereignty. The desired result takes place because the competitive price is, in a sense, an indemnity price—a price, the payment of which by the consumer covers the costs (including the so-called "opportunity costs") incurred by those persons who supply the required "factors of production." Under regulated monopoly, there is no such automatic adjustment. Instead, the adjustment must be brought about by the aid of policies of rate regulation expressly designed for this purpose. This involves a system of modified cost pricing. But, as we shall note later, there are formidable difficulties in the way of a successful application of such a system—among them, the difficulty arising when rates sufficient to cover the incremental costs of the different services, considered separately, would fail to cover all the overhead costs.

The prominence here given to the function of utility rates in restricting the demand for service to those demands for which consumers are ready to cover costs of rendition may seem shocking to persons familiar with the emphasis placed by modern rate-making practice on so-called "promotional rates"—on rates expressly designed to stimulate increased use of public utility services including, particularly, use of electric energy. This abundant-

resistance thereto, Mr. Harry E. Jordan, Secretary of the American Water Works Association, Inc., has declared that "when cities have gone from an unmetered to a fully metered basis the records show a reduction from one-third to one-half of the prior use per person." But he adds: "No one has suffered. No citizen has had to 'go dirty.' He simply stopped letting faucets drip and water run unnecessarily." Letter to New York *Times*, Oct. 2, 1954.

use philosophy of rate making has won qualified acceptance among the private utility companies within the past twenty years. But it has been accepted with greatest enthusiasm by some of the spokesmen for public power, who have been charged by critics with carrying their enthusiasm to an irrational excess.

The merits of this criticism could hardly be appraised intelligently without reference to specific claims on behalf of the public power advocates. But in any case, there is no inconsistency between the view that rate schedules should be promotional in character and the view that an important function of rates is that of preventing wasteful consumption.[15] A promotional rate schedule is simply a schedule that lowers the price barriers against incremental blocks of consumption. Only if these barriers are lowered to the point at which incremental use fails to cover incremental cost is there a violation of the consumer-rationing standard of rate making as here construed. In fact, a failure to lower the barriers to this point with respect to those utility services for which the demand is highly elastic indicates a departure from the consumer-rationing standard.[16]

4. THE INCOME-DISTRIBUTIVE FUNCTION

All three of the functions of public utility rates mentioned up to this point are tasks designed cooperatively to serve one com-

[15] A public utility may promote the use of a specific type of service, not only by lowering the rate charged for this service but also by raising the rate charged for alternative types of service, thus taking advantage of what economists have called "cross-elasticity of demand." For example, off-peak consumption of the service may be encouraged by a reduction in rates for off-peak uses combined with an increase in rates for peak-time use.

[16] The problem of making utility rates perform most effectively a consumer-rationing role is rendered extremely difficult by the paucity of reliable data on the price-elasticity of demand for different kinds of utility service. This problem is complicated by the influence on demand of variables other than the rates charged for the services in question: by changes in demand resulting from changes in consumers' income, changes in the price of substitute or complementary services and commodities, and increases or decreases in advertising and in high-pressure salesmanship. For example, in determining the extent to which electricity will be put to various household uses, the form and level of the rates of charge can exercise only a participating influence along with the availability and the pricing of electrical equipment and along with the promotional activities of the company's commercial department. Some years ago, the manager of the Winnipeg Municipal Electric Plant told me that the effect of a drastic decrease in rates had proved very disappointing until the household use of electricity was spurred on by an intensive advertising program and by resort to an expensive and appealing showroom of electric appliances, located on Winnipeg's main shopping street.

mon goal of rate-making policy: the provision of the community with adequate kinds and amounts of public utility services, produced in an economical manner. Thus, the rates charged for its services by a given public utility enterprise are designed (a) to determine the types and amounts of service that the enterprise must undertake to render, (b) to enable the enterprise to pay operating expenses and to attract the capital needed in order to render this service, and (c) to encourage the management of the enterprise to meet the demand for service at a minimum cost of production.

But a fourth rate-making function must now be distinguished: that of transferring, in the form of a cash payment, a desirable amount of purchasing power from buyer to seller, from consumer to producer, from the persons who receive the benefits to the persons who bear the burdens. This return flow of cash is here regarded neither as a mere means of stimulating production nor as a mere means of limiting consumption, but rather as a means of income distribution or redistribution.

In order to distinguish this income-distributive function from the other three, let us assume, contrary to experience, that the demand for electric energy by domestic consumers would be unaffected by any change in rates within a wide range—say, a range between 2¢ and 10¢ per kilowatt-hour. At either extreme or at any intermediate point, consumption per customer would stay fixed at 200 kilowatt-hours per month. Let us also assume that even the 2¢ rate would yield revenues quite adequate to enable the company to attract needed capital and hence that any higher rate, at least up to 10¢ per kilowatt-hour, would also yield revenues adequate for this purpose—in fact, *more* than adequate.

Under these assumptions, none of the rate-making functions so far discussed offers us a criterion of reasonable rates by which to prefer any rate between 2¢ and 10¢ over any other rate within the same wide range. For any one of these rates will meet equally well both the capital-attraction test and the consumer-rationing test. As to the efficiency-incentive test, its relevance is doubtful to the problem here under review. We must therefore look to a fourth standard of reasonable rates, the income-distributive standard. Under this standard, the question how much currency should be transferred from the consumer to the producer must be decided "on its own merits" and not merely because of the effect of the contemplated transfer on producer or consumer motivation.

An answer to the question just raised calls for a relevant standard of income distribution in the sale of public utility services. But, here, two quite different standards are suggested in the rate cases and in the literature on public utility pricing. They may be called the compensation standard and the quasi-tax, or ability-to-pay, standard.

Under the compensation standard, the outflow of cash from consumer to producer is designed merely to offset or counterbalance the cost incurred by the producer in supplying the service. The consumer "accounts" to society, through the producer, for a draft on its limited resources, thereby reducing his own opportunity to purchase other commodities or services with a given income. In this way public utility rates participate with other prices in the task of making the relative *money* incomes of individuals determine their relative purchasing power, their relative "*real* income." But these rates are not designed to offset inequalities in money incomes as among different consumers, or as between consumers and producers.

Under the quasi-tax version of the income-distributive function of public utility rates, the prices that consumers pay for public utility services are not necessarily set at amounts designed to transfer the cost of supply from producer to beneficiary; nor are they designed to offset the value of the service to the beneficiary. Instead, the price mechanism is here used as an occasion to make some slight correction for unequal distributions of money income between producers and consumers, or among different classes of consumers. Here, for the first time, ability to pay becomes directly relevant to the determination of a "reasonable" rate.[17] But the ability-to-pay principle cannot be carried beyond severe limits, since any attempt to do so would lead to a breakdown in the other functions of utility rates.

In our hypothetical example of domestic service for which the demand is completely unresponsive to changes in price within a range from 2¢ to 10¢ per kilowatt-hour, the compensation

[17] One should be on guard not to confuse an ability-to-pay standard of reasonable rates with a value-of-the-service principle. For a comment on the distinction, see p. 111, n. 5, *infra*.

version of the income-distributive function of rates would require that the rate be set at 2¢, since this price suffices to cover cost including a capital-attracting rate of return. Even under a quasi-tax version the same answer may be given if we assume (a) that the corporate stockholders are richer than the domestic consumers and (b) that a uniform rate must apply to all of these consumers. But if the stockholders are poor and the consumers rich, then a 10¢ rate would be suggested by ability-to-pay considerations of reasonable rates.

Very wisely, in my opinion, the standard literature on utility and railroad rate theory expressly or implicitly supports the compensation version rather than the quasi-tax version of the income-distributive function of utility rates. That is to say, these rates are not regarded as instruments for an attempted correction of personal-income disparities, save in highly special cases, and then only as makeshift devices.[18] Taken as a whole, moreover, the actual practices of American rate regulation comport with this position of modern rate theory. But there is also no doubt that, in actual practice, ability-to-pay considerations have material influence in rate-making practice, although the extent of their influence is impossible to assess because their lack of a recognized, official standing in the law of rate making leads rate makers or rate regulators to mix them in vaguely with other considerations. In the regulation of the rates of private utility companies, which are held entitled to an opportunity to enjoy a "fair return" regardless of the income status of their customers,[19] the influence of ability-to-pay factors is most potent in the design of the rate schedules as distinct from the rate levels. It accounts for the failure of many rate schedules to recover full out-of-pocket costs for small consumptions of service; for the allowance by railroads and local transit companies of special fares (or, in extreme cases, free service) to classes of patrons with reputedly low incomes combined with reputedly high merits (clergymen); and for a part of the public hostility toward proposals by economists for the imposition of higher local-transit fares in metropolitan cities

[18] See on this point, pp. 111–112, 116–117, *infra.*

[19] But during the depression of the 1930s, income-distributive arguments were merged with "value-of-service" arguments in support of utility rate reductions that were not expected to yield standard or conventional "fair rates of return." See pp. 259–262, *infra.*

during rush hours and in rush directions—proposals quite justified on cost-of-service principles.[20]

In closing this exposition of the income-distributive function of utility rates, let me warn the reader that neither popular nor legal thinking on standards of reasonable rates draws the clear-cut distinction that I have attempted to draw between the "compensation" and the "quasi-tax" version of this function. Indeed, popular thinking betrays notions of "fair pricing," derived perhaps from the mediaeval conceptions of "just price" (*justum pretium*), which do not nicely fit either of these two versions. It reveals itself in a contention that uniform rates should be charged for services that are the same in some highly superficial or conventional sense, despite marked differences in cost of rendition: for example, uniform prices per kilowatt-hour of electric power regardless of time of delivery, or regardless of distance from the source of power, or regardless of density of the area to be served. A striking example of this point of view was given me, some years ago, by an official of the New York Farm Grange, who insisted that every farmer in the state of New York, however remote and isolated, "has an inherent right" to enjoy electric utility service on payment of the same rates paid by city dwellers. This contention, he felt, did not need to be proved valid on "economic" grounds since it followed from "the very nature of the public utility concept." Against such "natural-rights" points of view, scientific approaches to rate-making problems are impotent weapons.[21]

THE PROBLEM OF RECONCILING THE DIFFERENT FUNCTIONS OF PUBLIC UTILITY RATES

Under a system of unregulated but perfect competition as expounded by the classical treatises on economic theory, prices are supposed to perform all four of the aforesaid functions, in their competitive versions, with optimum efficiency. So far from being in conflict, the four functions are in perfect harmony. Indeed, no one function can be performed except in cooperation with the others.

[20] See p. 346, n. 7, *infra*.
[21] For a further discussion of this point of view, see Chap. VIII.

Even in the regulation of rates to be charged by a monopolistic public utility, there is a *large measure* of harmony among the correlative functions assigned to the rate-making machinery. As will be noted in the following chapter, this workable degree of harmony is attained by programs of rate making based on modifications of, and variations from, the principle of cost of service. Thus, rates based largely on cost may be designed, at one and the same time, (a) to make the enterprise self-supporting and to attract required new investment, (b) to restrict demand, and (c) to secure a return flow of cash consistent with the income-distributive function of prices in its more defensible version (the compensation version). Even the efficiency-incentive function of prices can be given some recognition, through flexible features in regulation of the type suggested in an earlier paragraph.

But in the determination of rate levels and rate structures for a regulated monopoly, a perfect harmony or perfect interaction among the various tasks assigned to these rates is impossible to secure. On the contrary, the functions are in partial conflict. By this I mean that the acceptance of rules or measures of reasonable rates designed exclusively to secure maximum efficiency in the performance of any single function would necessarily impair the efficiency of the same rates in the performance of some, at least, of its other functions. In consequence, the development of sound rate-making policies calls for a resort to wise compromise.[22] It may also call for consideration of proposals to minimize the conflict among the various tasks imposed upon the rate-making machinery by resort, on occasion, to alternative or ancillary forms of economic control including the use of tax-financed subsidies on the one hand and excess-profits taxation on the other hand.

The nature of these partial conflicts among the various purposes that public utility rates are designed to serve will be noted throughout the following chapters, which will also point to conflicts among subdivisions of each function and not just among the four primary functions noted in the present chapter. But the

[22] The fact that rate-making policy must seek to attain partly conflicting objectives and hence that it must resort to wise compromises is well recognized by the abler practical experts on utility rate making. L. R. Nash, *Public Utility Rate Structures* (New York, 1933), p. viii: "A basic purpose of this book is to portray rate making not as an exact, scientific procedure but as a skillful balancing of conflicting objectives."

fact that conflicts exist and that they create serious problems for the theory and practice of rate control may be illustrated here by two examples: first, that of a conflict between the capital-attraction and the efficiency-incentive functions of utility rates; and second, that of a conflict between the capital-attraction and the consumer-rationing functions.

The first type of conflict arises because, with a regulated monopoly such as an electric power company or a telephone company, financial experience strongly suggests that the type of regulation best designed to maintain corporate ability to raise capital is one which goes as far as feasible toward protecting the security holders against the risks of financial loss. In other words, maximum security rather than the sporting chance of high gains is the most effective inducement for this type of investment. This view is accepted by modern rate regulation, which undertakes to supply the financial security within limits. But if regulation goes too far in an attempt to supply this security, there arises the serious danger of a loss of managerial incentive toward efficient operation. And some writers believe that American rate regulation has gone well beyond this danger point in its readiness to give the capital-attraction function priority over the efficiency-incentive function of the price system.

The second type of conflict among functions of utility rates—that between the capital-attraction and the consumer-rationing functions—is of even more importance from the standpoint of rate theory. To be sure, performance of both of these functions calls for the acceptance of a cost-of-service standard of reasonable rates (or of reasonable *minimum* rates), with the result that there is a degree of harmony between the two functions. But the harmony is not complete, since the cost relevant to the capital-attraction standard is total experienced cost, whereas the cost most clearly relevant to the consumer-rationing standard is prospective incremental cost—the estimated additional cost of additional units of service. In consequence, the rates of charge that would be best designed to serve the one objective of rate making are not the most efficient rates for the attainment of the other objective. Standard public utility regulation is vaguely, though imperfectly, aware of this dilemma and attempts to meet it by discriminatory deviations from cost pricing of a "value-of-the-service" character.

On the other hand, some economists have supported another, more radical, escape from the dilemma, in the form of a proposal to base rates entirely on marginal or incremental costs, any resulting deficiencies in total revenue being made good by tax-financed subsidies.

But this chapter is not concerned with the merits of alternative ways by which to avoid or minimize serious conflicts among the different functions performed by public utility rates. Instead, it has been content merely to point to the existence of these conflicts and to suggest that the fact of this existence presents modern public utility rate theory with its most frustrating set of problems.

IV

COST OF SERVICE AS THE BASIC STANDARD OF REASONABLENESS

In stressing the fact that public utility rates, like other prices, are usually designed to perform multiple functions, the preceding chapter distinguished four functions as primary under modern rate regulation. But only casual attention was paid to the question what *measures* of reasonable rates can be expected to secure satisfactory performance of these functions, together with the question whether rates best designed to perform any one function are also best designed to perform the others. A discussion of these two related questions constitutes the main subject of the remaining chapters.

Most of the treatises on public utility economics, following the case law on rate regulation, divide the subject of rate determination into two parts. The first part is concerned with the measures of a reasonable rate level for any one company or group of companies. The second part is concerned with the principles of the rate structure or rate differentials. This distinction is essential for purposes of analysis, and it will be observed in Parts Two and Three of the present study. But it suffers the disadvantage of cutting across certain basic standards of reasonable rates, such as the cost standard, which apply, though with variations, both to rate levels and to rate relationships. In order to minimize this disadvantage, the present chapter and those that immediately follow will use a different breakdown of subject matter. First, separate chapters will be devoted to the two most frequently cited criteria of reasonable rates—cost of service and value of service. Then will come a chapter on the standard of hypothetical competitive price—a standard often thought to embody an ideal reconciliation of cost and value factors. In sharp contrast with the competitive-price norm is a group of related rate-making standards to be considered in Chapter VII, "Social Principles of Rate Making." Chapter VIII will reclassify the tests of reasonable rates by distinguishing between considerations of "fairness" or "equity" and considerations of "functional efficiency." Only in the last chapter of Part One, a transition chapter, will attention be turned to the distinction between rate-level standards and rate-structure standards.

THE WIDESPREAD ACCEPTANCE OF A COST-PRICE STANDARD

No writer whose views on public utility rates command respect purports to find a single yardstick by sole reference to which rates that are reasonable or socially desirable can be distinguished from rates that are unreasonable or adverse to the public interest. A complex of tests of acceptability is required, just as would be the case with the tests of a good automobile, a good income-tax law, or a good poem. Nevertheless, one standard of reasonable rates can fairly be said to outrank all others in the importance attached to it by experts and by public opinion alike—the standard of cost of service, often qualified by the stipulation that the relevant cost is *necessary* cost or cost reasonably or prudently incurred. True, other factors of rate making are potent and are sometimes controlling—especially the so-called value-of-service factor in the determination of the individual rate schedules. But the cost standard has the widest range of application. Rates found to be far in excess of cost are at least highly vulnerable to a charge of "unreasonableness." Rates found well below cost are likely to be tolerated, if at all, only as a necessary and temporary evil.

A cost standard of rate making has been most generally accepted in the regulation of the levels of rates charged by private utility companies. But even more significant is the widespread adherence to cost, or to some approximation of cost, as a basis of rate making under public ownership. Thus the great Hydro-Electric Power Commission of Ontario purports to apply the principle of "service at cost" in its charges for wholesale power supplied to the various municipal distribution systems of the province. And thus most of the Federal power projects in the United States, including the Tennessee Valley Authority, purport to sell electric power at rates designed to cover operating expenses plus a compensatory return on allocable capital investment—one form of a cost-of-service standard. To be sure, critics of these projects have insisted that, under proper accounting, revenues would be shown to fall short of full-cost coverage. But the mere fact that these allegations are generally denied by the responsible managements of the Federal agencies implies that these managements themselves concede the validity of a cost principle of rate making.

Lest the foregoing remarks be taken to imply an adherence to a cost standard more rigid than the facts would justify, let me at once note exceptions. In the first place, the principle is followed far more closely as a measure of general rate levels than as a measure of individual rate schedules. In the second place, it is deliberately violated by those municipal power plants, said to be fairly numerous, that use the sale of electricity as a source of large profits for the city treasury. And in the third place, it has been waived to a minor degree through the use of indirect subsidies in support of rural electrification in the United States; and waived to a major degree through the use of heavy subsidies for rural electrification in the province of Ontario. One may also note the huge deficits incurred in the operation of the Canadian National Railways, and the failure of most metropolitan transit systems, in recent years, to charge fares that cover operating expenses plus fixed charges.

Important, however, as are these and other deviations from a cost-price standard, they are generally treated as exceptions to the general rule of rate making. In Great Britain, even a Labor Government that went much farther than did this country in the direction of socialization, including socialized medicine, did not see fit to abandon the general criterion of service at cost when it nationalized its public utilities. Instead, it instructed the various boards, such as the British Electricity Authority, to undertake to realize total revenues sufficient to meet total outlays properly chargeable to revenue account, "taking one year with another." [1]

[1] The British statutes governing the rates to be charged by the nationalized public utilities and railroads do not expressly forbid sale of services at prices designed to yield revenues *in excess* of total costs. But they have been interpreted by British commentators as contemplating the provision of service "without mak-

THE THREEFOLD RATIONALE OF A COST-PRICE STANDARD

No doubt one of the reasons for the popularity of a cost-of-service standard of rate making lies in the flexibility of the standard itself. "Cost," like "value," is a word of many meanings, with the result that persons who disagree, not just on minor details but on major principles of rate-making policy, may all subscribe to some version of the principle of "service at cost." The best known, though not the most important, illustration of such a disagreement is that between supporters of an "original-cost" basis of rate making and supporters of a "reproduction-cost" basis.[2]

But before turning attention to alternative meanings of "cost of service," let us first ask ourselves what reasons are advanced in favor of *any* cost-price standard of rate making—of any standard under which an attempt is made to transfer the cost of supplying the service from the producer to the consumer, no more and no less. The answer is that there are at least three related, though not identical, reasons, each one associated with a different function of public utility rates.[3]

The first support for the cost-price standard is concerned with the consumer-rationing function when performed under the principle of consumer sovereignty. Under this principle, potential consumers should be free to enjoy whatever kinds of service, in whatever amounts, they desire as long as they are ready to indemnify the producers, and hence society in general, for the costs of rendition. Only in this way can the consumers be put in a position, as it were, to ration themselves by striking a balance between benefits received and sacrifices imposed. If the rates were set at less than cost, either

---

ing, so far as possible, either a deficit or a surplus." William A. Robson, ed., *Problems of Nationalized Industry* (New York, 1952), p. 335.

[2] A more important disagreement is that between those who identify "cost of service" with some kind of average or prorated total cost, and those who identify it with differential or marginal or out-of-pocket cost.

[3] A fourth possible defense of a cost-price standard of utility rates is suggested by the British economist Professor W. Arthur Lewis as applicable to the British public corporations which now operate the major utility and railway enterprises. These corporations, he declares, should make neither a loss nor a profit after meeting all capital charges, because "to do otherwise is to contribute either to inflation or to deflation." Robson, ed., *Problems of Nationalized Industry*, p. 181.

overt rationing would be necessary or else service would have to be supplied in wasteful amounts. If the rates were set at more than cost, use of the services thus priced would be unduly restricted.

But the pricing of public utility services at cost of production is supposed not only to bring about a proper control of demand; in addition, and at the same time, it is supposed to motivate and enable the producing company to supply the service in the amount demanded, thereby avoiding the need for resort to tax-financed subsidies. That is to say, the sale of the service at cost will supply the company with necessary revenues to pay operating expenses and capital charges. For this purpose, to be sure, cost must be given a broader definition than is customary in the language of accounting, since it must include allowance for a capital-attracting rate of return on investment. But a capital-attracting rate of profit is here considered a part of the necessary cost of service.

The third defense of the cost-price standard is related to the income-distributive function of rates in the more generally acceptable version that I have called, in Chapter III, the "compensation version." Under this version, an individual with a given income who decides to draw upon the producer, and hence on society, for a supply of public utility services should be made to "account" for this draft by the surrender of a cost-equivalent opportunity to use his cash income for the purchase of other things.[4]

Of the four primary functions of rates set forth in the preceding chapter, the only one that may require a deviation from *any* kind of a cost-price standard is the management-incentive function, which calls upon the price system to impose penalties for inefficient management, and to award special profits for superior performance. But even this function could be subserved, in theory at least, by a factitious definition of cost to mean, not necessarily the cost of service actually incurred under the existing management, but rather an estimate of the cost that *would be* incurred under a management of standard efficiency. The application of such a hypothetical-cost criterion presents formidable practical difficulties. But American

[4] Compare F. M. Taylor's discussion of price fixing in a socialist economy, in which Taylor declares that, under socialism, prices should equal cost of production in order to give meaning to any given income-allotment by the state to the individuals therein. "The Guidance of Production in a Socialist State," 19 *American Economic Review* 1–8 (1929).

regulation has attempted a very limited and crude application through the acceptance of the doctrine that, in public utility rate cases, companies may secure reimbursement only for "prudent" or "legitimate" or "reasonable" operating expenses and capital outlays. Some commissions, moreover, have purported to consider efficiency factors in their allowance of a fair rate of return.[5]

PARTIAL CONFLICT AMONG THE THREE OBJECTIVES OF A COST-PRICE STANDARD

In view of what has just been said, one might suppose that, with the possible exception of the management-efficiency function, all of the primary functions assigned to public utility rates could be performed in complete harmony, since all of them agree in calling for a cost-price measure of reasonable or optimum rates. That is to say, when utility rates are fixed at costs, they will simultaneously serve effectively (a) to keep the demand for the service within economic bounds without resort to overt rationing, (b) to enable and induce private capital to undergo the expenses of supplying whatever service is thus demanded, and (c) to transfer from the consumer beneficiaries to the producing enterprise, and hence to the suppliers of the "factors of production," compensatory amounts of purchasing power. This harmony of objectives under a cost-price system of rate regulation would thus be similar to the harmony of functions supposedly performed by unregulated prices when determined under the forces of market competition.

Unfortunately, however, the harmony attainable under a cost-price system of utility rates is far from complete. On the contrary, the functions are in partial conflict, sometimes seriously so. The major source of the conflict lies in the fact that a cost-price standard is subject to many different interpretations and that the interpretation which would best comport with any single objective of rate making is almost sure to be ill-adapted to the attainment of some of the other objectives. Some of the more serious dilemmas presented by different meanings of "cost of service" will be noted briefly in the following paragraphs.

[5] See Chap. XV.

ENTERPRISER COSTS VERSUS SOCIAL COSTS

A cost is always a cost *of* something *to* some individual or group of individuals. When cost of service [6] is offered as a measure of reasonable rates, the answer to the question, Cost to whom? may seem self-evident. Obviously, cost to the consumer is not the answer, since this cost is set by the very rates under review. Equally obviously, cost to the vendor would seem to be the relevant cost, since this is the only cost for which the vendor has a claim to compensation. Since the vendor is usually a private or public corporation, money costs to the corporation or "to the enterprise" would seem to be the proper criterion. Indeed, this is the interpretation generally placed upon a "cost-of-service" standard of rate making.

But if we view a cost-price standard of rate making, not merely as a means of enabling public utility companies to supply the required services at minimum prices, but also as a means of controlling the demand for these services—the consumer-rationing function—the acceptance of the enterpriser's cost as the only relevant cost is justified only under one important assumption: namely, that this cost reflects, with tolerable approximation, the sum total of costs or losses imposed upon the entire nation or community by the construction and operation of the utility plant.[7] If this assumption is valid, the consumers of the service, in indemnifying the public utility corporation or agency for money costs of production, are also indemnifying society for its total social costs. But if the assumption is invalid, then the theory of cost pricing faces a dilemma—that the payment of cost by the consumers to the corporate producer does not constitute full payment of cost or losses by the consumers to the whole community.

In the economic textbooks, the stock example of social costs of production not typically borne, at least not fully, by the producing

[6] This chapter will note only a few of the many distinctions between alternative meanings of "cost of service." For a much more exhaustive treatment, see John Maurice Clark, *Studies in the Economics of Overhead Costs* (Chicago, 1923). See also W. Arthur Lewis, *Overhead Costs* (London, 1949); Joel Dean, *Managerial Economics* (New York, 1951), Chap. 5, "Costs."

[7] The distinction, familiar to accounting, between a "cost" and a mere "loss" becomes blurred in the concept of social cost. By referring to the adverse consequences of purposive action, social costs seem to fall into the category of true costs. But in their inclusion of consequences that are adverse to "innocent bystanders," even though not to the actors, they seem to fall into the category of losses and not of costs.

firm or corporation, is the example of smoke, smell, and noise nuisances of plant construction and operation. Other unindemnified losses suffered by society as a whole or by "innocent bystanders" may be stream pollution, encroachment on the area's water levels, fire hazards, and consumption of limited natural resources (such as fuel oil or natural gas used in the generation of electric power), purchased by the producer at current prices that fail adequately to reflect their long-run scarcity values.[8]

But one must not assume that enterpriser costs always fall short of reflecting total social costs. Indeed, they may greatly exceed the *net* social costs of plant construction and operation, since they may include outlays by the public utility enterprise resulting in social benefits other than those derived from the use of the service. The most striking and important example of this possibility is that of a public utility plant constructed during a period of business depression, such as the deep depression which prevailed in this country during the 1930s. During this period, the money outlays for plant construction did not even roughly reflect comparable net social costs, since these outlays went largely to the employment of labor and plant that would otherwise have remained idle. This fact was well recognized by the national administration under President Roosevelt and by Congress, which granted heavy subsidies and loans for the construction of multiple-purpose dams and other public works projects.

The argument that enterpriser money costs are not even rough measures of the net social costs of producing public utility services, combined with the further argument that the prices which consumers will pay for services fail to measure the benefits received, forms the basis of those "social theories" of rate making to be discussed in Chapter VII. This argument is not accepted by most economists as justifying the general abandonment of a cost-price system of rate making. But it has special force in particular situations, or during a severe and prolonged business depression. When

[8] In the economic literature, the classical discussion of the distinction between social costs and private enterpriser costs is that by Professor A. C. Pigou, *Economics of Welfare*, 4th ed. (London, 1932), Chap. IX. For a recent monograph on the subject, see Professor K. William Kapp, *The Social Costs of Private Enterprise* (Cambridge, Mass., 1950). The analysis of this interesting book is somewhat limited by its failure to recognize that enterpriser cost, while it may fall short of total social cost, may also exceed net social cost.

these situations arise, they present rate theory with a serious dilemma. As far as possible, the dilemma should be avoided by efforts to bring enterpriser outlays into closer conformity with social costs —for example, by government action to compel public utility companies to minimize smoke nuisance even at considerable cost to them and hence to their consumers. Also to be considered is the possible use of a system of special utility taxation, designed to raise or lower the company's tax bills in recognition of social-cost differentials. But the practical difficulties of measuring social costs are formidable obstacles to their overt recognition as factors in rate determination.

TOTAL COSTS VERSUS COSTS OF SPECIFIC SERVICES

Let us now resolve the question, Costs to whom? by assuming that the relevant costs are the money costs to the public utility enterprise and turn to another question, Costs of what? Already this question is partly answered by the nature of the assignment, which is to measure the costs incurred in supplying public utility services. But any one company renders various kinds of service to many different consumers at many different rates of output and at different times. This circumstance gives rise to the question whether the significant costs are the particular costs of specific classes or units of service or whether the cost principle refers rather to a scheme of rate making designed to balance total revenues against total costs over some shorter or longer period of time.

In the regulation of private utility companies, and even in the rate-making practices of publicly owned plants, the determination of general rate levels is likely to take precedence over the determination of specific rate schedules; and there the most directly pertinent costs are the total costs, including the overhead costs. In other words, the cost principle is taken to mean that rates as a whole should cover costs as a whole. But even so, the specific rates are also based in part on cost calculations, and the problem of measuring the costs of separate classes and amounts of service is one of the most controversial problems in public utility rate theory.

We discuss this problem in Part Three, "The Rate Structure." For present purposes what is important to note is the fact that the sum of the costs specifically allocable to different classes and amounts of public utility service will not equate with the total costs of supplying the services jointly, save under conditions rarely met with in actual practice. To be sure, public utility cost analysts have sometimes *purported* to apportion the total costs among the various classes of services so as to bring about the appearance of an equation between the whole and the sum of the parts. But this accomplishment is a tour de force of cost accounting, for reasons developed at length in Chapter XVIII. Hence there is an unavoidable partial conflict between a cost-of-service principle of rate making as referring to total cost, and a cost-of-service principle as referring to the costs of specific classes and amounts of service.

Where the objective of rate regulation is that of securing the lowest level of rates consistent with the avoidance of a public subsidy—and this is the primary objective of regulated private ownership in America—the major emphasis of the cost criterion is on total cost rather than on specific cost. In consequence, the design of the rate structure is based only in part on specific costs or cost differentials. But some modern economists, who lay great stress on the consumer-rationing function of utility rate making, would choose the other horn of the dilemma. That is to say, they would base utility rates on those costs that can be specifically assigned to definite types and amounts of service by a process of differential or incremental cost analysis. If the rates thus fixed should fail to yield total revenues sufficient to cover total costs including a fair return, these economists would have the government make good the deficiency by tax-financed subsidies.

SUNK COSTS VERSUS ESCAPABLE COSTS; ORIGINAL COSTS VERSUS REPLACEMENT COSTS

In America, a long-standing controversy as to the proper interpretation of the cost-of-service principle of rate making has been that between supporters of an actual-cost or original-cost basis of rate control, and supporters of a reproduction-cost basis, often associated with a "fair value" rate base. According to the former doctrine, rate levels should be sufficient to cover actual operating expenses plus a "fair rate of return" on the actual, depreciated cost of the utility plant and equipment; according to the latter, rates should cover operating expenses plus a "fair rate of return" on the depreciated replacement or reproduction cost of the properties, calculated on the basis of current price levels.

Direct discussion of this controversy will be reserved for later chapters, since it is both involved and confused in its objectives. But what concerns us here is the related, but more fundamental, distinction between "sunk costs" and "escapable" or "avoidable" costs as measures of reasonable rates or prices. The distinction is of critical importance for rate theory in the light of the contention, advanced by some writers on price economics, that the only relevant costs for purposes of price determination are the escapable costs. "Sunk costs"—that is to say, costs which, having already been irretrievably incurred, can no longer be avoided or minimized through a curtailment of output—are asserted to form no basis for rational future economic action. "In commerce," wrote a famous British economist, "bygones are forever bygones." [9]

With a public utility plant, a large fraction of the costly action needed to supply any particular output of service or stream of services will have been taken, and irrevocably taken, long before the services are delivered and long before the rates of charge therefor can feasibly be set. These "historical costs" should be completely ignored according to the opponents of "sunk-cost pricing." But their place, so it is argued, should be taken by estimates of replacement cost, of the cost that would be incurred to supply the service by a new plant of modern design, constructed at prevailing levels of prices. In short, the only cost of public utility services relevant for rate-making purposes is hypothetical replacement cost, not actual historical cost.

For reasons to be discussed at length in the chapters on the competitive-price standard and on the rate base, I do not believe that the arguments on behalf of replacement-cost pricing of utility services are well taken. On the contrary, I am convinced that some version of an actual-cost basis of rate making, with its frank acceptance of a sunk-cost price philosophy, is preferable for practical reasons and is by no means inferior for reasons of price theory. But the deficiencies of a sunk-cost standard of rate making should be recognized; and they are especially serious in the determination of the individual rate schedules as distinct from the general rate levels.

What, then, are these deficiencies? They are revealed most clearly by reference to the consumer-rationing function of public utility

[9] W. Stanley Jevons, *The Theory of Political Economy*, 4th ed. (London, 1911), p. 164.

rates. In support of this function, rates should be made just high enough to deter potential consumers from demanding services of types and in amounts for which they are unwilling to defray the costs of rendition. But for this purpose of demand control the relevant costs are those costs that can still be avoided by a restriction of output—in short, the escapable costs rather than the sunk costs. In a celebrated article on public utility and railway rate theory,[10] Professor Harold Hotelling illustrates this point by the example of a bridge which, despite a capacity far in excess of the total use that would be made of it even if it were toll free, is operated on a toll basis high enough to cover total operating and capital costs. This action by the toll-bridge authority in covering all costs including sunk costs will result in an uneconomic curtailment of service, a curtailment that will have only a trivial effect in reducing total costs since the major part even of the maintenance expenses is a function of time rather than of use. Hotelling therefore contends that, as long as the bridge has excess capacity, it should be kept toll free even though the operating and capital costs must be met by taxation. This means the abandonment of a cost-of-service standard of pricing in one sense of the term, the popular sense. But it means a transfer to another standard of cost pricing—the standard of incremental or marginal cost which, in the extreme example of the underused bridge, happens to be almost zero. The merits of this proposal will receive detailed discussion in Chapter XX and incidental attention in earlier chapters.

SHORT-RUN COSTS VERSUS LONG-RUN COSTS

Intermixed with questions of choice between sunk costs and escapable costs is the further question as to the period of time over which a cost-price system should attempt to equate rates on the one hand with costs on the other. Should the rates to be charged today be designed to equal the costs incurred today, or should a longer period of time—say a year or a cycle of four or five years—be chosen in which to keep rates on the average in line with costs on the average? If rates (or, at least, *minimum* rates) are to be based on marginal or incremental costs, should these cost increments be based on short-run or long-run assumptions as to the feasibility of a change

[10] "The General Welfare in Relation to Problems of Taxation and of Railway and Utility Rates," 6 *Econometrica* 242–269 (July, 1938).

in plant capacity in response to a change in demand for the service?

These are formidable questions admitting of no simple answers. Among the reasons for their perplexing character is the fact that they present dilemmas of rate-making policy that cannot be resolved by theoretical analysis because they do not represent choices, the adverse and favorable consequences of which can be balanced against each other in quantitative, mathematical terms. Concededly, for example, utility rates ought not to be too volatile and too unpredictable, as they would be if the attempt were made to let them respond to every shift in unit production costs. Concededly, also, they should not be too unresponsive to shifts in cost factors and in demand factors. But the problem of securing a wise compromise between excessive volatility and excessive stability must rely largely on "common-sense" solutions, with only limited help from economic analysis. More will be said on this point in Chapters XVII and XX.

COST PRICE AS A STANDARD OF COMPENSATION TO THE PRODUCER VERSUS COST PRICE AS A STANDARD OF PAYMENT BY CONSUMERS

Most discussions of the principle of reasonable pricing, whether with respect to the prices of public utility services or with respect to the prices of ordinary commodities, tend to assume that the sale of the service or the commodity represents a two-way transaction in which the payment made by the buyer-consumer represents the very payment received by the vendor-producer. Under this assumed identity between price payable and price receivable, a "fair" price must be a price alike fair to the buyer and fair to the seller, just as would be the case in a barter economy in which so many arrows are offered in exchange for one bow.

This tendency to assume an identity between the prices that consumers should pay for commodities and services and the prices that producers should receive as compensation for outlays and efforts in supplying the products persists throughout most of the law and literature of public utility rate making. But there is no such necessary identity; and on occasion, even American public utility and railroad law has recognized distinctions between standards of reasonable compensation to producers of service and standards of reasonable payment by consumers. The possibility of drawing this distinction arises through the interposition of the government, or of some governmental agency, into the transactions of purchase and sale. On the one hand, the government may supplement the prices paid by consumers for utility services, thereby making the compensation to the producer larger than the burden on the consumers. On the other hand, the government may levy taxes on sales of service, or taxes closely related to such sales, thereby driving a wedge between the higher charges imposed on consumers and the lower rewards conferred on producers.[11] The first situation is illustrated by any of the various public utility subsidies, such as the airmail subsidies of domestic American air transport. The second situation is illustrated by the former "Recapture Clause" in the Transportation Act of 1920, under which railroads earning returns in excess of a certain standard percentage were required to turn over 50 per cent of the excess as a financial aid to "underprivileged" railroads. Under this law, Congress recognized that the rates which American railroads might reasonably exact from their customers might yield revenues in excess of a reasonable compensation for service rendered.[12]

[11] If the public utility taxes were designed merely to idemnify the public treasury for those costs of government, such as police and fire protection, that are imposed by the operation of the utility plant and business, the taxes could then be treated like any other cost of producing the service. But American public utility taxation does not have this limited objective. See pp. 402–406, *infra*.

[12] With the Recapture Clause, the ground for the distinction drawn by the Transportation Act between reasonable rates for consumers to pay, and reasonable compensation for producers to receive, lay in the practical necessity for uniform tariffs among the competing railroads in any one transportation area. In 1920 it was believed that tariffs high enough to maintain the credit of the higher-cost carriers would be higher than necessary to maintain the credit of their lower-cost rivals. In upholding the constitutionality of this clause, the Supreme Court, speaking through Chief Justice Taft, held that rates may be reasonable from the standpoint of the shipper even "though their net product furnishes more than a fair return for the carrier." Dayton-Goose Creek R. Co. v. U.S., 263 U.S. 456, 484 (1924). For an interesting commentary on this case and on the economic problem which it illustrates, see Robert L. Hale, *Freedom through Law* (New York, 1952), pp. 523–528.

Proposals to distinguish between reasonable prices for consumers to pay and reasonable compensation for producers to receive have been made as a solution of the farm price-control problem, alternative to the prevailing system of parity pricing. Such proposals, which have been adopted in England, were embodied

The merits of a policy of rate making which distinguishes between the standards of reasonable rates of charge for utility service and the standards of reasonable compensation to the producers of such service will be discussed in later chapters rather than in this preliminary chapter on cost-price standards. But the possible distinction between the two standards is noted here because of its bearing on the relevant definition of cost of service as a basis of rate making. Insofar as a cost-price standard of rate making is designed to yield revenues adequate, but no more than adequate, to permit a company to recover its necessary operating expenses and to maintain sound credit, it must be interpreted in a manner consistent with this objective of rate-making policy. This means, among other things, (a) that the relevant costs are enterpriser costs and not social costs, (b) that the important equation is between total revenues and total costs rather than between the rate for any specific service and the cost of rendering this particular service, and (c) that "cost of service" must be interpreted somewhat broadly to include an allowance for a capital-attracting rate of return. It also means that a strong, even if not conclusive, case can be made for an original-cost basis of rate making rather than a replacement-cost or an escapable-cost basis—a subject reserved for later discussion.

But insofar as the primary function of public utility rates is deemed to be that of optimum consumer rationing rather than that of producer motivation, a different conception of a cost-price standard of rate making is suggested. Among other things, (a) enterpriser cost is not a good price determinant unless it is acceptable as a rough measure of social cost, (b) the most obviously relevant costs are incremental or marginal costs rather than average total costs, and (c) costs that are "sunk" in the sense that they cannot be avoided or even minimized by denying service to potential customers should

in the "Brannan Plan" for the United States. Under the plan, farm products were to be priced at free-market prices, but the income of farmers was to be stabilized and, if necessary, fortified, by taxation.

Henry George's single-tax doctrine, under which the annual rental values of unimproved land would be completely or largely expropriated by government instead of by landlords, rests on the same essential distinction between the prices that consumers should pay for commodities or services and the proceeds of sale that the owners of the "factors of production" should be permitted to retain. See Paul A. Samuelson, *Economics*, 3d ed. (New York, 1955), Chap. 27, especially pp. 505–507. See also Kenneth E. Boulding, *Economic Analysis*, 3d ed. (New York, 1955), section on "The Attack on Economic Rent," pp. 724–725.

be excluded from consideration if their inclusion in the rates of charge would seriously restrict demand for the service.

Rightly or wrongly, American rate-making policy has made only sparing use of the machinery of taxation and of bounty in an effort to reduce the conflict between rates designed to yield adequate revenues to producers and rates designed to secure an optimum use of public utility service. Hence, regulation is compelled to approve rates that will be tolerably acceptable for both of the above-noted functions of pricing (producer motivating and consumer rationing) as well as for other functions. The prevailing American principles of rate making show the effects of this preference for compromise by a lack of sharpness in their details and by the acceptance of blurred concepts and measures of cost of service.

With this brief introduction to the nature and rationale of a cost-of-service criterion of public utility rates, we turn in the next chapter to the counterpart criterion, popularly called "value of service."

V

VALUE OF SERVICE AS AN ANCILLARY STANDARD

Despite a flexibility of definition which permits many variations in the meaning of "cost of service" designed to fit the special purposes of rate determination, the very nature of a cost standard gives it limitations that preclude its acceptance as the sole measure of reasonable public utility rates. Among these limitations is the failure of the cost principle to give direct weight to the value of the service to the consumers as distinct from the cost of production to the producers. While the relevance of cost is generally conceded, the contention is made that value should also be taken into account. This proposition is asserted to apply no less to the fairness of the prices charged for electric power, telephone service, and railroad transport, than to the fairness of the price charged by a tailor for a suit of clothes or by a department store for a lady's hat.

Writers on public utility rates who stress this position are not uniform as to how they would incorporate both cost and value components in the determination of specific rates. But perhaps the most frequent position is that a "reasonable" rate is one intermediate between cost of production as the lower limit and "value" of the service as the upper limit, the precise point being set by "practical" considerations rather than by any scientific rule of rate making.[1]

[1] Propositions of this character are plausible only if "cost" is narrowly defined so as to limit it to out-of-pocket cost or marginal cost, the receipt of any additional revenues being justified as reasonable compensation for overhead costs including or plus a fair rate of profits. For statements of the position of leading spokesmen for the public utility companies, see L. R. Nash, *Economics of Public Utilities*, 2d ed. (New York, 1931), pp. 121–122; Alexander Dow, *Some Public Service Papers* (Detroit, 1927), p. 211; Henry L. Doherty, in National Electric Light As-

EXAMPLES OF VALUE-OF-SERVICE PRINCIPLES IN RAILROAD AND UTILITY RATE MAKING

Postponing for later discussion the formidable problem of defining "value of service" so as to qualify it as a definite standard of rate making, let us first note examples of the actual use of a standard which, at least, goes under that name.[2] The most prominent American examples are those of railway freight rates. While both the class rates and the special, "commodity" rates for freight shipments are designed to reflect cost differences to a limited extent, they are notoriously and openly based, for the most part, on the so-called value-of-service principle. In line with this philosophy, the charges per 100 pounds for the transportation of the more valuable commodities between any given points are higher, often several times higher, than the comparable charges for the transportation of commodities with a low value relative to their weight or bulk. These differentials are admittedly much greater than any which could be justified by the higher costs of handling and insuring the more valuable shipments.

Among the nontransport utilities, perhaps the most overt resort to a value-of-service standard is that by the telephone industry, led by the Bell System. Under policies of rate making acquiesced in by most state public service commissions, a single telephone company serving an entire state will adopt a state-wide system of rates designed to yield adequate total revenues through the levy of higher charges on subscribers in larger cities than on subscribers for comparable service in smaller communities. In any given city, moreover, higher rates are quoted for business use than for residential use. To be sure, both of these types of rate differentials can be, and have been, defended to some degree on strictly cost grounds—the community-size differential on the ground that telephone switch-

sociation, *Proceedings*, May, 1910, pp. 291–321, reprinted in Edison Illuminating Company of Detroit, *The Development of Scientific Rates for Electricity Supply* (Detroit, 1915).

[2] In the literature of public utility rates, three terms, all referring to noncost standards of rate making, are often used interchangeably or with only minor differences in meaning: "value of service"; "charging what (or in proportion to what) the traffic will bear"; and "rate discrimination." The latter term is in favor among economists as a synonym for "value-of-service" pricing. But defenders of the practice in actual rate cases are loath to characterize it by the invidious term "discrimination." See Chap. XIX.

ing costs increase more than in direct proportion to any increase in the number of subscribers; the kind-of-use distinction mainly on the ground that business telephone calls create more serious peak loads than residential calls. But the telephone companies themselves do not rely on this defense alone in rate cases. On the contrary, and particularly with respect to size-of-community differentials, their major emphasis has been on "value" considerations. In the larger communities, so the argument goes, telephone service is "worth more" than in smaller communities, since each subscriber has potential toll-free connections with more people.

As to the electric and gas utilities at the present time, the deficiencies and paucity of published cost analyses make it impossible to determine the effect of "value-of-service" principles on typical rate structures.[3] Few persons, however, would deny that these principles have influence in the direction of rate differentials not completely supported by cost differentials. The major influence is probably to be found in what, from a cost standpoint, are excessively heavy discounts for large quantities of electricity or gas—discounts offered even when the unit costs of supplying the larger quantities are not reduced by more favorable load factors.

All of the examples of value-of-service pricing so far offered are those concerned with the *structure* of the rates, with the rate differentials rather than directly with the general rate levels. Indeed, at least with respect to the relatively monopolistic nontransport utilities, rate-level determination has adhered far more closely to the principle of service at cost in the sense of total cost coverage plus or including a "fair rate of return." [4]

But with the railroads, the Interstate Commerce Commission has

[3] A recent monograph by Ralph K. Davidson entitled *Price Discrimination in Selling Gas and Electricity* (Baltimore, 1955) concludes that current rate differentials are by no means justified on a cost basis. Other experts have expressed the belief that Davidson grossly exaggerates the extent of the deviations of rates from costs. However this may be, one should not assume that a failure of relative rates to correspond closely to relative costs means that they must therefore necessarily be based on value-of-service factors. Most rate patterns are to a material extent the outgrowth of historical developments, of political pressures on public service commissions, of popular prejudice, and of questionable theories of cost analysis. This statement applies alike to publicly and privately owned utility enterprises. It is only fair to add my conviction that the electric utilities, in recent years, have taken the lead in closer approaches to a rational system of rate making, the greatest lag being that on the part of the railroads.

[4] But so-called value-of-service considerations have probably influenced public service commissions in their findings as to what constitutes a "fair rate of return." See p. 138, *infra*.

sometimes declined to sanction the full rate-level increases requested by the carriers, not on the ground that lower rates would suffice to yield a "fair rate of return," but on the ground of serious doubt whether the higher rates would yield higher net income, at least over the longer run.[5] A refusal on these grounds is based on value-of-service considerations in one sense of this highly ambiguous term.

A significant though vain attempt to invoke a "value" standard of reasonable rate levels in the interest of consumers was made by the Wisconsin Public Service Commission during the depression of the 1930s. In 1934 the Commission ordered the Wisconsin Telephone Company to reduce its rates, not on the ground that they were yielding a rate of return that would be judged excessive by tests applicable to normal years, but rather on the ground that some decrease in rates was called for in response to the reduction in general price levels and in consumer incomes.[6] In an opinion speaking for the Commission, Chairman Lilienthal made the point that the depression had resulted in a decrease in the (money) *value* of telephone service. But the Commission's order was overruled by the Wisconsin Supreme Court, which held that even a severe business depression did not deprive the utility company of a constitutional right to the enjoyment of an opportunity to earn a "fair rate of return."

In thus using value-of-service language in partial defense of the rate-reduction order, Chairman Lilienthal doubtless hoped to secure support from an early, cryptic dictum by the Supreme Court in the famous case of *Smyth v. Ames*.[7] In this dictum, the Court de-

[5] Sharfman, *The Interstate Commerce Commission*, Vol. III B (New York, 1936), Chap. XIV. Railways have complained bitterly against refusals to grant rate increases on the ground that the increase would be ill-advised even from the applicant's point of view. The subject is an important one from the standpoint of sound regulatory practice. I am deliberately slighting it, however, in this book, since it involves an issue of legal jurisdiction rather than an issue of rate theory.

[6] See p. 261, *infra*.

[7] Smyth v. Ames, 169 U.S. 466 at 546–547 (1898). For discussions of the possible meaning of the dictum, see my *Valuation of Property* (New York, 1937), pp. 1108–1110; Robert L. Hale, "Non-Cost Standards in Rate Making," 36 *Yale Law Journal* 56–67 (1926); Eleanor Heyman, "Value of the Service; Its Various Meanings and Uses," 9 *Journal of Land and Public Utility Economics* 252–265 (1933); Henry H. Edgerton, "Value of the Service as a Factor in Rate Making," 32 *Harvard Law Review* 516–556 (1919); Hillyer Brown, "Value of the Service as a Ceiling on Public Utility Rates," 33 *California Law Review* 283–297 (1945); Clyde O. Fisher, "Value of the Service and Public Utility Rates," 12 *Journal of Land and Public Utility Economics* 76–83 (1935).

clared in effect that a company supplying a public service might not be denied an opportunity to charge rates that would yield a "reasonable return on the fair value" of its property devoted to the public service. But it qualified this statement by declaring that the consuming public has an overriding right to enjoy service at rates no higher than "what the service is reasonably worth."

Although this qualification of the "fair value" rule was repeated by the Supreme Court in later cases, its import remains a mystery. The mystery lies in the question how the Court could have assumed that any company, however greedy and however possessed of monopoly powers, could coerce free men and women to buy its products or services at more than they are worth. Obviously, "reasonably worth" must have been used in a normative, not in a positive, sense; but the nature of the norm has gone undefined to this day.

THE LACK OF A QUANTITATIVE DEFINITION OF VALUE OF SERVICE

Before considering the merits of a value standard as a basis of rate making one must first ask just what the standard means. This is no easy question, since the term has been used loosely, without benefit of any formal definition. One reason for the looseness is that, except when applied in special situations, "value" does not lend itself to quantitative expression as a measure of reasonable rates. Under a *cost* standard, the price per unit of service is supposed to be made equal to the cost per unit. But under a *value* standard, as ordinarily construed, one cannot first find that the value of the service is, say, 2¢ per kilowatt-hour or 10¢ per telephone call and then set the rate at this value.[8]

What has just been said is obviously true if "value" is taken to mean selling price or market value. For the price at which the service may be sold is the very point at issue in a rate case. But we are also in trouble, though of a different kind, if "value" here refers to the worth of the service to consumers as measured by the prices

[8] Referring to the use of the standard in railroad rate regulation, Professor Sharfman writes: "The 'value of service' principle, as a basis for rate-making, provides at best a vague and indeterminate formula, rather easily construed as justifying any system of rates found expedient by the carrier. Taking the words in their most obvious sense, no rate can exceed the value of the service and still continue to be paid by the shipper." Sharfman, *The Interstate Commerce Commission*, Vol. III B, pp. 321–322.

that they would pay rather than go without. Thus defined, the value of any given class of service will be different for different customers, and also different for various amounts of service delivered to any one customer. This variability in the values placed by different people on different amounts of any commodity or service at any one time is represented in the economic textbooks by "demand curves," which start with the relatively high prices that would be bid for small amounts of the product and which slope down to zero bid prices for redundant amounts. Of course, one might arbitrarily define "the value" of the service as the highest price for which *any* amount of service, however small, could be sold. But this definition would be useless for rate-making purposes. On the other hand, no specific intermediate price between the peak price and zero can be singled out as alone reflecting the value of the service,[9] no more than can one intermediate point on the side of a hill be picked out as reflecting the height of the side.

The fact that "value of the service" does not usually lend itself to quantitative expression as a measure of reasonable rates will be appreciated by anyone familiar with the use made of this type of standard in actual rate cases. Telephone companies, for example, insist that their service is "worth more" to subscribers in big cities than in little cities. But, to the best of my knowledge, they offer no evidence of the amounts of these values, city by city. The relative "values," if such they can be called, would seem to have merely ordinal, not cardinal, magnitudes. And the same statement would appear to apply to value-of-service pricing by the other utility companies and by the railways.

THREE DIFFERENT MEANINGS OF A VALUE-OF-SERVICE STANDARD

The foregoing remarks justify the conclusion, reached by writers who are meticulous in their use of economic terms, that "value of the service," when invoked as a criterion of reasonable rates or even

[9] Except in a case in which the maximum rate of output is predetermined by limited plant capacity, in which case the market-clearing price, as measured by the marginal bid, might be said to represent "the value of the service." But in most cases in which rates are said to be fixed on a value principle, the rate of output, so far from being predetermined, will depend on the amount of service demanded at the price to be fixed.

of maximum rates, is an inaccurate expression, and a misleading one unless the inaccuracy is clearly recognized. What, then, does the term really mean? The answer is that it has been used in a wide variety of senses.[10] But aside from its not infrequent use as a mere synonym for "fair price" judged by *any* standard or standards of fairness, the three alternative meanings that seem most significant are those set forth below.

1. MOST PROFITABLE RATE OF CHARGE

An earlier paragraph has suggested that, while "value of the service" *might* be defined as the highest price which any consumer would bid, if necessary, for any quantity of service, this definition would be useless for rate-making purposes. But a really significant price would be whatever price for a given service would yield maximum profits to the public utility company—possibly an exorbitant price but also possibly only slightly above out-of-pocket costs depending upon the product's cost functions and on the price elasticity of demand for the service ("what the traffic will bear"). In economics, such a price is called "the monopoly price" on the assumption that the objective of a monopolist is to maximize his net earnings. To be sure, this price does not represent the value of the product, in a nice sense of the word "value," any more than a variety of alternative prices at which the monopolist might be able to sell different quantities of his product. But business lingo lends some sanction to this use of words. Consistently with this usage, a public utility company which sets its rates, whether deliberately or through miscalculation, at levels higher than enough to yield maximum profits could be said to be charging "more than its services are worth."

The very fact that public utility rates are subject to regulation clearly disqualifies this version of the value-of-service standard as a *general* criterion of reasonable rates. But if, for reasons of competition or of a decline in the prosperity of the community, a utility company is unable to earn a "fair rate of return" under any feasible level and pattern of rates, a plausible though far from conclusive

[10] In the article cited in footnote 7, Professor Robert L. Hale, after a study of the language of commissioners and judges in rate cases, concludes that the term has been used on occasion to refer to *any* noncost criterion of rate making. I may add that it has also been used to refer to some standards more appropriately considered special types of a cost standard.

argument can be made for permission to exact whatever rates of charge will minimize the deficiencies in total revenues. This situation, which presents the rate regulator with a serious dilemma, will receive attention in Chapter IX.

2. RATE DISCRIMINATION

At least in the literature of economics, the "value-of-the-service" principle is taken most frequently to mean that principle of rate design under which the differences in the prices charged by a given enterprise for its various products are based, not just on differences in the costs of production but also in part on differences in the relative "price elasticities of demand." Products for which the demand will not be seriously curtailed by relatively high prices will be made to bear these prices. Products for which the demand will vanish or fade if the prices are set far above out-of-pocket or marginal costs will be priced nearer to these costs. This practice is called "price discrimination" in economics; and it is widely prevalent among public utilities and railroads despite the legal restrictions imposed by the rules against "unjust discrimination."

Viewed from the standpoint of the public interest, the restrained use of this version of a "value-of-the-service" principle is defensible chiefly on the ground that it is a *relatively* harmless means of making good the deficiency in total revenues that would result from the sale of all public utility services at mere marginal or out-of-pocket costs. To be sure, sale of utility services at some kind of average costs is a possible alternative method of full-cost coverage, in some respects a superior method. But it is inferior in other respects since it will sometimes preclude the consumption of services for which consumers would gladly pay more than incremental costs. For a further discussion of the subject, the reader is referred to Chapter XIX.

3. MARKET-CLEARING PRICE

Public utility rate theory suggests a third possible meaning of the value-of-service standard, that of a market-clearing price—a price just sufficient to secure full but unrationed use of service in whatever amounts are temporarily available in view of limited plant capacity. Such a standard would be applied by a toll-bridge authority if it were to set tolls just high enough to hold traffic within limits of safety and without inordinate delays. Similarly, it would be applied by a gas distribution company if, on facing a rapid growth of demand for natural-gas heating beyond the present capacity of its sources of supply, it were to raise its rates for service up to the point at which no demand at these rates would long go unsatisfied.[11]

Readers familiar with the textbook expositions of the theory of a competitive price will recognize the similarity of this possible market-clearing-price standard of rate regulation to the price-determining forces of an unregulated market, operating in "short-run" periods. But in actual practice public utility regulation makes very sparing use of the price system as a means of securing short-run adjustments between plant capacity and demand for service. More will be said on this point in the next chapter.

This chapter has discussed briefly a number of possible noncost standards of reasonable rates that often go under the name of value-of-service standards because of their direct reference to the benefit or demand side of public utility service as distinguished from the cost or sacrifice side. More will be said about them in the chapters of Part Three, on specific rate schedules. But enough has been said here to suggest that, in one of several versions—especially the second one noted above—they play important though subordinate roles in the modern theory and practice of rate regulation.

Why is the role subordinate to that of cost? The short answer is that the task of regulation is one of fixing values, not of finding them—of bringing the prices of public utility services into line with the prices of other products by relating the former prices to the costs of production. But this is an incomplete answer, and supplementary comments are called for.[12] Three such comments will serve to close the chapter.

[11] While this practice would constitute a violation of cost pricing in any usual sense of the term, it would be consistent with a cost principle in a special sense. For as long as there is an absolute limit to the number of customers who can be supplied, the granting of service to any one customer involves a social cost in the resulting necessity of denying the service to some other customer. What this other customer would pay for the service may be said to reflect the cost of the denial. Needless to say, however, it would not constitute a cost to the producer. See p. 393, *infra*.

[12] From the standpoint of economic theory, the major inadequacy of this reply lies in its failure to answer the question why the adjustment of public utility prices to costs of producing the services should not take place, as it is supposed

First, cost of service is itself affected by the values that individuals place thereon and hence by the demands that they make for services of different kinds and in different amounts. By way of illustration, consider the practice of electric and gas utilities in charging higher rates for service supplied at peak-time hours or seasons than for otherwise identical service at off-peak periods. Differential pricing of this nature is quite justified—indeed, it is required—by strict cost principles since increments in the output of service during peak periods impose additional capital outlays not imposed by increments in output produced at times when the plant is partly idle. But these cost differentials would not exist except for differences in the consumer demand for service at different periods of time.

Secondly, the lack of a coordinate status for cost and "value" factors in rate determination is also partly attributable to the lack of symmetry between the position of the producer and that of the consumer of public utility services. Unlike the situation prevailing in unregulated business, the producer is under a duty (qualified, to be sure) to supply services to all applicants in its franchise territory at the approved rates of charge and in whatever amounts the applicants are ready to pay for. On the other hand, the consumer of public utility service, like the purchaser of ordinary commodities, has the option to take as little or as much service as he desires to take at the filed rates. Having this option, he will presumably refrain from buying any service at rates in excess of its value to him. Under these conditions, any attempt to abandon a cost-of-service standard in favor of a value or benefit standard of rate making would have absurd results. Consumers to whom the service is of little value might then have the right to demand that the company supply it to them at this low value even though the cost of production might be ten times as high!

Thirdly and finally, in actual rate cases the cost principle is always given modified interpretations which, while not converting it into a value principle, take indirect account of the effectiveness of the cost incurrence in contributing to the benefit of the consumers. Thus, in principle at least, costs are subject to compensation only

to take place without regulation but under competition, *indirectly* through increases and decreases in the output of the services rather than by directly through price control. In short, why should not regulation substitute output control for direct price control? On this point, see pp. 99 and 335.

if prudently incurred; and thus, with some exceptions, a "fair rate of return" is allowed only on those capital outlays still embodied in properties "used and useful in the public service." [13] The refusal of public utility law to guarantee against losses, combined with the allowance of an opportunity to earn a reasonable profit for successful risk taking, are also related, though in a very crude way, to the principle of payment for benefit received and not merely for costs sustained. Finally, the tendency of commissions to base the allowance for these capital outlays on the original construction costs of the plant and equipment, even if the properties have later been sold to the present accounting company at higher or lower prices, rests on the principle that the only capital entitled to compensation is the capital usefully devoted to the service of the public.

[13] See pp. 173 and 214, *infra*. For a brilliant exposition of the relationship between the cost principle of rate making and a value, or service-rendered, principle, see Justice Jackson's dissenting opinion in Federal Power Commission v. Hope Natural Gas Co., 320 U.S. 591, 628–660 (1944).

VI

COMPETITIVE PRICE AS A NORM OF RATE REGULATION

Before turning in the next chapter to those unorthodox principles of rate making often called "social theories," we may consider the merits of a general standard of reasonable rates that has received at least verbal support both from public service commissions [1] and from public utility spokesmen.[2] This is the standard of the hypothetical competitive price. Regulation, it is said, is a substitute for competition. Hence its objective should be to compel a regulated enterprise, despite its possession of complete or partial monopoly, to charge rates approximating those which it would charge if free from regulation but subject to the market forces of competition. In short, regulation should be not only a substitute for competition, but a closely imitative substitute:

[1] "The purpose of regulatory policy, in the protection which it is designed to afford the consumer, is to simulate and substitute the effects of competition and give the consumer the benefit which he would derive from a system of competition." From National Association of Railroad and Utility Commissioners, Report of the Committee on Progress in Public Utility Regulation, *Proceedings of the 53d Convention* (Washington, D.C., 1942), p. 369. Many commission opinions and a number of judicial dicta could be cited to the same effect.

[2] "The justification for the regulation of utility rates being the absence of competition, regulation should aim to produce the same results as competition would do if competition were economically desirable and feasible—neither more nor less." Brief for the New York Telephone Company in a case insisting upon its right, under the New York statutes, to receive the benefit of a "fair value" rate base that gives weight to reproduction costs. New York Telephone Co. v. Public Service Commission, 142 N.Y. Supp. 569 (1956). The brief cites, as supporting a competitive-price standard of reasonable pricing, Walter A. Morton, "Rate of Return and the Value of Money in Public Utilities," 28 *Land Economics* 91–131 (1952) and Arthur H. Dean, "Provision for Capital Exhaustion under Changing Price Levels," 65 *Harvard Law Review* 1339 (1954). See also C. P. Guercken, "An Economic Appraisal of the Private Electric Utility Industry," 55 *Public Utilities Fortnightly* 739–754 (1955).

This is a most intriguing proposition in view of the contention, familiar to economists, that competitive prices are optimum prices. One of its possible virtues is that it may offer definite answers to two formidable sets of questions raised in the preceding chapters: first, questions as to the relevant definitions of "cost of service" and "value of service"; and second, questions as to the respective roles of cost factors and of value or demand factors in price determination. Should "cost," for example, be taken to mean original cost or replacement cost, marginal cost or average cost, sunk cost or escapable cost? Let these and similar questions be resolved by a study of the types of costs that govern competitive-price determination. Should differences in rates of charge for different classes of service be based entirely on cost differences or should they depend in part on "value" differences (differences in the price elasticity of demand for the respective services)? Again, let the answer depend on the question whether firms producing multiple products under competition can and do practice price discrimination. And so on with respect to all of the other debated issues of rate-making policy.

During the postwar years of inflated price levels, the defense of a competitive-price standard has come largely from spokesmen for investor interests or for the public utility companies, who object to an original-cost rule of rate making on the ground that it unfairly deprives utility stockholders of the hedges against inflation said to be enjoyed by the owners of equities in unregulated enterprise. This is a forcible objection, the merits of which will be discussed in the chapters on the rate base and the "fair rate of return." But one may surmise that the alternative of a competitive-price norm would lose its charm for many of these writers were they to face the full implications of its adoption. In a dynamic economy, unrestrained competition is supposed to be a pretty tough game, often leading to individual or corporate bankruptcy.[3] "Fairness," in the sense of protection against the loss of hard-earned savings, is not one of its many virtues. Be that as it may, the view that a regulated monopoly should be induced or coerced to charge whatever rates would prevail under competition is so frequently urged, and so plausible in its appeal to economists, that it deserves

[3] "No war, no strike, no depression, can so completely destroy an established business or its profits, as new and better methods, equipment and materials in the hands of an enlightened competitor." From a statement by the Society for the Advancement of Management, repeatedly quoted in the issues of the magazine *Systems*.

sharp analysis. By way of introduction, let us therefore assume its acceptance and ask what rules of rate control would be required in order to put it into effect.

ASSOCIATION OF COMPETITIVE PRICE WITH REPLACEMENT COST

The popular conception of a competitive-price standard of rate making is much more primitive than any of its more sophisticated modern versions. But it is worth a brief exposition, since it probably still reflects the views of most practical rate makers who invoke the standard. It is derived from a simple exposition of the forces of competitive price determination to be found in the earlier textbooks on economics.

Under competition, then, prices are supposed to be determined proximately by the forces of supply and demand, and at a point set at the intersection of the curves representing these two offsetting forces. At any given time, the competitive market price may be much higher or much lower than the unit cost of production, and especially the unit *normal* cost. But any such discrepancy indicates a state of disequilibrium, which will tend to be corrected, even if the rate of demand for the product remains the same, by the expansion or contraction of output on the part of producers, acting under the attraction of profits or under the discouragement of deficits. Thus, there is a tendency for a competitive price to come into correspondence with normal cost of production; and this normal cost is sometimes said to represent the "normal value" of the product as distinct from its current market value.

Under the competitive-price standard of rate regulation, in its popular form, the object of regulation is deemed to be that of making the prices charged for public utility services conform to normal costs or normal values. But under regulation, just as under competition, these normal costs are not set by the actual or historical costs incurred by any given company by virtue of the prices paid in the past for the construction or acquisition of plant and equipment. On the contrary, the relevant costs are supposed to be the current and prospective replacement costs, since these are the only costs that would guide the action of competitive producers in their future decisions to expand or contract their output.

Thus, the competitive-price norm is brought to the support of a special version of the cost-price principle of rate making—a replacement-cost version that came to be closely associated, even though not completely identified, with the "fair-value" rule of rate regulation discussed at length in Part Two of this book. In the academic field, its most vigorous supporter has been Professor Harry Gunnison Brown of the University of Missouri, who for years waged a valiant fight on its behalf in the economic journals and against Justice Brandeis's increasingly popular "prudent-investment" principle.[4]

Properly qualified, the above-noted view of the nature of a competitive price, which led Brown and others to associate it with a price measured by replacement cost, cannot be said to have been discredited by modern price theory. But this theory has undergone developments that throw grave doubt on the relevance of a replacement-cost basis of rate control in its familiar sense.

Among these developments has been one which distinguishes important *types* of competition and which therefore destroys the earlier, simple distinctions between competitive and monopolistic pricing. The modern defense of any competitive-price norm of rate making therefore faces the necessity of defining the competition which it seeks to emulate. Is the proposed model that of "pure" or "perfect" competition? Or is it one of those mixed kinds of practical or "workable" competition typical of large-scale industry in this country?

Although few writers outside of academic economics have yet recognized the necessity of answering this question, those public utility representatives and public service commissioners who express approval of a competitive standard almost certainly have in mind a type of competition associated with fairly large-scale industrial companies, and not a type approximating the economist's concept of pure competition. Nevertheless, it will be worth while first to consider the implications of the standard of *pure* competition—competition bereft of monopoly elements—since this type is the only one that has been claimed by economists to offer a model of optimum pricing.

[4] See pp. 226–236, *infra*.

THE STANDARD OF PURE OR STRICT COMPETITION

For present purposes, pure competition may be defined as competition under which there exists no collusion among producers and under which no single producer controls a portion of the potential supply sufficiently large to give him any appreciable influence over the market price of his product.[5] It is "competition among the many." Here, unlike the situation of "competition among the few," each producer must accept the market price of his product as "given"; he has no opportunity to practice a "price policy."

The principles of price determination under assumed conditions of pure competition, as set forth in the modern textbooks,[6] are fairly complex, and I shall not here undertake even a brief summary. Instead, I propose to consider what rate-making criteria seem to be required by an attempt to make the rates that are in fact charged by a monopolized enterprise behave as if they were determined solely by the market forces of competitive supply and demand.

RATES SHOULD CORRESPOND TO PRODUCTION COSTS ONLY UNDER CONDITIONS OF EQUILIBRIUM

It is an elementary principle of competitive-price theory that the influence on price of cost of production is indirect and that the cost-price relationship states merely a condition of static equilibrium. At any given time, prices are set by the offsetting forces

[5] The terms "pure" and "perfect" competition are used sometimes as mere synonyms but sometimes with distinctions. Those who draw a distinction use the former term to denote any competition completely devoid of monopoly elements, reserving the latter for competition which, in addition to its purity, has other attributes of "perfection," notably, perfect two-way mobility of the factors of production (including contractability and expansibility of plant capacity in immediate response to changes in demand), and perfect foresight (itself a dubious concept). See Edward Chamberlin, *The Theory of Monopolistic Competition,* 6th ed. (Cambridge, Mass., 1950), p. 6; John Maurice Clark, "Toward a Concept of Workable Competition," 30 *American Economic Review* 241–256 (1940). As a norm of regulated monopoly pricing, "perfect competition" would make no sense. The objective of sound competitive rate-making policy should be to make the best use of whatever plant happens to be available in view of its noncollapsibility. And there is no point in assuming a greater degree of foresight than intelligent people can hope to enjoy at the time of a rate case.

[6] See, e.g., George J. Stigler, *The Theory of Price,* rev. ed. (New York, 1952), Part I.

of supply and demand and are likely to be materially above or below their normal production costs.[7] While they are *tending* in the direction of these costs, the adjustment of prices takes time and synchronizes with an adjustment of output. This lag in the adjustment process is not thought of as a defect of competition. On the contrary, a temporary disparity between cost and price is an essential device whereby the forthcoming supply of the commodity is brought into harmony with the demand.

It follows that a revision of the orthodox principles of rate regulation would be called for in any strict application of a competitive-price philosophy. Instead of the principle that utility rate levels should be raised or lowered so as to yield operating expenses plus a normal rate of profit, year after year, there must be substituted some rule of rate making which more closely emulates the competitive forces of supply and demand—forces under which rates will yield highly abnormal profits when there is a shortage of existing plant capacity, and under which rates will fall to mere short-run incremental or marginal costs of service when plant capacity is temporarily redundant. This objective can no more be accomplished under a replacement-cost or "fair-value" rule of rate making in its traditional form than under an original-cost or prudent-investment rule.

The drastic import of this aspect of a strictly competitive-price standard of public utility regulation will be apparent if we consider its probable consequences, first, in a period of severe business depression and, secondly, in a period in which the growth of plant capacity has failed to keep pace with the demand for the service forthcoming at "normal" rates of charge. During a depression, the forces of competitive supply and demand, if operating under conditions of strict competition, would soon bring rates down to temporary, marginal costs so low that the resulting revenues would probably bankrupt companies capitalized in the manner typical of public utility capital structures. On the other hand, during a period of plant shortage, rates might need to be raised drastically in order to preclude the necessity for rationing or in order to avoid a long waiting list of unsatisfied potential consumers.

[7] See, e.g., Alfred Marshall, *Principles of Economics*, 6th ed. (London, 1910), Book V. At pp. 401–402 of this great classic is an interesting comment on the invalidity of the assumption that, under competition, the market price at any given time will be likely to approximate reproduction cost.

IF MARKET-CLEARING RATES YIELD EXCESS PROFITS, A COMMISSION SHOULD COMPEL THE EXPEDITIOUS ENHANCEMENT OF PLANT CAPACITY

Under the prevailing rules of public utility regulation, decisions as to the proper policies of plant expansion are seldom made by the regulating agency. Instead, they are made by the management of the public utility company, motivated partly by the expectation that the investment in the larger plant will at least pay for itself, but partly by the legal obligation to supply, without unreasonable delay, all services demanded at the scheduled rates of charge.

This regulatory policy of leaving the responsibility for plant expansion in the hands of the corporate management would need to be abandoned, or at least modified, in an effort to make monopoly pricing behave like competitive pricing. For as long as plant capacity is inadequate, competitive rates will yield excess profits —profits that will fall to normal (or to zero under *some* definitions of profits) as soon as the plant has been enlarged to proper size. As trustees for investor interests, corporate managements would therefore be under impelling pressure to retard plant expansion in order to continue the sale of services at their high, market-clearing prices. Nor would the obligation to satisfy all prevailing demand suffice as an offsetting pressure; for this very demand will be kept from becoming embarrassingly heavy by the establishment of the market-clearing prices.

Thirty years ago, the point that the enforcement of a competitive standard of rate making must take place primarily through governmental control of the *investment* activities of a public utility company, rather than directly through control of rates under the rule of a "fair rate of return," was developed by Professor Bruce W. Knight in an article entitled "Why Not Regulate Investment Instead of Return?" [8] In a commentary on this article, published in the same periodical, I took issue with Knight's contention that such a change in the rules of regulation would be desirable and feasible. But there can be little doubt that control of investment would be required by a strict, competitive-price rule of rate making.

[8] 6 *Public Utilities Fortnightly* 406–419 (1930). Compare the proposals by Lerner and others for competitive-price simulation under outright socialism. A. P. Lerner, *The Economics of Control* (New York, 1947), especially Chaps. 5 to 7.

RATES WOULD BE REQUIRED TO EQUAL BOTH AVERAGE COSTS AND MARGINAL COSTS

Under conditions of competition, both the prices and the outputs of commodities are supposed to tend to reach an equilibrium position in which the prices are equal to average unit costs of production. But, in this same position, they are also supposed to equal their marginal or incremental costs—the additional unit costs of producing them at enhanced rates of output. That is to say, under competitive equilibria, average costs and marginal costs coincide.[9]

But this requirement of correspondence of prices alike to average cost and to marginal cost presents a dilemma under monopoly-price regulation. For, if the monopoly is operating under conditions of decreasing cost with increasing size—a condition often assumed to be characteristic of the public utility industries—it will be impossible to bring rates into accord with the average costs of the service without making *some* of these rates, at least, higher than marginal costs. One might as well attempt to draw a square circle! Hence, the very type of cost behavior that precludes the maintenance of *actual* competition in the public utility industries may also preclude the application of a standard of *hypothetical* competition.

THE RELEVANT COSTS WOULD BE FUTURE COSTS, NOT "SUNK" COSTS

Here we have the basis for the popular assumption that the competitive-price standard of public utility regulation calls for the fixation of rates at replacement costs rather than at levels that will yield a fair rate of return on the original costs of the utility assets. Under the assumed conditions of pure competition, the only costs that govern the actions of competing producers in their decisions to increase or decrease output are those costs that are still

[9] Under the leadership of the great British economist Alfred Marshall, competitive-price theory has developed the concept of multiple equilibria, under which the earlier simple division between short-run (or virtually instantaneous) and long-run price determinants has been superseded by a recognition of a series of short, longer, and still longer runs, depending on the time required for more or less complete readjustments of plant capacity and of rate of plant output designed to meet prevailing demands. This complication will not be introduced in the present chapter, although it is important for utility-rate theory, especially so in raising the question whether the relevant marginal costs are long-run or short-run marginal costs. See Chap. XVII.

under the producers' control. A competitive price has no tendency to rise and fall in such a way as to cover costs of production already irretrievably incurred—not, that is, unless these sunk costs happen to coincide with anticipated escapable costs.

However, this principle of competitive pricing, characterized by the aphorism that "in commerce bygones are forever bygones," by no means supports the replacement-cost or "fair-value" basis of rate making in its orthodox, American sense. For, in the first place, replacement cost has here been identified with the estimated cost of a substantially identical plant rather than with the estimated cost of replacing the service by the most economical modern substitute. And in the second place, the replacement-cost or "fair-value" principle of rate making, in undertaking to make total corporate revenues equal total replacement costs of service including a fair rate of return, ignores the other requirement of competitive-price equilibria—that the specific rates should equal the marginal costs of the specific services.

ALL RATE DISCRIMINATION WOULD BE OUTLAWED

According to the treatises on price theory, the practice of price differentiation based on "value" or demand-elasticity differences rather than on cost differences would be impossible under conditions of strict or pure competition. This practice constitutes discrimination; and the power of a seller to discriminate, with profit to himself, is held to be limited to sellers possessing at least some degree of monopoly power.

Even under existing regulation of railways and public utilities, the law places limits on the right of a company to practice rate discrimination. But what the law forbids is merely certain types or degrees of discrimination which, for one reason or another, are deemed adverse to the public interest; for example, so-called "personal discrimination." No such distinction between just and unjust forms of discrimination could persist in a thoroughgoing attempt to apply competitive-price theory to railway and utility rates. The whole practice of rate fixing based even in part on "what the traffic will bear" would have to be outlawed.

THE RATES OF RETURN SHOULD CORRESPOND TO THE PROFIT-AND-LOSS DIFFERENTIALS OF A COMPETITIVE ECONOMY

In a dynamic economy, the function of competition is by no means limited to that of bringing about a more or less gradual adjustment of prices to costs of production. An even more important function is that of stimulating innovations and improvements in products and in techniques of production.[10] To this end, competitive business is not business conducted strictly at cost, nor even at cost plus a "fair" profit. Instead, it is business transacted at prices temporarily yielding very high rates of profit under efficient or lucky promotion and management and yielding substandard returns or even operating deficits under inefficient or unfortunate operation.

Under any regulation designed to make the prices charged by regulated monopolies perform in the manner in which competitive prices are supposed to perform over the years, public service commissions would face the problem of setting these prices so as to approach the differential profit-and-loss status of competitive industry and so as to break away from the fairly standardized "normal profit" status of orthodox regulation. This is a truly formidable assignment, and one which, to the best of my knowledge, has never been accomplished effectively. To be sure, schemes of differential returns, designed to reward highly efficient or highly successful performance, have been tried out from time to time, occasionally with fair success during periods of stable price levels. But even if experience with these schemes had been such as to warrant general adoption, their acceptance would provide a very weak imitation of the behavior of actual competition in a period of rapid technological development. Something far more radical would be imposed by the standard of simulated competition.

[10] During his chairmanship of the Federal Power Commission, Leland Olds stressed this function of a competitive price and suggested that commissions should undertake to emulate it. "Regulation," he wrote, "if it is to be a worthy substitute for competition, must similarly be able continuously to make it impossible for a public utility to charge prices higher than it could charge if an efficient and economic competitor could reasonably be expected to enter the field and capture the market." *20th Annual Report of the Federal Power Commission* (1940), p. 13. As a means toward the accomplishment of this objective, Mr. Olds has looked sympathetically on the use of Tennessee Valley Authority rates and other public-plant rates as "yardstick rates."

Incomplete as is the foregoing summary of the major implications of a purely competitive-price standard of public utility rate making, it should suffice to indicate what drastic changes in the established principles of rate regulation would be called for by any thoroughgoing attempt to embody the standard in practice. Certainly the standard does not comport with an original-cost or net-investment basis of rate control. But neither does it comport with the use of a fair-value or replacement-cost rate base in any familiar or legally accepted sense of these terms. The popular assumption that a strictly competitive price is a price equal to replacement cost must be rejected, even as a rough approximation.

But what about a possible contention that all of our orthodox systems of regulation, whether of the original-cost or of the replacement-cost variety, should be discarded in favor of a scheme of rate control designed to simulate the forces of a competitive market? Such a contention has been made; but it must face formidable objections. In the first place, there is the dilemma presented by the fact that a condition of competitive-price equilibrium is one in which the price is simultaneously equal to average cost and to marginal cost (not to mention long-run and short-run varieties of each of these costs). In the second place, there are the serious practical problems to be faced by a commission if required to dictate how far and how fast a company must go in expanding its plant—a necessity noted in an earlier paragraph. In the third place, there are the difficulties of corporation finance presented by any scheme of rate making yielding the highly variable rates of profit and loss characteristic of competitive industry in a dynamic economy. And in the fourth place, there are the problems that would be faced by consumers of public utility services in adapting themselves to the frequent and rapid changes in rates imposed by competitive responses to changes in current demand and supply.

What these difficulties suggest is that the very characteristics of a public utility business which rule out reliance on actual competition as an automatic price regulator also rule out attempts closely to emulate the behavior of competition in the control of monopoly prices. Indeed, so far as concerns pure or strict competition, which is the standard now under review, modern economists seem to be in general agreement that this "atomistic" type of competitive behavior is not even applicable to the large industrial companies. Here, the competition which must serve as an alternative to regulation or to public ownership is that of a mixed type, which Professor John Maurice Clark has called "workable competition." [11]

THE ALTERNATIVE STANDARD OF "WORKABLE" COMPETITION

Mindful of the serious objections to a competitive-price standard of rate control when defined as a standard of strict or pure competition, we may now turn to the alternative standard of "realistic" or "workable" competition—the type of competition that prevails in real life and that characterizes industries which, because of their size and their relatively heavy capital investments, would seem most nearly comparable to the large public utility companies. As economists have pointed out, competition of this nature is far from "pure," since all large industrial companies possess important attributes of monopoly status.

There is at least a fair prospect that, at some future time—say within the next two decades—standards of socially acceptable "workable competition," covering, among other things, rules of competitive price determination, may have been developed to the stage at which they can serve, with important qualifications, as norms of public service regulation. This stage, however, has not yet been reached, nor does its attainment appear to be just around the corner. There are three related reasons for this cautious conclusion.

In the first place, too little is known today as to the nature of price determination by unregulated industrial companies. It is generally agreed that the "administered prices" of large-scale in-

[11] "Toward a Concept of Workable Competition," 30 *American Economic Review* 241–256 (1940). Paradoxical as it may seem, there is good reason to contend that, both with the public utilities and with heavy industry, a closer approach to strictly competitive price determination could be attained under outright public ownership than under private ownership. Indeed, one school of socialists, the so-called "market socialists," has defended government ownership as making possible the realization of a system of "optimum prices" similar to that which economists have associated with pure competition. See, e.g., A. P. Lerner, *The Economics of Control,* Chap. 7. These contentions, however, involve a dubious assumption: that the ideology of a socialist state and the attitudes of the aggressive types of people who would probably control such a state would be friendly to the principle of consumer sovereignty.

dustry are far less volatile and far less responsive to the forces of short-run supply and demand than would be their behavior under pure competition. But this is a negative conclusion; as to the positive factors in price determination, the subject remains highly controversial.

In the second place, enough is known about modern industrial price policies to belie the assumption that these policies can be reflected by the adoption of any simple, feasibly administered, rule of rate making such as the rule that rates should be based on replacement cost of service. Indeed, these policies do not conform to any single theory of price determination, coordinate with pricing under pure competition. They are the outcome of a whole range of inter-firm relationships intermediate between strict competition and strict monopoly. When modern economists, in attempting to rationalize the price strategy of "competition among the few," are led into an elaborate mathematical analysis called "the theory of games," [12] their findings do not offer very promising material for decisions in rate cases!

And in the third place, since the competition of the type supposed to govern unregulated industrial pricing has no claim for recognition as resulting in the socially optimum prices, emulation on the part of a regulating commission would be of doubtful wisdom even if a fair job of mimicry were feasible.[13] One must remember that the attempts by current economists to develop standards of workable competition for the purpose of antitrust law administration are not attempts to create standards of optimum pricing. On the contrary, they are attempts to secure types of com-

[12] John Von Neumann and Oskar Morgenstern, *The Theory of Games* (Princeton, 1944); J. D. Williams, *The Compleat Strategyst* (New York, 1954). The different price policies prevailing in different types of unregulated industry are presented and compared in Professor Walton Hamilton's book, *Price and Price Policies* (New York, 1938). In his *Managerial Economics* (New York, 1951), pp. 400–401, Professor Joel Dean notes important differences in price policies, not just among different companies or industries but even in the same company with different products. He cites the pricing of the various DuPont products as a conspicuous example. The highly controversial nature of the theory of imperfect or impure forms of competition is illustrated in the discussion of "Concepts of Competition and Monopoly" by Messrs. Clark, Weintraub, Machlup, Gordon, and Ackley at the Dec., 1954, annual meeting of the American Economic Association, published in 45 *American Economic Review, Proceedings* 450–490 (May, 1955).

[13] On the other hand, as long as nonutility prices fail to represent *pure* competitive prices, the fixation of public utility rates at such prices could not be claimed to yield optimum results.

petitive pricing good enough to render price control unnecessary.

Since the competitive-price standard of rate regulation has so often been identified with the acceptance of a replacement-cost or "fair-value" principle of rate control, one may raise the question to what extent the types of competition characteristic of large-scale industrial companies have actually brought prices into rough correspondence with current costs of production plus a normal rate of profit on the depreciated replacement costs of plant and equipment. This question is unanswerable in the absence of wide-scale and careful appraisals of industrial plant and equipment comparable to the tremendously expensive "physical valuations" of the American railroads made by the Interstate Commerce Commission under the Valuation Act of 1913. I think it almost certain, however, that the correspondence would not be close.[14]

Lest the reader of this chapter gain the impression that it is intended to deny the relevance of any tests of reasonable rates derived from the theory or the behavior of competitive prices, let me state my conviction that no such conclusion would be warranted. On the contrary, a study of price behavior both under assumed conditions of pure competition and under actual conditions of mixed competition is essential to the development of sound principles

[14] During the years since the Second World War, prior to the time of the recent stock-market boom, the stocks of many of the best-known industrial companies sold at market prices below their book values, values in turn presumably well below depreciated replacement costs. The steel industry offers a conspicuous example. In testimony before the Senate Banking and Currency Committee in 1955, Mr. Benjamin F. Fairless, chairman of the United States Steel Corporation, was reported to have stated that the current cost of building fully integrated steel-plant capacity from mines to finished product was on the order of $300 per ton, whereas the investor valuation of the plants, as measured by current security prices, was only $56 per ton for the ten largest steel companies, on the average. In its 1954 Annual Report to Stockholders, the Marquette Cement Manufacturing Company stated that, in 1953, it had earned 9.6 per cent on its "original-cost value" but that these earnings represented a return of only 3.6 per cent on estimated reproduction cost (after adjustments for additional depreciation charges on this higher cost). No doubt these and other examples of substandard returns based on replacement-cost tests could be matched by examples of superstandard returns.

The Feb., 1955, issue of *The Exchange*, a monthly publication of the New York Stock Exchange, reported that a study of 1,053 listed common stocks disclosed that 42 per cent were selling at less than their latest available book values. At the extremes among the separately noted industrial stocks, Armour and Company common was selling at 68 per cent below book value, whereas International Business Machines common was selling at 588 per cent above book value.

of utility rate control. Not only that: any good program of public utility rate making must go a certain distance in accepting competitive-price principles as guides to monopoly pricing. For rate regulation must necessarily try to accomplish the major objectives that unregulated competition is designed to accomplish; and the similarity of purpose calls for a considerable degree of similarity of price behavior.

Regulation, then, as I conceive it, is indeed a substitute for competition; and it is even a partly imitative substitute. But so is a Diesel locomotive a partly imitative substitute for a steam locomotive, and so is a telephone message a partly imitative substitute for a telegraph message. What I am trying to emphasize by these crude analogies is that the very nature of a monopolistic public utility is such as to preclude an attempt to make the emulation of competition very close. The fact, for example, that theories of pure competition leave no room for rate discrimination, while suggesting a reason for viewing the practice with skepticism, does not prove that discrimination should be outlawed. And a similar statement would apply alike to the use of an original-cost or a fair-value rate base, *neither* of which is defensible under the theory or practice of competitive pricing.

This chapter has been written under the assumption that the utility subject to regulation enjoys a monopoly, so that any emulation of competitive-price behavior would have to be imposed by governmental authority or adopted as a matter of policy. But this assumption is never strictly valid; and in the field of intercity transport, the degree of railroad monopoly has now become so limited because of road, water, and air competition, that the acceptance of a competitive-price standard of rate control, in some sense of competition, would cease to be the acceptance of a mere make-believe. While the complete abandonment of rate regulation is even here out of the question, the development of new and less rigid standards of rate control seems necessary. In this development, more is to be said for standards suggested by modern ideas of "workable competition" than can be claimed for such standards with the more nearly monopolized utility companies.

So far as concerns the electric power utilities, competition in the sense of rate making by a comparison of the performance of other utility enterprises, including public "yardstick" plants, has been favored by spokesmen for consumer interests. This is not competition as the term is used in economics; but it has promising possibilities for limited and cautious use, and both the promises and the limitations will be noted in Chapter XV, on the fair rate of return.

VII

SOCIAL PRINCIPLES OF RATE MAKING

Despite the failure of the familiar rules of public utility regulation to result in the same rates that might be expected to emerge "automatically" under actual competition, this failure does not necessarily imply a fundamental difference between the objectives of rate regulation and the recognized functions of a competitive price. In large measure, at least, the different results are imposed by technical obstacles in the way of any attempt to compel natural monopolies to behave contrary to their nature.[1] Regulation can still be regarded as a substitute for competition—probably as an inferior substitute.

But the statement that regulation is a substitute for competition would be accepted only with qualifications by any writers aware of its full implications, whereas it would be rejected sharply by a minority of writers on the ground that "public policy," and not merely technological or administrative difficulties, justifies deliberate departures from "commercial" standards of reasonable utility rates.[2] Both the qualifications and the wholesale rejection are based

[1] But regulation is deterred by notions of *fair* prices and *fair* profits from going even as far as technical difficulties would permit it to go in emulating the somewhat ruthless forces of competition.

[2] For a strong defense of this minority position, see the article by Professor Horace M. Gray entitled "The Passing of the Public Utility Concept," noted on pp. 24–25, *supra*. A similar point of view was expressed by Mr. Louis P. Goldberg, former member of the New York City Council, in a letter to the New York *Times* opposing further increases in subway fares designed to make the riders pay the full costs of transit. *Times*, Feb. 8, 1952. Subway service, he contended, is a "social service"—even more completely so than education, health, housing, libraries, etc. The costs of supplying this service, he concluded, should therefore be apportioned on social principles.

I do not include among the advocates of "social" rate making those economists who contend that public utility rates should be set at *marginal* costs even if the resulting revenues would fail to cover total costs (see Chap. XX). For these econo-

on a variety of reasons often said to rest on "social" principles of rate making rather than on "economic" principles. Elsewhere I have questioned the validity of such an antithesis between the adjective "economic" and the adjective "social."[3] But we may here ignore this verbal issue in an effort to discover the intended meaning of, and the reasons for, the distinction.

THE MEANING OF SO-CALLED "SOCIAL" PRINCIPLES OF RATE MAKING

Earlier chapters have cited a number of examples of "social" principles of rate making, and other examples will be noted presently. Obviously, the term is used loosely and in a variety of senses. By and large, however, it refers to any policy of rate control designed to make the supply of utility services responsive to social needs and social costs, and rejecting as even tolerable measures of these needs and these costs the prices that consumers are able and willing to pay for the services and the money costs that the enterprise must incur in their production.[4] An extreme form of this policy is that which views the welfare of the nation as an end in itself rather than as an expression for the welfare of the people who compose the nation. But, at least in America and Western Europe, such a point of view is not typical even of those who lay most stress on the distinctions between money values and social values.

A study of the different statutes setting forth criteria of rate-making policy, together with a study of the rate cases coming before public service commissions and courts, would reveal many deliberate departures from cost-price standards based on considerations often called "social," although critics might be unkind enough to call them "political." Most of these departures are defended on either or both of two different grounds: on the ground of "ability to pay," and on the ground of "diffusion of community benefits."

---

mists accept the logic of the principle that competitive prices are optimum prices. Their heresy is not based on those distinctions between money values and social values implicit in a "social" theory of public utility rates. I have emphasized this point in an earlier paper: "Major Controversies as to the Criteria of Reasonable Public Utility Rates," 31 *American Economic Review, Proceedings* 379–389 (1941).

[3] Pp. 31–33, *supra*.

[4] See Hans Staudinger, "Social Rates in Electricity," *Social Research*, Aug., 1936, pp. 259–281; also his "Planning in Electricity," *Social Research*, Nov., 1937, pp. 417–439. But many of the principles cited by Staudinger as examples of social rate making seem to me to be justified on sound "business" grounds.

As a rule, the two grounds are merged in discussions of the actual practice of rate making. But they may be distinguished here for purposes of analysis.

THE ABILITY-TO-PAY PRINCIPLE

The "ability-to-pay" principle[5] is, of course, a transfer of a familiar canon of taxation to the realm of fair pricing. It refers to the contention that the rates of charge for public utility services should depend, in part, on the personal-income status of the consumers. This contention has had its most widespread influence on the rate structure rather than on the entire rate level of a public utility enterprise. But the practical difficulties in the administration of rate schedules which impose higher rates on rich people than on poor people, combined with a wholesome tradition against "personal discrimination," have sometimes led to the acceptance of below-cost rate levels, available alike to rich and poor consumers, and defended on the ground that the service is of a general type urgently needed by persons of low income. This defense was made for the long delay on the part of the City of New York in raising the subway fare above 5¢ and is still made as one reason against any attempt to make the subway system self-sustaining by going above the present 15¢ fare. As to rate *differentials* based on ability-to-pay factors, one may note the special railroad passenger fares favoring deserving persons of reputedly low income (for example, ministers of the gospel); the rule on the Paris Metro permitting wounded veterans and parents of more than one child to ride at special, low fares; and, in some American jurisdictions, the special electric rates for charitable establishments, hospitals, and public housing developments.[6] Among the electric and gas utilities, the

[5] An ability-to-pay principle of rate making should not be confused with a value-of-the-service principle. To be sure, both principles involve forms of price discrimination. But under the former principle, lower rates for the same kind of service should be charged to individuals of lower income regardless of the question whether or not the lower income corresponds to a higher "price elasticity of demand" for the service; whereas, under the latter principle, relative incomes of different classes of customers are taken into account only for their possible bearing on relative elasticities of demand. Indeed, since wealthy consumers may have feasible alternatives not available to poorer consumers, the prices that they could be made to pay for a utility service might well be even lower than the prices that could be imposed on poorer consumers.

[6] The City of New York, which at first opposed a petition by the Consolidated Edison Company asking leave to withdraw its special low-rate classifications for gas and electricity supplied to public housing projects, was reported to have

tendency of commissions to hold the charge for very small rates of consumption to amounts that fail to cover even directly traceable costs has been defended in part on the assertion, denied by company spokesmen, that small customers are typically poor customers.[7]

THE DIFFUSION-OF-BENEFITS PRINCIPLE

Taken by itself, the ability-to-pay principle of rate making rests merely on the contention (a) that public utility services are essentials rather than luxuries and (b) that persons of low income should not be deprived of essentials by any inability to pay the full costs of rendition. But what may be called the "diffusion-of-benefit" principle rests on a different rationale. According to this principle, the benefits derived by the community from public utility service are by no means limited to those persons who pay for the service either directly as consumers or indirectly as the purchasers of products made by the aid of their services. On the contrary, encouragement of full use of the service is in the interest of the entire community or of the entire nation. Moreover, the maximum community benefits are by no means necessarily secured by sale of different types and amounts of service at relative money costs. Some kinds of service may properly be sold at less than cost, and other kinds at more than cost, for the purpose of attaining social benefits or avoiding social costs not attainable under the principle of "service at cost" in an ordinary sense.

Diffusion-of-benefit arguments for breaches of a cost-price system of rate making are perhaps more frequently advanced in the field of transportation. Thus, in New York City and other metropolitan areas, the operation of the public transit systems on a subsidized basis has been defended partly as a means of relieving congestion of street traffic by underloaded private automobiles. And thus, rapid-transit fares that disregard distance of travel have been defended as a means of encouraging decentralization of population. A similar argument has been made in support of the establishment of uniform, "blanket" electric rates over wide areas of an entire

discontinued its opposition. New York *Times*, March 12, 1956. But in a case, still pending as this footnote is written, in which the Consolidated Edison Company has sought leave from the Public Service Commission to abandon the practice of "conjunctional billing" of adjacent buildings held in common ownership, strong opposition has been forthcoming from the New York City Housing Authority.

[7] See Hubert Havlik, *Service Charges in Gas and Electric Rates* (New York, 1938), Chap. 5.

state or region. Related to rate policies of this nature has been the tendency of the Interstate Commerce Commission to take intermediate positions between one establishing railroad freight-rate relationships which preserve for nearer sources of market supply the natural advantages of lower transport costs, and one designed to put the potentially competing sources more nearly on a par [8]—as, for example, with citrus fruits from Florida and California to the Northeastern markets. In a somewhat similar category belongs the same Commission's policy, much criticized by the railroads and by some economists, of forbidding railroads to make special rate reductions to meet road or water competition beyond whatever limited reductions will permit both types of carriers to maintain a "fair share" of the business—a noncompetitive notion of fair competition! [9]

Population-decentralization or industry-decentralization arguments for deviations from cost standards have often been associated with national-defense arguments, which have also been accepted as reasons for special subsidies (the merchant marine, the airlines, the early railroad land grants, etc.), justified on the ground that the national defense requires the construction of plants and the continuous operation of services that could not or should not be made to pay their full way by charges for peacetime consumption. Far less justifiable from the standpoint of an economist has been the use of national-defense arguments in support of requirements, imposed especially on railroads, to maintain noncost-recovering services, the burden of the resulting deficits falling either on the stockholders or else on the users of those other services that can be made to stand the loss. This practice amounts to a distorted use of social principles of rate making and would be criticized by many of the supporters of these principles.[10]

Other benefit-diffusion arguments have had more or less influ-

[8] I. L. Sharfman, *The Interstate Commerce Commission*, Vol. III B (New York, 1936), pp. 656–693.

[9] See "Problems of the Railroads," *Report of the Subcommittee on Surface Transportation of the Committee on Interstate and Foreign Commerce*, April 30, 1958. U. S. Government Printing Office, 1958. The Transportation Act of 1958 contains a provision adverse to the above-noted policy.

[10] "Common carriers cannot be expected to recover from peacetime freight revenues the cost of maintaining their equipment or facilities in readiness for defense needs; this should be financed by government from general taxation for defense, and not borne by carriers or shippers." Statement by D. E. G. Plowman, Vice President, Traffic, United States Steel Corporation, quoted by *Railway Age*, March 7, 1955, p. 10.

ence on American rate-making policies. Among these we may note the Interstate Commerce Commission's reluctance to let railroads quote cost-justified discounts for full-trainload shipments—a reluctance based on a desire to favor smaller-scale industrial companies; the familiar objections, based on health and sanitation grounds, against metered charges for city water supply; and a number of plausible arguments in favor of rates designed to encourage the use of electricity rather than alternative forms of energy,[11] one such argument running in terms of relative fire hazards. The more ardent supporters of low-priced public power, both in this country and in England, have gone to great lengths in their enthusiasm for the bountiful use of electricity, regarding the growth in power consumption as the prime mark of industrial progress and of modern living.[12]

For obvious reasons, the most unrestrained advocates of social theories of rate making are not to be found among the representatives of the private utility companies. Yet even these representatives sometimes adduce social-benefit arguments on behalf of the rate-making policies which they favor. At a conference on telephone rates which I attended several years ago, an official of the Bell System used such an argument in defense of the System's policy of charging rates higher in larger cities than in smaller communities, even beyond the point at which this policy could be justified strictly on grounds of relative costs. The benefit conferred by the

[11] See Hubert Havlik, *Service Charges in Gas and Electric Rates*, pp. 115–119, referring to the "public policy" that all buildings in a populated district should have electric service. In the first edition of his *Electrical Engineering Economics* (London, 1928), pp. 195–196, D. J. Bolton wrote: "The actual physical need for electricity, with its absence from fumes, etc., is greatest in the dark and overcrowded rooms of those who are at present least able to afford it."

[12] In an article on "The Price of Fuel," 6 *N.S. Oxford Economic Papers* 226–242 (1954), D. L. Munby refers to a memorandum by the British Electricity Authority ending "with a paean of praise of electricity as 'one of the most powerful tools yet devised by man to raise his standard of life.'" Munby states that, while the same Authority defended its rate-making practices as in harmony with the principle of consumer choice, the British Gas Council denied the existence of such a harmony in the relationship between rates for electricity and rates for gas.

In recent public statements, Mr. Philip Sporn, one of America's outstanding private-utility operators, has denied that the industrial progress of a nation should be identified with the development of cheap electric power. See his "Observations on Private Versus Public Power," 53 *Public Utilities Fortnightly* 717–733 (1954). But even if the cheap-power enthusiasts are right, what is important is that strenuous efforts be made to minimize the costs of producing and distributing electricity, not that the sale prices be held below costs by tax-financed subsidies.

installation of any one telephone station, he noted, is not limited to the particular subscriber. Indeed, the interest of the whole community demands the widest feasible access of families and of businesses to the interchange of messages. This objective, he urged, can best be attained by a system of differential rates of charge that do not place too much emphasis on relative costs of service.[13]

THE LIMITED RECOGNITION OF "SOCIAL" PRINCIPLES OF PUBLIC UTILITY RATES

Many other examples of "social" principles of rate making could be found by an extensive study of the rate cases. Transportation rates, especially, would provide a rich field for exploration. Yet I think it fair to say, as was said in the introductory chapter of this book, that American rate making has adhered in the main to the standard of service at cost, and that even most of the departures therefrom have been due to administrative, historical, and business reasons rather than to "social" reasons.[14] Indeed, if the social considerations were to become dominant, the enterprises to which they apply would cease to be public utilities in the accepted sense of the term. They would then become "socialized," like the public schools, the tax-financed or endowed universities, and (to a greater degree) the police, the courts, the navy, and the city-street departments.

[13] From the standpoint of the private telephone companies, this point of view has dangerous implications. For it seems to justify subsidized telephone service, and arguments for subsidies are almost sure to be associated with arguments for public ownership.

[14] Pp. 22–24, *supra*. The same general statement would apply to the rate-making practices of the nationalized British utilities. See J. F. Sleeman, *British Public Utilities* (London, 1953), who cites, as exceptions, numerous examples of "social" rate making. In a 1956 report on electric rates by a government committee, the British Electricity Authority was criticized for deviations from "business principles" including acceptance of "national-policy" reasons for its decision to carry insurance with British insurance companies instead of practicing self-insurance. *Report of the Committee of Inquiry into the Electricity Supply Industry*, Jan., 1956, Cmd. 9672, pp. 7, 57, 96–97. The Committee declared that, if "political" objectives are to govern electric-power policy, these objectives are the responsibility of the Ministry and of the Parliament, not of the Authority. It insisted that, if there were any need for a rural-electrification subsidy (which it doubted), the subsidy should be provided by Parliamentary grant, not by charges against non-rural consumers. Professor R. H. Coase notes the different philosophy of the earlier Committee on Land Utilization in Rural Areas (the "Scott Committee"), which insisted that the supply of electricity is an "essential service" and that it should be available at the same price to country people as to city people. 15 *The Manchester School* 139 (1947).

But why should not the public utility services be socialized in the sense just mentioned, as has been urged by a minority of writers? The short answer is that the reasons which justify their price regulation are not reasons which justify their socialization.[15] But a brief exposition of this answer is in order. For this purpose, it will be convenient to discuss separately the ability-to-pay factor in social-pricing theory, and the benefit-diffusion factor, even though in practice the two factors are often intermixed.

CRITIQUE OF THE ABILITY-TO-PAY PRINCIPLE

Taken by itself, the ability-to-pay argument for departures from a cost-price or competitive-price principle of utility rates stands on the same footing as would an argument for the differential pricing of *all* commodities and services of comparable necessity. What, then, are the reasons for its general rejection, even by those who support the analogous principle of taxation?

Aside from formidable difficulties of administration, the objections are twofold. In the first place, to the extent that the ability principle of pricing were to accomplish its purpose, to that extent it would convert unequal distributions of cash income into equal distributions of real income—a conversion that could not feasibly be confined to those cash-income differentials deemed undeserved or otherwise socially indefensible. But in the second place, if public policy favors a reduction in income discrepancies, this objective can be accomplished by alternative instruments of social control, including progressive taxation, social security, and free public education, more feasibly and less harmfully than by a system of discriminatory prices. In short, the more promising attack on the maldistribution of cash incomes lies in a more direct attack on the maldistribution or on its causes, not in the administration of antidotes.[16]

[15] This position is taken by Professor I. L. Sharfman with respect to transportation rates. See his *The Interstate Commerce Commission*, Vol. III B (New York, 1936), pp. 690–693. Sharfman insists that transportation costs should enter the prices of commodities and should motivate the location of industries just as much as any other production costs.

[16] This point of view, familiar to modern welfare economics and modern tax theory, is developed at length by A. P. Lerner in his *Economics of Control* (New York, 1947), especially Chap. 4. Lerner stresses the point that if the prices of commodities and services are made to deviate upward and downward from their (marginal) costs, in an effort to offset differences in individual incomes, the economic organization of society will lose efficiency because individuals will be under

To be sure, the above-noted objections to price discrimination in favor of low-income recipients have not deterred modern capitalism from putting it into limited application in public housing, in hospital charges, in medical fees, in educational scholarships, and elsewhere. But in most of these situations, the benefit-diffusion consideration has also been prominent. With public housing there is the further point that rentals constitute a very large fraction of a low-income family budget, unlike the typical situation with public utility services. But if standard rates of charge for these latter services are beyond the means of the occupants of subsidized housing, a persuasive case can be made for the assumption of a part of the charges by the housing authorities rather than for the concession by the utility enterprise of substandard rates, to be offset by surcharges imposed upon other customers.

CRITIQUE OF THE DIFFUSION-OF-BENEFITS PRINCIPLE

Taken alone, the case for the acceptance of ability-to-pay standards of public utility rates is a very weak case save as a jury-rudder expedient under emergency conditions. A far more persuasive case can be made for departures from cost pricing with respect to services, the performance of which serves important community needs in addition to those needs for which it is feasible to exact a price. The extreme examples, of course, are those of government services financed entirely by taxation rather than by sale, and therefore ruled out of the public-utility category by very definition. But intermediate examples can be found inside those enterprises traditionally called public utilities—examples in which a plausible case can be made for relieving the regular consumers from the full payment of those costs not incurred entirely for their benefit. Examples of this possibility have already been cited. But we may add another example: that of a national-defense reason for a rule re-

---

an inducement to buy things that are offered to them at below-cost prices, even though they would prefer other, higher-priced things that could be produced at no higher social costs.

But some recent economists have suggested that the limited power of a state to effect a direct redistribution of income by taxation—a limit made serious by heavy revenue requirements of a modern state—may justify a partial resort to the price system to take care of the underprivileged. See, e.g., Marcel Boiteux, 58 *Revue générale de l'électricité* 321 (1949). In his *Welfare Economics and the Theory of the State* (Cambridge, Mass., 1952), p. 127, Professor William J. Baumol cites several references on the question whether rationing of incomes is better than rationing of commodities.

quiring an electric utility company to maintain, say, a 20 per cent stand-by reserve even if a 10 per cent reserve would be deemed quite adequate for peacetime purposes. Here a forcible argument can be made for a national-defense subsidy sufficient to offset the special, national-defense cost.[17] Such a subsidy would not conflict with the old legal maxim, "Let the beneficiary bear the burden." On the contrary, it would be in harmony with this maxim.

Although social principles of rate making are usually advanced in support of *lower* public utility rates than would otherwise be warranted, the logic of these principles may work also in the other direction. For, as was noted in Chapter IV, the money costs of producing the service—the so-called enterpriser costs—may not cover losses imposed on innocent bystanders.[18] Moreover, the act of supplying service to any given consumer may impose indirect burdens on other consumers. Striking examples of this are traffic congestion on streets and highways and passenger congestion in local buses or subway trains. When any one passenger crowds into a subway train at rush hours, he adds to the misery of the other passengers in addition to becoming a fellow sufferer. Under the logic of a social theory of rate making, under which the rates should cover the social costs, this cost-diffusion factor might be taken into account through the imposition of higher fares than would be warranted on cost principles in the ordinary sense.[19] But popular opinion would not support any such use of the price system to relieve rush-hour congestion.

[17] See Benjamin S. Loeb, "Electric Power Supply and National Security; a Study in Public Policy," Ph.D. Dissertation, typescript, Columbia University, 1959. Reproduced by U.S. Atomic Energy Commission, Washington, D.C.

[18] Referring to Great Britain, D. L. Munby writes: "The recent interim report of the Committee on Air Pollution has estimated that the annual cost of pollution may be more than one hundred million pounds per year, apart from waste of fuel." 6 *N.S. Oxford Economic Papers* 239 (1954), citing Cmd. 9011, pp. 20–21. Compare a lower estimate for an earlier year cited by I. M. D. Little in *The Price of Fuel* (Oxford, 1953), p. 79. On social versus private costs in the development of natural resources, see Joseph I. Fisher, "Natural Resources and Technological Change," 29 *Land Economics* 57–71 (1953).

[19] Here, however, the "social" theory of rates, while justifying a departure from the cost-price principle in its usual sense, is in harmony with a strict competitive-price theory for reasons suggested by Professor Frank H. Knight in a now famous article taking issue with a position once taken by Professor A. C. Pigou in the early editions of *The Economics of Welfare*. That is to say, if transit could be, and were, supplied by unregulated companies under conditions of strict competition, their profit-maximizing objectives would lead them to charge differential fares that would preclude any excessive rush-hour use caused by the failure of each passenger to feel the hardship that his presence imposes on all other passengers. See Knight, "Some Fallacies in the Interpretation of Social Cost," 38 *Quarterly Journal of Economics* 582–606 (1923/24).

Neither the diffusion-of-benefits argument for deviations from ordinary cost pricing, nor its counterpart, the diffusion-of-costs argument, can be ruled out on principle in public utility rate making, as can the simple ability-to-pay argument except as a makeshift device. To what extent, however, such deviations are justified in actual practice is quite another question. The reasons for caution and skepticism in use are indeed forcible. First, there is the extreme difficulty of prophesying and measuring indirect social benefits and social costs. Secondly, and in the absence of objective tests, there is the certainty that exaggerated claims of community benefits will be put forward by pressure groups. And thirdly, there is the question whether the indirect benefits from the production of any given public utility service will be greater than those that would result from the alternative production of other commodities and services offered for sale at market prices that do not take social benefits into account. Unless this question can be answered in the affirmative, a reduction of utility rates only, made in order to allow for social benefits, would result in a distortion between these rates and other prices.[20]

If the tone of this chapter is somewhat less than enthusiastic for those views of rate-making policy said to be based on social principles, the coolness is not due to a belief that these views have no place in the theory or practice of public utility rate making. It is due, rather, to a conviction that those services now called public utility services belong in that great class of economic products, including both commodities and services, that can best be offered for sale instead of being supplied without charge, and that can typically best be sold on the general principle of service at cost rather than at prices designed by a legislature or public service commission to accomplish some specific objective deemed by it to be in the public welfare.[21] This conviction is supported by a recognition of the greater importance of socialized services in *other* sectors of a modern economy. Even aside from the terrific burdens

[20] In his *Welfare Economics*, p. 32, William J. Baumol makes the related point that, assuming a given level of output, the divergence between marginal social cost and marginal private cost need not cause competition to result in a non-optimum allocation of resources so long as the divergence is of equal strength in all industries.

[21] "The Court never before has confided to our regulatory body the reshaping of our national economy." Justice Jackson, dissenting, New York v. United States, 331 U.S. 284, 362 (1947).

of national defense, the Federal, state, and local governments will be called upon to devote increasing shares of the total national income to the production of services, such as those in the fields of health, education, and recreation, that cannot feasibly be distributed by the mechanism of the price system. In order to finance these needs for truly collective services, the tax system will be put under heavy strain, as indeed it is put today. Only for compelling reasons should the strain be enhanced by the inclusion in governmental budgets of vast appropriations for subsidized utility and transportation services.

But the viewpoint just set forth merely expresses a general attitude, which is that of a *rebuttable presumption* in favor of so-called "business principles" of rate making.[22] Even if modern rate-making policy continues to accept this presumption in the future, as it has in the past, it still faces the task of setting up conditions for successful rebuttal. This constitutes one of the most formidable assignments of modern welfare economics—that of developing principles of social valuation and of social-cost determination.

[22] This is a very different attitude from that of those extremely "conservative" writers who appear to treat the presumption as conclusive. Thus a Task Force of the "Second Hoover Commission," in its report on Federal power policy, criticized the Federal administration for having used water resources and power-development projects, "which should be undertaken exclusively for economic purposes, to accomplish indirect social and political ends." Commission on Organization of the Executive Branch of the Government (1953–55), *Task Force Report on Water Resources and Power*, Vol. I (Washington, D.C., 1955), p. 13.

VIII

FAIRNESS VERSUS FUNCTIONAL EFFICIENCY AS OBJECTIVES OF RATE-MAKING POLICY

Cutting right across that classification of rate-making criteria suggested by the titles of the last four chapters—cost of service, value of service, competitive price, and "social" principles—is a vitally important distinction between two general points of view from which a choice may be made in favor of any particular version of any one criterion. Looked at from the first point of view, the choice should depend on considerations of "fairness" or "equity" or "justice"—fairness as between investors and consumers when the general *level* of rates is at issue; fairness as among the different classes of consumers when the rate *relationships* are under inquiry. Looked at from the second point of view, the choice should depend on a variety of "practical" or "economic" considerations only indirectly related to the equities of the specific parties—considerations perhaps best characterized as those of functional efficiency.[1]

Thus, the reasonableness of a company's request for approval of an across-the-board increase in its rate schedules may be judged by reference to the question whether or not the increase is called for in simple fairness to the corporate investors, who have committed their capital to the enterprise in reliance on an opportunity to

[1] The first point of view is said to be derived from the medieval conceptions of a "just price," *justum pretium*. See the article "Just Price," by Edgar Salin, in the *Encyclopaedia of the Social Sciences*, Vol. 8 (New York, 1932). He writes: "The just price, on the other hand, as a conception and as a doctrine is basically ethical rather than economic." While not completely devoid of economic content, "it recognizes no validity for economic activity as such nor independent economic norms. Its law is derived from theological doctrines and from the philosophy of mediaeval class society."

enjoy a fair rate of return. But it may also be judged by reference to the question whether or not an enhancement in earning power, fair or unfair, is required in order to maintain corporate credit and hence in order to enable the company to supply adequate service. And thus, the question whether or to what degree rates for off-peak electric service should be lower than rates for service supplied partly on the system peak can be viewed as a problem of *equitable* cost apportionment between the two classes of consumers. But it can also be viewed as a problem of establishing whatever rate differentials are best designed to stimulate the optimum use of plant capacity as well as best designed to avoid the necessity of an uneconomical expansion of this capacity—a functional-efficiency problem.

In railroad and public utility regulation, most contested rate cases are argued by the opposing parties from both points of view. But many of the opinions of courts and commissions—the earlier opinions more so than some of the recent ones—might seem to support a popular assumption that the major concern of rate regulation is with the *fairness* of the rates. Indeed, fairness and reasonableness are often used as interchangeable terms. Forty years ago, a writer with an engineering background well reflected not only this use of language but also this implied mode of thinking when he gave to his entire book on public utility regulation the title, *What Is Fair?* [2]

Today, despite the persistent use of the words "fair" and "reasonable" as mere synonyms, no responsible writer or public service commissioner would attempt to develop or appraise rules for the determination of reasonable rates by sole reference to standards of fairness or equity. Yet there remain important differences in the relative emphasis placed by different persons on fairness considerations as compared to other considerations, together with even more important differences in the accepted standards of fairness. By and large, expert opinion has tended to place more and more emphasis on "practical" or "efficiency" factors, restricting the fairness issues to such problems as those concerned with retroactive rules of rate making.[3]

[2] William G. Raymond, *What Is Fair?—A Study of Some Problems of Public Utility Regulation* (New York, 1918).

[3] My colleague Professor Ernest L. Williams calls my attention to the shift in the Interstate Commerce Act from a standard of reasonable rate levels expressed

The academic economists have gone farthest in this direction, some of them taking the position that issues of fairness are beyond their professional competence and concern.[4] But even courts and commissions, which are naturally highly responsive to popular or traditional ideas of equity among the parties to a rate case, have shown a tendency, though a lagging one, to shift their emphasis. Here, as was noted in Chapter II, the lead was taken by the Supreme Court, which, in 1944, abandoned its earlier, legalistic test of fair rates in favor of the acceptance of a functional approach.

At the risk of being subject to the prejudices of my profession, I am convinced that the modern tendency to view fairness criteria of reasonable rates as secondary criteria, to be accepted primarily as constraints on the application of the so-called economic criteria, is a mark of progress in the development of rate-making policies designed to serve the public interest. But this means merely that fairness issues should be kept in their place. It does not mean that they should be cavalierly dismissed or even belittled. What these issues are, and what part they should play in the development of rate-making policies, are the questions discussed in the present chapter.

---

in terms of *fairness* (Transportation Act of 1920) to a standard which emphasizes the adequacy of the rates from the standpoint of maintaining an efficient railroad system (Act of 1933).

[4] For example, Ralph K. Davidson, whose monograph on rate discrimination insists that it is concerned solely with "economic" criteria and not with fairness. *Price Discrimination in Selling Gas and Electricity* (Baltimore, 1955), p. 111. Another economist, reviewing Davidson's book, comments adversely on lawyers "vainly seeking equity where efficiency should have been the objective." H. S. Houthakker, 46 *American Economic Review* 734–735 (1956).

The fairness of public utility rate levels was much disputed during the great depression of the 1930s, some writers insisting and others denying that, on grounds of equity, these rates should be reduced in harmony with the reduction in personal incomes and in other prices. While favoring flexibility, J. K. Galbraith, now Professor of Economics at Harvard University, took the position, characteristic of economists, that "fairness" was not the main issue. "No longer," he declared, "is public policy toward monopoly power concerned merely with fair prices. . . . It is also concerned with *harmonious* prices and *harmonious* price changes; the problem of prices which do not intensify and prolong fluctuations in the economy." "Monopoly Power and Price Rigidities," 50 *Quarterly Journal of Economics* 474 (1936).

Some modern "welfare economists," associating or identifying arguments about equity with arguments about income-distributive justice, have denied that the issues are subject to rational solution on the ground that they involve a hopeless attempt to make interpersonal comparisons of satisfactions. See Kenneth E. Boulding's article on "Welfare Economics" in Bernard F. Haley, ed., *A Survey of Contemporary Economics*, Vol. II (Homewood, Ill., 1952), pp. 1–34. But other economists disagree, e.g., A. P. Lerner, *Economics of Control* (New York, 1947), Chap. 3.

THE BABEL OF CONFUSION AS TO WHAT CONSTITUTES "FAIRNESS"

The broad distinction, noted above, between fairness standards and efficiency standards should not be taken to imply the existence of two alternative measures of reasonable rates, the one derived from considerations of fairness, the other derived from considerations of economics or of corporation finance. On the contrary, persons who think in terms of equity or justice are likely to be as hopelessly at odds with one another as they are with persons who think in terms of social efficiency.[5] A major part of the disagreement lies in the association of "fair" treatment of individuals with "equal" treatment. But, in the first place, equality in one respect means inequality in others.[6] And, in the second place, a decision on rates designed to put one set of persons on a par with another set will often make still others the victims of discrimination.

Consider, for example, the fairness aspect of the question whether, in a period of price inflation, permitted rates of return should be raised in such a way as to give utility common stockholders an offsetting increase in money income. Such a raise would put these stockholders more nearly on a par with the holders of stock in many industrial companies. But it would thereby give them an advantage over the recipients of incomes from investments in savings banks, in U.S. savings bonds, in annuities, and in life insurance policies, not to mention incomes from holdings in preferred stock and in limited-dividend housing projects. No wonder that, quite aside from practical considerations of corporation finance, opinions differ as to the fair claims of public utility stockholders.[7]

Consider, again, the fairness of a proposal for uniform electric

[5] "In the nature of their functions, value-symbols and moral injunctions must suffer from logical inconsistencies if they are to be part of working ideologies. If they were qualified to the point where they could be applied literally, they would lose their evocative force." Francis X. Sutton, *et al., The American Business Creed* (Cambridge, Mass., 1956), p. 264.

[6] One is reminded of the comments of Anatole France on the legal obligation of the poor under received ideas of social justice: "Ils y doivent travailler devant la majestueuse égalité des lois, qui interdit au riche comme au pauvre de coucher sous les ponts, de mendier dans les rues et de voler du pain." *Le Lys rouge* (Paris, 1894), pp. 117–118.

[7] Pp. 270–274, *infra*.

rates throughout both urban and rural service areas—a proposal made, let us say, in the face of undeniable evidence that the unit costs of supplying the rural service are very much higher than the unit costs of city service. Here, to be sure, there may be plausible *practical* grounds for rate uniformity if the costs of the rural service do not bulk too large in the total costs of supplying the whole area. But if the question is argued on grounds of fairness among customers, opinions are likely to be widely divided. On the one hand, there is a strong tradition in support of the fairness of rate differentials based on cost differentials.[8] But on the other hand, there is a widely held, conflicting belief in the "inherent fairness" of a rule of equal prices for services regarded as "the same" in some superficial sense and despite marked differences in cost of rendition. This popular preference for rate uniformity beyond the limits justified by the advantages of simple rate structures has been repeatedly noted by current economists, most of whom keenly regret its political appeal. Only under special circumstances can the demand for this spurious equality of treatment properly claim whatever merit can be claimed for ability-to-pay principles of pricing, or for those other "social" principles discussed in the preceding chapter.

Of all of the many problems of rate making that are bedeviled by unresolved disputes about issues of fairness, the one that deserves first rank for frustration is that concerned with the apportionment among different classes of consumers of the demand costs or capacity costs—those costs of service that are regarded as a function of required plant capacity and not of rate of output in kilowatt-hours, cubic feet of gas, ton-miles of traffic, etc. Should the capacity costs be assigned to the different consumers on the basis of system peak responsibility, or of coincidental class demand, or of any one of the other thirty-odd proposed bases of assignment to be found in the literature of rate theory? Here, notions of "fair apportion-

[8] "Man's inborn sense of fairness and equity seems strangely satisfied if he be assured that he is paying for a unit of energy only what it has cost to generate and deliver it." A. C. Marshall and H. A. Snow, "Distribution Costs—Residential Service," reprinted in Samuel Ferguson, *Public Utility Papers* (Hartford, Conn., 1947), I, 354–381 at 355. Compare Professor J. K. Galbraith's remark on farm subsidies: "In the Puritan ethos there is no such thing as a legitimate subsidy. If one must nevertheless be subsidized, how much more seemly to have it out of sight." "Farm Policy: The Current Position," 37 *Journal of Farm Economics* 292–304 at 300 (1955).

ment" are almost sure to conflict with economists' convictions as to the relevant cost allocations. But these notions are themselves neither stable nor uniform, although they reveal a general tendency in favor of a fairly wide spreading out of the costs, as butter would be spread over bread in a well-made sandwich. Awareness of these unresolved conflicts about "fair" cost apportionment has led the British economist Professor W. Arthur Lewis to exclaim that, in rate determination, "equity is the mother of confusion." [9]

THREE DIFFERENT TYPES OF FAIRNESS STANDARDS

What can be done, if not to resolve the confusion to which Professor Lewis refers, at least to determine its sources and to minimize its frustrating influence on the theory and practice of rate making? Perhaps the best way to begin is to ask what "fairness" means when used in the context of a rate case. Just what does a contestant intend to assert, for example, when he opposes any rate or rule of rate making, not on grounds of difficulty of administration or of ineffectiveness in attracting capital or of nonpromotional character, but simply on the ground that it is unfair? Unless one is ready to accept completely the cynic's position that a "fair rate" means whatever rate is in my self-interest,[10] this is no easy question. And the dictionaries are here of little help, since their relevant definitions of fairness merely put the reader into a dizzy merry-go-round of synonyms (equity, justice, etc.).

To be sure, we can go a certain distance toward an answer by noting that a finding that rates are fair, as distinct from a finding that they are otherwise socially acceptable, implies a proper balancing of the relative claims of the interested parties when these parties are viewed, not as mere economic agents or factors of production, but as human beings entitled to respect for their own interests. In

[9] *Overhead Costs* (London, 1951), p. 23.

[10] Referring to the Electricity Consultative Councils in the organization of the nationalized British electric power industry, a recent report by a departmental committee remarks: "They also have to face criticism from the public who, in general, are averse to any changes in electricity tariffs, unless of course it is to their benefit." *Report of the Committee of Inquiry into the Electricity Supply Industry*, Jan., 1956, Cmd. 9672, pp. 104–105. But in Sutton *et al., The American Business Creed*, cited in footnote 5, the authors conclude that a good part of the businessman's notions of "fairness" and "justice" cannot be accounted for as mere reflections of self-interest.

the majority of cases, moreover, we are also correct in associating the fairness issue with the income-distributive effect of the rates and not directly with their producer-incentive or consumer-motivation effects.[11] But we are still uninstructed not only as to tests of a fair income-distributive effect but even as to the very meaning of the requirement that the effect be "fair."

Any attempt to get much further with this mixed problem of semantics, social psychology, folklore, and ethics would go quite beyond the field of economics and even farther beyond the scope of a book on public utility rates. But one point is clearly revealed by a mere study of the rate cases: namely, that there are different conceptions or standards of fairness, only loosely related to one another and often in conflict with each other. Among these standards, three types may be distinguished here because of their relevance to the theory and practice of public utility regulation. They may be called, respectively, (1) good-faith standards, (2) income-distributive standards, and (3) notional-equality standards.[12]

1. GOOD-FAITH STANDARDS

The meaning of "fairness" in business transactions is most clearly definable when referring to a moral obligation, which may also be a legal obligation, to avoid deception and to live up to previous commitments, expressed or implied. If judged by this test alone, any rule of rate making would be fair to investors, whatever its merits or demerits on other grounds, if it conforms to the terms, on the faith of which the investment was originally made—fair no matter how onerous or how profitable these terms may prove to be in the light of hindsight.

In the history of American public utility regulation, the most

[11] But income-distributive aspects of economic policy are not *necessarily* viewed in terms of "fairness" or "justice." They may be viewed, e.g., hedonistically, as a problem of securing that income distribution which will maximize total happiness—viewed in this manner by persons who purport to take no interest in the question whether actions that they approve, as contributing to maximum happiness, are "fair" or "unfair." Or, in the more modern manner, they may be appraised from the standpoint of over-all economic or social effects on community welfare—e.g., from the standpoint of the possible effect of a more nearly equal distribution of income in reducing the severity of business depressions, or in contributing to the elimination of tuberculosis.

[12] In his book on *The Sense of Injustice* (New York, 1949), Edmond N. Cahn notes six popular demands viewed as demands for fairness or justice: equality; recognition of desert; respect for human dignity; conscientious adjudication; confinement of government to its proper functions; and fulfillment of common expectations.

conspicuous example of a line of cases in which this conception of fairness was put to the test (and, incidentally, found wanting) was that presented by the early 5¢ fare franchises granted to the street railways and to the rapid-transit companies. In the heydays of the "street car," operation under these franchises proved highly profitable. But as a result of the rise in prices and wages during and after the First World War, the 5¢ fare first became unprofitable and then ruinous. As might have been expected, the corporate managements made a determined effort to secure relief from these outmoded obligations; and usually the relief was granted, either by the courts whenever they could find the franchise to have been noncontractual in character, or else by obliging state legislatures. In large part, this forgiveness even of contract obligations was justified on the "practical" ground that the companies could not otherwise continue in service. But there can be little doubt that the action was also influenced by appeals to the "injustice" of any attempts to enforce the letter of the early franchises.

The 5¢ fare cases are exceptional in the history of American utility regulation. For the modern law of commission regulation does not predetermine the rates that a company must continue to charge in future years, nor does it even set up a precise formula by which the fairness of these rates shall be redetermined from year to year. But this very lack of a definite understanding as to the criteria of a fair return, or as to the circumstances under which the company may assert a right to the enjoyment of such a return, has given rise to countless disputes between company spokesmen and consumer interests as to the implicit understanding or "reasonable expectation" under which capital was committed to the enterprise. One such dispute concerns the claim, once frequently advanced by public utility representatives, that a retroactive application of a prudent-investment rule of rate making, without compensation for early investments made on the faith of the more generous but much criticized fair-value rule, would be unfair to existing investors. Twenty-five years ago, when the fair-value rule was still held to represent the constitutionally imposed "law of the land," this claim was conceded to have weight even by defenders of a prudent-investment rate base in principle. But the argument has now lost much of its force with the lapse of time and with the adaptation of the stock market to newer philosophies of rate making.

Good-faith standards of fairness have been invoked by consumers no less than by investors in support of rates that would otherwise be deemed indefensible. The appeal is likely to take the form of an insistence by a particular class of customers on the right to the continued enjoyment of low rates which, while originally justified by cost analyses or otherwise, have subsequently been made obsolete by changed conditions including, particularly, changes in load factors or the disappearance of competition. The argument runs to the effect that the consumers were induced to locate their factories, or to abandon their isolated generating plants, or to convert their furnaces from coal burners to gas burners, in contemplation of the low, "promotional" rate and on the faith that this rate would remain in effect for the indefinite future. As a matter of legal doctrine, such an argument has dubious standing in view of the generally accepted principle that public utility rates are subject to revision if and when they become "unreasonable." But in the politics of utility and railroad rate regulation, it is sometimes pressed with enough force to retard for years changes in rate structure otherwise clearly desirable.[13] Indeed, one of the strongest practical objections to the pricing of special classes of service at rates only slightly above immediate incremental costs, even if these rate concessions would be required in order to attract the service, is that the low rates may create a vested interest in their continuance after the growth of demand has brought the incremental costs to a point equal to, or even above, the average costs.[14]

[13] In electric and gas rate making, commissions have sometimes approved a makeshift solution by permitting old customers to continue service at the old rates, new customers being subject to the higher, revised rates. But this action runs against a rival standard of fairness, the generally accepted principle adverse to "personal" discrimination.

[14] Some years ago, a number of manufactured-gas companies found themselves in this position after they had stimulated widespread resort to residential space heating by the establishment of heating rates originally justified because of redundant plant capacity. In railroad rate regulation, a striking example of a similar situation was that of the low "commodity" rates that the transcontinental railroads put into effect for eastbound Pacific Coast lumber, at a time when the balance of railroad traffic was heavily in the westbound direction. Partly, at least, because of the low lumber rates, the balance shifted. But there was a long delay before the railroads succeeded in securing a corrective shift in their rate structures. See Marvin L. Fair and Ernest W. Williams, *Economics of Transportation*, 1st ed. (New York, 1950), pp. 430–433.

2. INCOME-DISTRIBUTIVE STANDARDS

Sometimes supporting, sometimes conflicting with, good-faith or reasonable-expectation standards of fairness are other standards concerned directly with the income-distributive, or income-transfer effect of the rates—with the fairness (rather than with the motivating and restraining efficiency) of the payment for the service due from the consumer and receivable by the producer. This is the effect distinguished in Chapter III as giving rise to one of the four major functions of public utility rates—the income-distribution function. There two rival standards for the effective performance of this function were set forth: the compensation standard, in which the payment is based on an indemnity principle, or at least on some measure of fair payment not too far removed from that of indemnity; and the "ability-to-pay" principle. Both of these principles, frequently combined in indefinite mixtures, have had their influence on rate-making law and on its administration, and each of them has been defended, sometimes as an instrument for the efficient performance of the work of the world, sometimes on grounds of "inherent fairness." The first principle supports a cost-price basis of rate making; the second would support whatever deviations from cost can feasibly be applied in order to minimize the burdens falling on those consumers with the lower incomes. The previous chapter has taken the position, held in common with many other economists, that the compensation principle of rate making is far preferable to an ability-to-pay principle, save under certain emergency conditions. But the other position has considerable popular support.

3. NOTIONAL-EQUALITY STANDARDS

By these standards I refer to a popular tendency, already noted in an earlier paragraph, to assert the fairness of uniform rates for the same type of service despite significant differences in cost of rendition and perhaps, also, in elasticity of demand. Possibly this tendency is really a distorted reflection of an income-distributive standard. But it certainly fails to accord with any of the more general theories of proper income distribution. Instead, it accepts a specious egalitarianism.

One example of the influence of this egalitarianism has already been noted—the popular support given in this country to uniformity of rates throughout a wide area quite apart from considerations of simple administration and from difficulties in the way of reliable cost allocation. In a report on the electricity supply industry issued in 1956,[15] a British departmental committee refers similarly to contentions by consumers that the nationalized power industry should levy the same charges throughout the country. From this position the Committee dissented, although it qualified its dissent by noting that, with a nationwide grid and with improvements in long-distance transmission, charges may become more nearly uniform even on a cost basis. In an article on "The Price of Fuel" written in 1954, the British economist D. L. Munby insisted on the need to get coal prices away from a false equalization, made in disregard of important cost differences.[16] And in a recent book on British transportation rates, A. M. Milne regrets the existence of notions of fairness favoring the acceptance of "postal tariffs" for railroads, or at least of uniform mileage charges regardless of important cost differentials.[17] In American railroad rate regulation, Professor Sharfman and others have noted the conflict within the Interstate Commerce Commission itself between concepts of "equal treatment" in terms of formal rate uniformity despite cost differences, and of "equal treatment" in the different sense of uniform adherence to cost standards.[18]

From the standpoint of public utility rate theory, the most frustrating of all of these egalitarian notions is the one which tends to identify a "fair" charge with a charge related only to the cus-

[15] Cited in footnote 10.

[16] 6 *N. S. Oxford Economic Papers* 231 (1954). Munby adds: "The politicians' argument that a group which has benefitted for many years at the expense of the rest of the community has a prescriptive right to continue to benefit would prevent the rational use of the price system at all." See also I. M. D. Little, *The Price of Fuel* (Oxford, 1953), p. 49.

[17] A. M. Milne, *The Economics of Inland Transport* (London, 1955), pp. 158–159, 176–177. Along with other economists, Milne points out that uniformity of charges for first-class mail throughout an entire country, and regardless of distances, does not constitute a serious violation of the cost principle, since the major expenses here are for collection and delivery. Convenience is always a respectable ground for a disregard of minor cost differentials in rate making. On occasion, it may even be a valid ground for the disregard of major cost differences.

[18] I. L. Sharfman, *The Interstate Commerce Commission*, Vol. III B (New York, 1936), pp. 538–693. Among other examples of "formal uniformity" in rate making, Sharfman cites rates for shipments to the New York Port area denying to New Jersey any benefit of freedom from lightering costs; and a number of group rate cases including the milkshed cases.

tomer's intake of service as measured by energy-consumption, passenger trips, passenger miles, etc. Variations in charges by reference to time of day, to "demand" charges, or to charges for "readiness to serve" are often resented as unfair even if they are entirely defensible—as is by no means always the case—on grounds of strict cost responsibility. A British rate expert, D. J. Bolton, has this point in mind when he writes that "many people's feelings of equity will be outraged by a method which, under any circumstances, makes no charge for plant used, even in a purely off-peak period; or which makes a very large change in allocation for a very small change in load." [19] And in an earlier book on electric rates, G. P. Watkins had the same point in mind when he suggested that the so-called "Wright" form of rate, which embodies a demand-cost factor in the energy charges, "is probably more acceptable to the public . . . because it appears to be merely a modified kilowatt-hour charge." [20]

Enough has been said in this brief review of "fairness" aspects of rate regulation to suggest why their influence has been found so frustrating by economists and by many rate experts. In the first place, there are no uniformly accepted, measurable standards of fairness. "Equity," writes Professor Lewis, referring particularly to demand-cost allocations, "will support anything." [21] And Commissioner Connole of the Federal Power Commission takes a similar position in expressing a preference for the relatively objective, capital-attraction test of adequate rates as against a test based on some subjective consideration of fairness. "I emphasize 'objective standard,' " he writes, "because any attempt subjectively to demonstrate concepts so abstract as justness and reasonableness suffers all too often from the legendary abuses of the old English equity courts, summarized in the familiar bromide that equity justice varies with the length of the chancellor's foot." [22]

But in the second place, some of the more widely held considerations of equity clash seriously, at times, with those criteria of

[19] Bolton, *Electrical Engineering Economics*, Vol. 2: *Costs and Tariffs in Electricity Supply*, 2d ed. (London, 1951), p. 140. Similar popular notions of equity have influenced the Federal Power Commission in its choice of arbitrary methods of capacity-cost apportionment in natural-gas rate cases. See p. 354, n. 16, *infra*.

[20] G. P. Watkins, *Electrical Rates* (New York, 1921), p. 53.

[21] W. Arthur Lewis, *Overhead Costs* (London, 1949), p. 47.

[22] "Current Thoughts on Natural Gas Regulation," a talk before the New York Society of Security Analysts, Dec. 26, 1956. Mimeographed, p. 6.

functional efficiency which have been the primary concern of the economists and which ought to be one of the major concerns of regulation. Even the good-faith considerations clash at times, despite the impelling reasons for giving them due recognition. But more serious are those deterrents not only to rational practice but even to rational thinking set up by what I have called the notional-equality standards of equity.

What, if any, contributions can public utility rate theory make toward the reduction in these conflicts among rival standards of fairness and between fairness standards and efficiency standards? At least at the present time, the major opportunity for contribution lies in a disclosure of the presence of these conflicts throughout the field of public utility regulation, in a careful analysis of their nature, and in an effort to bring them sharply to the attention of the interested public in general and of persons concerned with rate making or rate regulation in particular. I say "at least at the present time" because, even today, there persists a fairly general unawareness of the extent and significance of these conflicts, and a widely held tendency to assume an identity between the question whether rates are reasonable and the question whether they are fair.

Whether or not the rate theorist should go further and take part, in his professional capacity, in controversies about rival standards of fairness is another question. The answer usually given by economists is in the negative, on the ground that the question, being one of ethics, goes beyond their professional competence. But while this answer is a forcible one and justifies a warning by an economist that he is out of his special field in expressing views on equity, it does not carry the complete conviction suggested by its implied distinction between means and ends. For even standards of fairness are properly judged, at least in part, by their utilitarian values,[23] and this statement applies alike to good-faith standards and to

[23] Compare John Maurice Clark's observation that, under the modern workmen's compensation laws, principles of incentives have modified older conceptions of fairness as to employers' liability for accidents. *The Economics of Overhead Costs* (Chicago, 1923), pp. 33–34. Despite economists' frequent denial that considerations of fairness fall within their professional concern, a study of their expressed or implied attitudes on concrete issues of fairness makes it clear that these attitudes are decidedly influenced by their profession. What distinguishes non-Marxian economists, at least (I am not competent to comment on the Marxians), is not a readiness to take positions on economic policy in ruthless disregard of considerations of equity, but rather a tendency to tailor their own conceptions of equity so as to bring them more nearly into harmony with their own conceptions of economic efficiency.

standards of fair income-distribution. The really serious obstacle to rational solutions of questions of economic policy involving issues of fairness lies, not in the philosophical insolubility of any dispute about ultimate ends, but rather in human inability to predict remote consequences. But in making attempts at a prediction, an economist is entitled to use whatever training and experience he may have in doing his fumbling best.

In any case, one conclusion as to the proper role of fairness standards in the determination of reasonable rates seems to me to be clearly justified: namely, that this role, though essential, should be a subordinate one, in that considerations of fairness or equity, when calling for separate recognition, must be regarded as restraints against the unqualified acceptance of general principles of rate making based on considerations of maximum economic or social efficiency.[24]

[24] I doubt whether this conclusion would be justified if carried over from the field of price policy to the field of tax policy, in which income-redistributive objectives properly receive much greater emphasis. See C. Lowell Harriss, "Sources of Injustice in Death Taxation," 7 *National Tax Journal* 289–308 (1954).

IX

RATE-LEVEL STANDARDS AND RATE-STRUCTURE STANDARDS

Problems as complex as those raised by modern price theory and rate regulation are best analyzed by the technique of cross-classification. Following this technique, the last chapter has cut across the previously discussed list of alternative criteria of reasonable rates in order to draw a distinction, familiar in the economic literature but not so often recognized in practice, between fairness conceptions of reasonable rates and economic-efficiency or social-efficiency conceptions. It is now time to turn to another distinction: that between the determination of a company's general *level* of rates, and the determination of *specific* rates or rate relationships. In the words of the late Chief Justice Stone, speaking for the Supreme Court in *Federal Power Commission v. Natural Gas Pipeline Company*,[1]

> The establishment of a rate for a regulated industry often involves two steps of different character, one of which may appropriately precede the other. The first is the adjustment of a general revenue level to the demands of a fair return. The second is the adjustment of a rate schedule conforming to that level, so as to eliminate discriminations and unfairness from its details.

Following the order of precedence thus suggested by Justice Stone, the chapters of Part Two will be concerned with rate-level determination under the standard of a "fair return," while the chapters of Part Three will discuss the far more complex problems of rate-structure determination. But the distinction between rate-level and rate-structure problems is one of convenience rather than one of analytical logic. The really basic distinction is that between

[1] 315 U.S. 575, 584 (1941).

an adequate-revenue or fair-return standard of reasonable rates and all other standards including those implied by rules against undue discrimination. Because the tests of a fair or capital-attracting rate of return are tests of adequate revenues from a public utility business in gross, or at least from a major division of that business, they are naturally associated with general rate "levels" rather than with specific rate differentials. The levels must suffice to make rates as a whole cover costs as a whole, including (or plus) a proper allowance for interest and profits. But the ability of a company to secure adequate over-all revenues depends on the structure of the rates as well as on their average height, as every management knows when it undertakes to improve its earnings by means of promotional rate reductions. Hence, even a so-called "revenue case," which involves a proposal for a fairly general increase or decrease in rate schedules designed to balance total revenues against financial requirements, raises at least incidental questions about proper rate relationships.

THE ASSUMED PRIORITY OF THE FAIR-RETURN STANDARD

Writers on the principles of public utility rates naturally dislike to deal with fuzzy rules or measures of rate determination. They are therefore under pressure to state these rules with a precision and definiteness that belie their nature in application or even in carefully expressed legal doctrine. This tendency is sometimes revealed in comments on the "fair-return" standard of reasonable utility rates. While all informed writers recognize that public utility companies are not *guaranteed* the enjoyment of a fair return, some of them have implied that neither statute laws nor public service commissions may impose rules of rate making which would deny a company whatever opportunity to enjoy such a return may be open to it in view of the potential demand for its services. Also, though less frequently, some writers have implied that the rates charged by any public utility must be deemed excessive, and hence subject to prompt reduction by voluntary action or by commission fiat, if the revenues therefrom are yielding more than a fair return and give promise of so doing in the future.[2]

[2] The fair-return standard probably has a firmer legal standing as a test of minimum rates than as a limitation on rate levels. Indeed, the standard was first developed in this country, in the form of the doctrine of a "reasonable return on fair value," as setting constitutionally imposed limits beyond which a legislature

This twofold assumption is not only convenient but almost essential for any simple exposition of rate theory developed under the constraint of a "budgetary" or fair-return standard. It fixes the status of this standard by giving it not just major importance but overriding priority. Strictly construed, it would predetermine not only the level but the pattern of a company's rates in the event that only one set of rates would yield the required net return, or in the event that one set would clearly come closer to this objective than any other set. Only if the company has redundant potential earning power, so that it could earn its allowed fair rate of return by any one of a variety of rate structures, would there then be room for the imposition of rules of rate making that contravene the objective of maximum profits or minimum deficits.

Not only is the above-noted assumption convenient for purposes of analysis. In addition, its acceptance reflects, as a first approximation, the actual practice of American rate regulation as applied to those public utilities enjoying a high degree of monopoly power—notably, the electric, gas, and telephone utilities. But it is never strictly valid either in legal doctrine or in practical application. And one need only look to the railroads for a situation in which it is far removed from the facts of life. To be sure, the railroads are now faced with such severe competition from road, water, and air carriers that most of them would probably fail to earn a conventional "fair rate of return" even if freed from all rate regulation. But one can hardly doubt that their earning power is further restricted by the application of commission-enforced rules of reason-

---

or legislative agency might not go in restricting a public utility's opportunity to charge compensatory rates. In some of its earlier opinions, the Supreme Court stated that, under ordinary conditions, "reasonable" rates might well be higher than those sufficient to yield a "fair rate of return." But the Court's views as to what return was fair were so liberal that public utility representatives at times publicly expressed their satisfaction with earnings insufficient to yield the constitutional minimum! See New York State, *Report of Commission on the Revision of the Public Service Commission Laws* (Albany, 1930), Minority Report of Messrs. Walsh, Bonbright, and Adie, N.Y. Legislative Doc. No. 75 (1930), pp. 52–55.

Modern public utility statutes, in setting forth general standards of rate control for observance by commissions, often include fair-return tests as factors to be "considered." Thus the New York State law governing gas and electric rates permits the Commission to "consider all facts which in its judgment have any bearing upon a proper determination of the question although not set forth in the complaint and not within the allegations contained therein, with due regard among other things to a reasonable average return upon capital actually expended and to the necessity of making reservations of income for surplus and contingencies." L. 1910, Chap. 480, Sec. 72, as amended L. 1934, Chap. 212.

able or nondiscriminatory rates—rules that do not yield priority to any assumed fair-return standard.

The full extent to which fair-return measures of reasonable rates may be made to give way, in practice, to other criteria of reasonableness is doubtless obscured by the very flexibility of the measures themselves. Persons who have made a close study of the rate cases will almost surely recall cases in which commissions seem to have reasoned backward, from their convictions as to what rates are reasonable to their findings as to what rate of return (or what rate base) is fair. But we may pass this point in taking note of *open* conflicts between the principle of a fair return and the principles governing specific rates or rate relationships—conflicts in which the latter principles sometimes emerge victorious.

Let us first recall that the primary verbal rule of public utility law in America, so far as concerns the rates of charge for service, is the twofold rule that rates must be both "reasonable" in themselves and not "unjustly discriminatory" in relation to other rates charged by the same enterprise. Only indirectly may this rule invoke a test of reasonable rate *levels*. In its immediate application, it is a rule for the determination of specific rates or rate schedules; and at least as a matter of formal doctrine, it is not subject to the proviso that a company may charge "unreasonable rates," or rates that involve "undue discrimination," in the event that these otherwise forbidden practices are an essential means of yielding adequate corporate revenues.[3]

The most obvious possible conflict with the fair-return objective is that raised by various restrictions against rate discrimination, a subject reserved for discussion in Chapter XIX. While the law does not forbid all forms of discrimination, despite careless statements to the contrary in much of the literature, and while commissions may tolerate, as necessary for the maintenance of sound corporate credit, forms or degrees of discriminatory rate making that they might otherwise forbid, there is a limit to this tolerance in that the discrimination must not be of a type which seriously injures customers in their competition with other customers. Dis-

[3] The Supreme Court may have had this point in mind in the early history of rate regulation when it qualified its famous fair-value rule of rate making by the statement that a public utility may not exact from its customers charges higher than the services "are reasonably worth." Smyth v. Ames, 169 U.S. 466 at 546–547 (1898). See pp. 85–86, *supra*.

crimination of this character will be deemed "undue" or "unjust" and presumably would not be excused even as a necessary means of forestalling corporate insolvency. The same statement applies a fortiori to certain types of discrimination expressly outlawed without qualification by statute, notably so-called "personal discrimination."

As to the possible conflict between the fair-return standard and the requirement that each rate must be "reasonable per se," without reference to any complaint of undue discrimination, the subject presents more difficulties from the standpoint of rate theory.[4] But the railroad rate cases supply examples of such a conflict, in the form of decisions by the Interstate Commerce Commission restricting the opportunity of railroads to make up deficiencies in the revenues received from highly competitive business by attempts to secure "unduly" high returns from noncompetitive or less competitive business.[5] In other words, the right of the carriers to enjoy a fair return on their entire business is subject to their ability to secure these returns by means of rate schedules that are not "unreasonable" in the extent to which they share in the coverage of overhead costs, and not unjustly discriminatory in their relation to other rates available to competing shippers.

SUBORDINATE POSITION OF THE FAIR-RETURN STANDARD IN THE REGULATION OF RAILROAD RATES

The warning, expressed in the previous section, that the fair-return standard is not an overriding rule of rate making, applies in principle to public utility industries in general, including the so-called municipal utilities as well as the railroads. All of these utilities are under a mandate to charge specific rates that are both

[4] The major theoretical difficulty is that of reconciling commission decisions restricting the opportunity of railroads or other public utilities to charge "disproportionately" high rates for noncompetitive service, with the acceptance by the same commissions of value-of-service principles of rate classification—principles which recognize that the types of service for which there is a more inelastic demand may properly be called upon to pay larger shares of the overhead costs. It would seem that a rate may be found "unreasonable" on the ground that it carries the value-of-service principle to excess. But I doubt the existence of any definite rule by which to draw the line of tolerance.

[5] In support of the position that rate-of-return standards must sometimes bow to specific-rate standards, not alone on the count of unjust discrimination but also on the requirement of "reasonableness per se," see the "Lignite Case," Northern Pacific R. Co. v. North Dakota, 236 U.S. 585 (1915).

"reasonable" and not "unjustly discriminatory"; and none of them can adduce either a constitutional or a statute-supported claim to a fair over-all return as exempting them from this twofold legal obligation.

But in the actual practice of rate regulation, the fair-return standard comes much closer to dominance with respect to those utilities that enjoy a high degree of monopoly status than it does with respect to the transportation companies, including the railroads. Indeed, in the regulation of the railroads during the past quarter of a century, the fair-return standard has been so submerged under a combination of competitive conditions and statutory or commission-enforced rules governing the rate structure, that it can hardly any longer be deemed an operating principle of rate making. Only in occasional years have the American railroads, taken as a whole, succeeded in earning on their invested capital or on their property "values" rates of income that would approximate a "fair rate of return" under traditional tests of fairness.[6]

This striking difference between the railroad situation and the local-utility situation is not due to any fundamental differences in the legal rules of rate making. It is due rather to the difference in the impact of these rules. With the local utilities, the fair-return standard has been actually applied, not because it here enjoys a legal priority over all other rules but rather because its application has not often imposed an intolerable constraint on the pattern of the individual rates. But with the railroads, the conflict between the fair-return standard of aggregate rates and the accepted criteria of proper specific rates and rate relationships has been so severe that the former standard has had to give way, not just occasionally but chronically.

Two circumstances in combination account for this difference between the dominance of the fair-return standard in the regulation of the electric, gas, and telephone utilities, and the submerged status of

[6] During the ten-year period 1949–1958, the reported ratios of net railway operating income to depreciated property investment for the line-haul, Class I railroads of the United States averaged 3.68 per cent per annum, with a minimum of 2.76 per cent in 1958 and a maximum of 4.28 per cent in 1950. Association of American Railroads, Bureau of Railway Economics, *A Review of Railway Operations in 1958* (Washington, D.C., 1959). What the rate of return might have been if applied to a "rate base" measured by current replacement costs but subject to full deductions for the obsolescence of railway transport as a whole is an unanswerable question.

the same standard in railroad regulation: first, the far more widespread and severe competition faced by the railroads; and secondly, the tendency of freight shippers to attach more importance to the relationship between their rates and rates paid by competing shippers than they attach to their absolute rates.[7] The first factor has put the railroads under special pressures to quote rates little higher than out-of-pocket costs for their highly competitive traffic and hence to make up for the resulting revenue deficiencies by disproportionately high rates for their less competitive traffic. But the second factor has put the Interstate Commerce Commission under special pressure to resist this tendency by the application of rules against undue discrimination. So serious has this situation become that a thorough overhaul of railroad rate regulation under the Interstate Commerce Act has been demanded, not just by the railroads but by many nonpartisan experts. A move in this direction was made by the Cabinet Committee Report of 1955, the "Weeks Report," which proposed to give the railroads much greater freedom in setting competitive rates. Under its recommendations, the Interstate Commerce Commission would be deprived of its present powers to set the exact railroad tariffs and would be limited to the setting of upper and lower limits of "reasonable" rates, although remaining under the difficult duty of enforcing rules against "undue" discrimination.

For the reasons just suggested, any author of a modern treatise on transportation rates in general and railroad rates in particular is under impelling pressure to put the fair-return standard of reasonable rates in a subordinate, or at least in an uncomfortably unsettled, position.[8] No longer can he usefully assume, even as a rough approximation, that the determination of individual rates or

[7] A possible third, contributory reason is suggested by the recent contentions of rate experts, led by Mr. John Alden Bliss, that public utility rate structures are far better designed than are railroad rate structures to meet competition without sacrificing adequate returns. The superiority of the utility rates is believed to lie in their resort to multi-part pricing based on multi-part cost functions. See the articles cited in Chap. XVI, footnote 22.

[8] The later editions of Professor D. Philip Locklin's standard treatise on the *Economics of Transportation*, departing from the chapter sequence of earlier editions, put the chapter on fair return in the middle of the section on tariffs and without implying a dominant status in rate regulation. In their *Economics of Transportation*, rev. ed. (New York, 1959), Professors M. L. Fair and E. W. Williams relegate their chapter "Railroad Rate Level and Financial Return" to the end of their study of railroad rates.

rate differentials, however important, is nevertheless a strictly secondary problem, the solution of which is subject to a prior principle that rates as a whole shall be designed, as far as possible, to yield a fair return. Instead, one of his major problems—a frustrating one at that—is to suggest possible resolutions of the conflict between fair-return standards and other rate-making criteria, or else to propose alternative forms of rate regulation that completely reject the fair-return philosophy. The alternative criteria include not only nonrevenue considerations but also revenue standards of a short-run or emergency character distinguished from those of a fair return in its usual sense. For example, in a recent general rate-level case brought by the railroads, *Ex Parte 175* (1951),[9] the Interstate Commerce Commission's grave concern for the carriers' financial needs led it to grant substantially the full requested proportionate rate increases despite forcible objections from certain classes of shippers, including the coal shippers, that were being forced to make good deficiencies in other parts of the railroad business, notably in the passenger business. But the financial needs emphasized by the Commission were specific, short-run needs for equipment and plant, not revenue requirements as measured by the traditional standards of fairness or even of long-run financial health and vigor.

But only incidentally is the present book concerned with transportation rates, and then mainly for purposes of comparison. Instead, its primary interest is in rate control as applied to the more nearly monopolized public utility industries. Here, the convenience of the assumption of priority for the fair-return standard outweighs its inaccuracy. Hence the chapters on fair return will come first, in Part Two, and all but one of the later chapters on rate structure (Part Three) will assume that this structure must

[9] 297 I.C.C. 17 (1951). Compare the following statement by Mr. J. C. Kauffman, Vice President and General Counsel of the Chesapeake and Ohio Railway Company: "Within my span of about thirty-five years of railroad experience, I have never seen a charge that the railroad industry was making more than a fair return. Our problem is to find enough income to live on, to sustain the railroads in their day-to-day operation and to provide funds for capital expenditures. We don't have to worry about their rate of return in the language of the public utilities, when we are earning about three and a half per cent on any fair evaluation of our property." Appearance for the Association of American Railroads before the Interstate Commerce Commission in hearings on a proposal for the creation of tax-equalization reserves for defense amortizations, Stenographer's Minutes, June 26, 1956.

be designed under the constraint of a fair-return rule. The exception is Chapter XX, which will discuss proposals by an important group of economists that public utility rates be based on marginal costs, regardless of the resulting effect on total revenues.

PART TWO

Fair-Return Standards of a Reasonable Rate Level

X

CRITERIA OF A FAIR RETURN

As already indicated, a fair-return or fair-profit standard of reasonable rate levels is generally accepted throughout the United States as a controlling basis of rate regulation with respect to those privately owned public utilities that have excessive potential earning power by virtue of their protected, monopoly status. But even when a fair-return standard is accepted in principle, and even if it were enforced in practice with much greater rigor than has been customary,[1] it would still be far from supplying a definite basis of rate control. For a mandate that the sought-for return shall be fair carries with it no instructions as to the criteria of fairness. The history of American rate regulation is in large measure a history of attempts by courts, legislatures, commissions, and economists to supply these criteria. But even today the subject remains highly controversial.[2]

[1] Commissions have seldom attempted a rigorous, mechanical application of the fair-return standard through regular, periodic reviews of rate levels. Instead, they have tended to let the existing rate levels stand, subject to minor revisions in the rate pattern, until there appears to be an impelling reason for a new general rate case. The impulsion may come from the company in the form of a petition for authority to increase the rates, or else from consumer spokesmen in the form of a complaint that the existing rates are too high. But the ablest commissions will themselves take the initiative on behalf of consumers if their expert staffs are convinced that a review is called for. Even with these commissions, however, one should not assume that a rate hearing will at once be instituted if actual corporate earnings materially exceed the "fair rate of return" accepted as the basis of the last rate order. Quite aside from the recognized undesirability of too frequent rate revisions, commissions recognize the "regulatory lag" as a practical means of reducing the tendency of a fixed-profit standard to discourage efficient management.

[2] A complete treatise on public utility rate regulation would not limit its discussion of revenue requirements to the two components of these requirements considered in this book—the fair return and the related problem of the annual allowance for depreciation. In addition, a review and critique of commission allowances for all operating expenses would be called for. See, e.g., Eli W. Clemens, *Economics*

Until a few years ago, any discussion of the subject from the standpoint of practical application necessarily included elaborate commentaries on the partly indeterminate, partly conflicting legal precedents established by the Supreme Court in the form of constitutional restrictions against the power of government to fix prices that would impair the profit-making opportunities of individual or corporate owners of private property. While these owners were held subject to rate regulation if engaged in a business "affected with the public interest," they were also held entitled to protection against regulation so stringent as to violate the injunctions of the Fifth and Fourteenth Amendments against the taking of property "without just compensation" or "without due process of law." In 1898, in its famous dictum in *Smyth v. Ames*,[3] the Supreme Court established the fair-value rule as setting the normal limit below which rates, if imposed by legislative or commission fiat, would be held "confiscatory." But since "fair value," used in this context, is an ambiguous and confusing term, and since the norm of a "reasonable" or "fair" rate of return on this value is also indefinite, long articles and even large books were written by lawyers, economists, and valuation experts on problems of "interpretation."[4]

Fortunately for the cause of continued private ownership of American public utilities, this submergence of important practical issues of effective regulation by esoteric legal issues no longer prevails throughout the country in view of the notable change in the philosophy of the Supreme Court as reflected not only by its decision, but by all of the five separate opinions of the justices, in the *Hope Natural Gas* case.[5] While judicial precedents on the measurement of a valid rate base and of an adequate rate of return are still

and Public Utilities (New York, 1950), Chaps. 6 and 22; Emery Troxel, *Economics of Public Utilities* (New York, 1947), Chap. 11.

[3] 169 U.S. 466 (1898). See pp. 163–166, *infra*.

[4] A summary of the early rulings and opinions on "fair value" is contained in Chaps. 30 and 31 of my *Valuation of Property* (New York, 1937). John Bauer and Nathaniel Gold published a monograph on the subject as a part of a series of studies of legal valuation prepared under my editorship. *Public Utility Valuation for Purposes of Rate Control* (New York, 1934). In the more elaborate early treatises, entire chapters were devoted to single components or "elements" of value, such as "going value," land valuation, and the hypothetical replacement costs of pavements over mains. See Robert H. Whitten, *Valuation of Public Service Corporations*, rev. ed. by Delos F. Wilcox, 2 vol. (New York, 1928).

[5] Federal Power Commission v. Hope Natural Gas Co., 320 U.S. 591 (1944).

important, and while they are controlling in any one jurisdiction when set forth in definite terms, they no longer constitute an elaborate "law of the land" as did the earlier rulings made under the influence of the fair-value doctrine. Moreover, since they are for the most part interpretive of statute law rather than of constitutional mandate, they are subject to change by statutory amendment. Hence, they will receive only incidental attention in these chapters, which discuss the criteria of an adequate return from the standpoint of the economic objectives of rate control and by reference to common-sense considerations of fairness between investors and consumers.

THE RELEVANT MEANING OF A "RETURN"

Under the usual forms of American public utility regulation, a "fair return" represents the entire excess in operating revenues, over and above current operating deductions, for which a commission will make provision in a rate case as a component of the company's annual revenue requirements. The operating deductions include allowances for depreciation and for nearly all taxes, not even excepting corporate income taxes. But the return in excess of these annual deductions does not coincide with corporate "net profit" or "net income" in an accounting sense, since it covers allowances for interest charges as well as for earnings on the stock equity. The published corporate earnings statements do not set forth the amount of return that a commission may have found "fair," nor even the precise return actually realized as measured by rate-making technique. But the reported "net operating income" is often an approximation of this latter figure.

In orthodox practice—and most regulation remains orthodox in this respect regardless of its choice between "original-cost" and "fair-value" principles of rate making—the allowed-for return is arrived at as a multiple of two factors: the rate base, and the "reasonable" or "fair" rate of return thereon.[6] The rate base, or "valu-

[6] In this country the most significant recent departure from a rate-base measurement of a fair return has been that of the "operating-ratio standard," applied to the passenger-bus and motor-truck industries by the Interstate Commerce Commission and by some state commissions. See Alan Wright, "Operating Ratio—A Regulatory Tool," 51 *Public Utilities Fortnightly* 24–29 (1953); Laurence S. Knappen, "Transit Operating Ratio—Another View," 51 *Public Utilities Fortnightly* 485–497 (1953); Charles W. Knapp, "Economics of Transit Operating Ratio," 56

ation" as it was called in former years, represents the total quantum of invested capital or of property "values" on which the company is entitled to a reasonable rate of compensation. The "fair rate of return" reflects whatever annual, percentage rate is found appropriate in the light both of historical conditions and of conditions prevailing or anticipated at the time of a rate case. Thus, the acceptance by a commission of a sum of $100,000,000 as the rate base, combined with a finding that 6 per cent constitutes a fair rate of return, will result in an allowance of $6,000,000 per annum as the fair-return component of the company's revenue requirements. But since general rate levels are seldom revised annually and are likely to stay put for the indefinite future, the tariffs approved by the commission may be designed to yield a reasonable rate of return on the average, over the next few years. Under this objective, account may be taken of future growth of capital investment and of net operating income as the company's plant and business expand.[7]

Public Utilities Fortnightly 467–479 (1953); National Association of Railroad and Utilities Commissioners, Report of the Special Committee to Study Principles of Rate Regulation in the Motor Bus Industry, *Proceedings*, 64th Annual Convention (1952), pp. 461–470; and commission and court opinions noted in these citations. The affected industries have favored the newer technique since, as usually administered, it has conceded much higher profit-making opportunities than would the orthodox technique if applied under traditional standards of a fair percentage rate of return. But, to the best of my knowledge, the supporters have not yet succeeded in finding a convincing rationale of an operating-ratio standard, nor have they yet effectively answered adverse criticisms such as those advanced by Mr. Knappen in the article noted above. The situation now calls for a thorough and objective reexamination, made in the light of actual experience in attempted application.

[7] Commission orders approving a rate-level increase or requiring a decrease are usually based on findings that, in the light of recent realized earnings, the existing rates would probably yield a deficient, or an excessive, rate of return in the near future. As a guide to such a finding, a commission may first determine the return realized during some twelve-month period taken as a "test year." In estimating the rate of return that may be earned during the next year, or during some other future period, the commission will accept convincing evidence of changes in operating expenses and in other operating deductions. Indeed, even in its finding of the rate of return already earned during the test year, a commission may substitute pro forma operating revenues and operating deductions designed to reflect current financial conditions. Payroll expenses, e.g., may be restated in terms of changed rates of wages. As to the rate base applicable to the test year, the usual procedure has been to accept some midyear rate base or some other average rate base in taking account of growth of plant investment during the very year under review. But in recent rate cases, some commissions that accept an actual-cost measure of the rate base have conceded a year-end rate base, presumably as a crude means of allowing for the so-called attrition factor during a period of price inflation. Discussion of these and other procedural details of rate control is beyond the scope of the present book.

Obviously, any assumption of an approximate offset between these two variables is likely to be belied by actual experience. Partly in view of this likelihood, partly because changes in financial conditions may justify changes in the allowed rates of return, companies and consumer groups are given the opportunity to institute a new rate case on a showing of plausible grounds therefor.

FIVE MAJOR CRITERIA OF A FAIR RETURN

In line with the usual procedure of American rate regulation, the four following chapters will consider the measurement of the rate base, leaving for a later chapter a discussion of the tests of a reasonable percentage rate of return on whatever rate base is accepted in a given jurisdiction. But by way of introduction, the present chapter will discuss briefly those major criteria of a fair or reasonable return by reference to which one may judge the merits of alternative measures alike of the rate base and of the rate of return thereon. The main purpose of this introduction is to guard against a tendency, to which some of the early Supreme Court opinions unfortunately lent support, to treat rate-base measurement and rate-of-return measurement as if they were governed by different standards of fairness instead of being merely two steps by which to determine corporate revenue requirements.

Let it be noted that the criteria under review are relevant, not just to the determination of the fair return in a given rate case but to the choice of an entire system of rate regulation applied with reasonable consistency over an extended period of years. What makes an allowed return fair in any specific rate case must be its fair conformity with the general principles applicable also in other years and under different conditions. Apart from such conformity, the fairness of the allowance at any one time is simply indeterminate. In an extreme case, for example, the allowance of a zero return or even of an operating deficit during a period of business depression may be quite "fair" and quite consistent with the need for capital attraction if imposed under a scheme of regulation that concedes to investors adequate opportunities to enjoy what would otherwise be excessive profits in times of prosperity. On the other hand, an allowed return of 25 per cent on the cost or "value" of the corporate properties might be penurious if rate-making policy should threaten to bring all profits to an end after the next several years.

1. THE CAPITAL-ATTRACTION CRITERION

Recalling, then, that we are concerned with the *principles* by which to measure a fair return rather than with details of application to any one case, what criteria of fairness or effectiveness should govern the choice of these principles? Among these criteria, high place, perhaps first place, must be given to that of capital-attracting efficiency. Judged by this test alone, choice should rest with whatever principles of rate control are best designed to permit well-managed, soundly financed public utility companies to attract needed capital.[8]

In view of the obvious plausibility of the capital-attraction standard of a fair return, one may raise the question why its recognition has not gone farther than it has to date in resolving the more serious controversies about principles of measurement. A partial answer is that, under favorable conditions, any of various systems of rate control can be so administered as to permit adequate capital financing.[9] Material scope for choice lies in the possibility of making alternative schemes attractive to investors by the introduction into each scheme of special attractions designed to offset special restrictions. Thus, any adverse effects of a system of regulation

[8] "Rates which enable the company to operate successfully, to maintain its financial integrity, to attract capital, and to compensate its investors for the risks assumed certainly cannot be condemned as invalid, even though they might produce only a meager return on the so-called 'fair-value' rate base." Federal Power Commission v. Hope Natural Gas Co., 320 U.S. 591, 605 (1944). Elsewhere I have discussed the significance and limitations of this famous pronouncement: "Utility Rate Control Reconsidered in the Light of the Hope Natural Gas Case," 38 *American Economic Review, Proceedings* 465–482 (May, 1948).

[9] The extent of the room for choice among alternative principles of rate control, any one of which might be expected to permit well-managed utility companies to attract needed capital, depends partly on the readiness of regulation, assisted by general corporation laws, to enforce compulsory financial reorganization as an effective means by which to bring the contractual interest and preferred-dividend requirements of public utility companies into conformity with the currently applied rules for the measurement of a fair return. For example, under a strictly enforced, fair-value rule of rate making, permitted dollar earnings might well fall to such low levels, during a period of price deflation, that even well-managed companies would suffer a serious impairment in credit and hence in their ability to serve the public. This impairment might be cured by a sufficiently timely and drastic reorganization involving the enforced readjustment of the contract rights of the senior security holders. But no one familiar with the prevailing law and practices of corporation reorganization could look with complacency on rules of rate making likely to give rise to the need for this kind of cure.

which denies to public utility stockholders direct protection against future price inflation may be offset by the concession of nominal rates of return more liberal than would otherwise be necessary. And thus, the deterrent effect of a rule of rate making which denies to a company the opportunity to amortize, through charges against consumers, the costs of property rendered useless by extraordinary obsolescence, may be counterbalanced by the allowance of returns on useful properties so liberal that investors are ready to take their chances of suffering for an uncompensated erosion of surplus, if and when the erosion takes place.

2. THE MANAGEMENT-EFFICIENCY CRITERION

In its analysis of the role of public utility rates, Chapter III distinguished between the function of rates in enabling a public utility to secure the capital required for the supply of the service, and the function of the same rates in stimulating managerial efficiency. Both functions may be at issue in the determination of a fair return. For the amount of the allowed return may be designed, not just to enable a company to attract capital but also to reward efficiency and discourage inefficiency of management.

More will be said about the management-efficiency criterion in Chapter XV. Here we merely note that an incentive standard of a fair return may come into conflict with other standards, especially with that of capital attraction. Such a conflict will become acute with respect to companies threatened with insolvency because of substandard earning power or of top-heavy capital structure for which the existing management, or some earlier management, has been at least partly to blame. In these situations, should commissions be especially lenient in their application of principles of rate control including allowances of a "fair return"? Or should they be rigorous in the enforcement of the general principles, possibly concluding that a "fair" rate of return for a poorly managed company is a very low return or even one expressed in red letters? Questions of this nature present serious dilemmas of rate-making policy.

3. THE CRITERION OF RATE-LEVEL STABILITY

If public utilities were required to raise and lower their rates year by year, with the object of maintaining a fixed annual rate of return, the resulting necessary changes in rate schedules would prove inconvenient alike to the consumers and to the corporate managements.[10] Even more serious would be the countercyclical direction of the change in rate levels required by an attempt to offset a depression-created decline in the demand for the service by an increase in the unit rates of charge. From an economic point of view, the most reassuring thing to be said about such an attempt is that it would be likely to fail!

The points just suggested raise one of the most difficult and most incompletely solved problems of fair-return determination. Most writers would probably agree that, during periods of prosperity, rates should be designed to yield profits sufficiently liberal to avoid the need for a countercyclical rate increase during a depression. But whether or not an attempt to secure cyclical flexibility in the right direction is desirable and feasible remains a highly controversial question. More will be said on this question in the chapter on rate of return; but the conclusions will be disappointingly uncertain.

4. THE CONSUMER-RATIONING CRITERION

Earlier chapters have called repeated attention to a possible conflict between a capital-attraction standard of reasonable rate levels and the "consumer-rationing" function of specific rates or rate differentials. Under the first standard, rates as a whole should cover costs as a whole, including the so-called costs of capital. Under the second standard, each rate should be designed to encourage all consumption for which consumers are ready to pay escapable, marginal costs, and so as to deter any consumption for which consumers are not prepared to pay these costs.

Economists, I think, would generally agree that no complete harmony between these two standards of rate making is possible. But those writers—and they are in the great majority—who do not accept proposals by Hotelling and other economists to abandon a full-cost or fair-return principle would tend to favor whatever feasible measures of a fair return may result in the least serious discord with standards of optimum specific rates and rate relationships.

[10] Compare the laws governing the rate-making practices of the nationalized British utilities, which call for the establishment of rates that will meet revenue requirements, including fixed charges, "taking one year with another." See W. A. Robson, ed., *Problems of Nationalized Industry* (New York, 1952), pp. 334–338.

In earlier years several economists, led by Professor Harry Gunnison Brown, defended the use of a reproduction-cost rate base as coming closest to a reconciliation between the principle that rate levels should be based on total cost of service and the principle that specific prices should be governed by specific costs or marginal costs, as they are supposed to be governed under competition. This line of defense is no longer as widely voiced as was formerly the case, for reasons stated in other chapters.[11] But it at least suggests one sought-for attribute of a fair-return program of rate control—the attribute of tolerable harmony with the demand-control or consumer-rationing objective of rate making.

5. THE CRITERION OF FAIRNESS TO INVESTORS

All four of the criteria of a fair return so far suggested might be classified as criteria designed primarily in the interest of the consuming public. But the very term "fair return" implies a standard of equity to investors not necessarily governed by considerations of consumer self-interests. Certainly this implication is borne out by the history of rate-making law. For the traditional rules of fair-return determination were originally developed by the courts as a means of protecting the private owners of public utility properties against "confiscatory" regulation. And even in those jurisdictions which today deviate from some of the older rules, public utility companies have an undeniable right to challenge the fairness of the newer practices.

But granting, as we must, that a "fair return" must be fair, we may still raise the question whether considerations of fairness to investors require the importation of *special* criteria, in addition to those criteria which would be justified solely in the long-run interests of the consumers themselves. A negative answer would, of course, greatly simplify the solution of problems arising in a rate case. And such an answer would at least be plausible, since it rests on the ground that most public utility companies, in order to render good service, must be able repeatedly to attract new capital

[11] See pp. 101, *supra*, and 224–237, *infra*. Compare the contentions of some British economists that the rates of the nationalized public utilities should be designed to cover, not the fixed charges imposed by the securities issued as compensation to the investors in the expropriated private companies, but rather computed interest on the appraised "values" of plant and equipment. References cited in Chap. XIV, footnote 16.

from investors who are free to commit their funds to any alternative investments including the purchase of stocks in unregulated corporate enterprise. Market acceptability may thus be thought to become, at one and the same time, the test of fairness and of corporate financial need.

Stated in terms of a very broad generalization, the point of view just suggested may be accepted. But, at least under the prevailing American types of rate regulation, the principle is subject to serious qualifications—so much so that disputes about equity intermingle in a most confused way with disputes about functional efficiency. The trouble arises because of the likelihood that, after a public utility corporation has already become established as a going concern, with possession of a large plant and with a fixed security structure, rates of earning that will permit it to attract new capital on whatever terms the current market may require do not necessarily correspond to rates of earnings that are fair to existing investors. Yet, in any given rate case, the only return *directly* at issue is, not the return necessary to attract new capital but rather the return necessary to compensate the existing investors for capital already attracted or for the use of properties already acquired with this capital.[12]

The situation would be quite different if the principles of American rate regulation were so unchanging and so determinate in their measurement of a fair return that all public utility investors could be deemed to have been put on constructive notice of the restrictive character of the regulation prior to their commitment of capital. But neither of these assumed conditions has prevailed down to date. In the first place, there have occurred important retroactive shifts in the "rules of the game"—shifts such as that from a "fair-value" rate base to an "actual-cost" rate base, or from retirement-expense accounting to accrued-depreciation accounting. And in the second place, the rules themselves, at any given time and in any single jurisdiction, are too indefinite to predetermine the reasonable expectancies of the purchasers of utility securities except within fairly wide limits. These purchasers are not in

[12] This is not to say that a capital-attracting rate of anticipated income on new investment has no bearing on a fair allowed rate of return on capital already invested. But the bearing must depend on the indirect argument that existing, captive investments are fairly compensated if permitted to receive whatever rates of return would currently induce free investments.

the happy position of the old lady who said that she always ordered hash in a public restaurant because she knew just what she was getting!

The conclusion that indeterminate rules of rate making are largely responsible for disputes about the fairness of commission actions is suggested by the relative freedom from similar disputes enjoyed by publicly owned utility enterprises financed without pledge of government credit by the issuance of revenue bonds. Here, the income claims of the investors are definitely established in the contracts, which also contain mandatory provisions for whatever increases in utility rates may be required to protect the debt service. Similar relative freedom from fairness questions has prevailed under private ownership with respect to companies that have issued fixed-dividend common stock, in the manner of the earlier British gas and electric companies.

But in this country, regulatory policy, following common-law traditions, has insisted on retaining material flexibility as to the measurement of a fair return. Investors in utility securities, notably in common stocks, must therefore take their chances as to the effect of future rate cases, or even of future amendments to regulatory law, on the earning power of the companies in which they invest. Today, for example, they may hold stock in a company which, under a statute law as interpreted by the courts, is entitled to charge rates designed to yield a "reasonable" rate of return, not on the cost of its properties but on their present "fair value." But no one can assure them that the fair-value rule will be in effect five years from now. And even if that assurance were forthcoming, not even an expert could tell them how this vague, ambiguous rule will be interpreted by a new commission or by a new set of appellate judges.[13]

In the present, unsettled stage in the development of public utility rate theory and practice, this American policy of flexibility

[13] The unfeasibility of any attempt to give to prospective purchasers of public utility securities definite and reliable information about the governing rules of rate regulation is apparently recognized by the (Federal) Securities and Exchange Commission. For the registration statements and prospectuses of public offerings of utility securities say little or nothing about the nature of these rules. Thus, a recent prospectus (Commonwealth Edison Company, April 15, 1958, $50,000,000 issue of First Mortgage 3¾ per cent Bonds), in referring to regulation, merely states the names of the commissions having jurisdiction over the company or its affiliates together with the *types* of regulation to which the company is subject (rates, issuance of securities, etc.).

and indeterminacy may be defended on the ground that a policy which makes no positive commitments is better than one which makes the wrong commitments. But as long as the indeterminacy exists, issues of *fairness*, both to investors and to consumers, cannot be ruled out of controversies about the meaning and measurement of a fair return. The best prospect of minimizing these frustrating issues lies in the gradual development of more definite rules, which may become more and more firmly established as their application stands the test of experience.

The points discussed in this introductory chapter on the criteria of a fair return repeat, with variations, the main theme, or "leit-motif," of this entire book on standards and measures of reasonable public utility rates. This theme runs to the effect that the standards of reasonable rates are multiple standards, and that all of them at times come into conflict with one another. It applies both to the criteria of reasonable rate levels and to the criteria of reasonable rate structure or rate differentials. In the determination of the rate levels, capital-attracting adequacy is properly considered a basic test of a fair return. But other criteria of reasonableness, notably the four others summarized in the preceding paragraphs, must also be taken into account, alike in the design of the rate base and in the determination of formulas or principles for the measurement of a fair allowed rate of return on whatever rate base may be accepted.

XI

THE RATE BASE: COST OR VALUE

The various objectives of rate-making policy reviewed in the preceding chapter are relevant to the entire determination of a fair return in terms of dollars per annum. But the amount of this return is usually calculated through the application of a percentage rate to a so-called rate base. We now turn to the measurement of this rate base—the most widely disputed legal issue in the history of American public utility regulation. But, for the most part, the discussion will be analytical rather than legal or historical, and the reader must turn elsewhere for a study of the case law and for a review of the actual American experience with regulation under the old "fair-value" doctrine of the Supreme Court—an experience so dismal that, in my opinion, it constituted a serious threat to the long-continued survival of regulated private ownership.[1]

The general principles of measurement are usually subject to only minor disputes with respect to plant newly constructed or equipment newly acquired in arms' length transactions. Here, actual, legitimate cost of plant and equipment, with "reasonable" allowances for interest during construction and for working capital, is the governing basis of calculation. But with the lapse of time between the dates of acquisition and the date of a rate case, the room for controversy widens. There now arise numerous plausible grounds for claims by opposing parties that the original-cost entries should be either disregarded or subject to revision: because the

[1] Most of the case law referred to in this chapter is reviewed in Chaps. 30–31 of my *Valuation of Property* (New York, 1937). See also the other references in footnote 4 of Chap. X. The adverse opinion expressed above about the actual experience of regulation under the fair-value doctrine is substantially the one that I expressed in 1930 in the Minority Report of the *Report of the Commission on the Revision of the Public Service Commission Laws*, a report signed by Commissioners Walsh, Bonbright, and Adie, N.Y. Legislative Doc. No. 75 (1930), pp. 334–410.

plant has been transferred to another company at a higher or lower "acquisition cost"; because the assets have ceased to be "used and useful" in the public service; because these assets, even though still useful, have undergone depreciation in efficiency or in life expectancy; because a part of their costs has already been recouped by the company through amortization or depreciation charges allowed as operating deductions; because current replacement costs would be higher or lower than historical costs; because increases or decreases have taken place in the "general price level" and hence in "the value of the dollar"—in short, because the original costs have lost their original economic significance.[2]

Within recent years, at least, the major division of practice and opinion on these issues of asset "revaluation" has been between the position that, as long as the assets remain "used and useful" for their intended purposes, they should stay in the rate base at their original costs, subject only to systematic annual deductions for physical and functional depreciation; and the position that the costs should be written up or down so as to take account of major changes in con-

[2] There is an obvious similarity between these problems of rate-base determination and problems of fixed-asset valuation in financial accounting. The similarity is much closer under an actual-cost or net-investment principle of rate control than under a reproduction-cost or "fair-value" principle. Indeed, the shift from a fair-value to a net-investment principle is often characterized, aptly though not with complete accuracy, as a shift from the realm of the appraisal engineer to the realm of the accountant. Hence, appraisal engineers may not welcome the shift! But even under a net-investment standard of rate control, the similarity between rules of accounting "valuation" and rules of rate-making "valuation" does not reach the point of identity, although it is being gradually increased through a shift in the accounting rules themselves as prescribed by commissions for application to public utility and railroad companies. The shifts are designed, sometimes to tighten, sometimes to depart from, those "generally accepted principles of accounting" applied to unregulated enterprises, so that the regularly recorded book values of the public utility assets can be used with minimum revisions as components of a rate base.

In disputes between company representatives and commission staff experts as to the proper measure of the rate base, both sides have sometimes appealed to "generally accepted principles of accounting." Thus, companies seeking inclusion in the rate base of any price at which they may have acquired public utility plants from other companies, however much in excess of original construction costs, have produced expert accounting witnesses to support their contention that arms' length acquisition costs are the relevant costs for accounting recordation by the acquiring enterprise. (See p. 175, *infra.*) And thus, some commission staff experts, in seeking to deny or restrict rate-base allowances for interest on work under construction, have pointed to the scant recognition given to such allowances not only in standard financial accounting but also in accounting for income-tax purposes. (See p. 178, *infra.*) What these arguments overlook is the fact that, in unregulated business, accounting valuations do not impose legal restrictions on corporate earning power.

struction costs or in general price levels. The first position is that of the "original-cost" or "net-investment" principle of rate making —a principle strongly espoused by the Federal Power Commission and now accepted, with or without qualifications, in the majority of jurisdictions. The second position is most frequently referred to as the "fair-value" principle—a principle no longer held legally mandatory throughout this country but still accepted in a considerable minority of states.

But the prevailing tendency to identify *any* departure from a net-investment or actual-cost standard of rate making with a "fair value" standard, at least whenever the departure takes account of current construction costs or current price levels, is both inaccurate and confusing. For it seems to imply the existence of only two alternatives, the one calling for a rate base measured by a historical record of depreciated cost, the other calling for a rate base designed to reflect, not the costs of the assets in *any* sense but rather what they are "fairly" or "really" worth at the time of a rate case. This loose and indiscriminate use of the ambiguous term "fair value" as an antonym for original cost conceals the presence of other significant alternatives, such as the adherence to an original-cost figure restated in terms of dollars of equivalent current purchasing power. It also obscures the prevalent use of "fair value" to mean, not a definitive measure of the rate base but rather an eclectic or compromise standard, which somehow splits the difference between conflicting measures.

In the hope of minimizing this confusion, the present chapter will consider the basic distinction between cost and value factors in rate-base determination and will stress the point that value considerations are by no means absent even under the so-called original-cost or net-investment principles of rate control.

THE RATE BASE AS REPRESENTING CONTRIBUTED CAPITAL VERSUS THE RATE BASE AS REFLECTING THE VALUE OF THE ASSETS

By way of noting the interplay between cost and value considerations in rate-base determination, let us return to those occurrences, already mentioned, that may be claimed to have impaired the "economic significance" of the unadjusted, original-cost data. A review of these claims suggests two, quite different conceptions of the nature of the alleged impairment. According to the first conception, the costs have lost significance because they no longer reflect, in terms of dollars of current purchasing power, the net financial sacrifice for which investors are still entitled to fair compensation under a cost-of-service principle of rate control. According to the second conception, these same costs have lost significance because they no longer reflect the current values of the assets devoted to the public service and hence because an allowed annual return thereon would not even crudely measure the annual service value or rental value of the assets to the consuming public. Thus, the first of these notions invokes an investor-sacrifice or cost-recoupment theory of a "fair return," whereas the second imports a somewhat roundabout version of a value-of-the-service standard.

In actual rate cases, these two significant factors in rate-base determination are more often than not intermixed instead of receiving separate recognition. But they can be distinguished readily for purposes of analysis. Consider, for example, the now uniformly accepted position that, even under an actual-cost type of rate base, a deduction should be made for the depreciation of depreciable assets. If the deduction is defended as fair to investors on the ground that it represents a portion of capital investment which has already been recouped by operating deductions charged, in effect, against former consumers, the defense rests on an investor-sacrifice or cost-recoupment principle of a fair return. But if the deduction is defended on the ground that the assets have "in fact" suffered physical or functional depreciation, regardless of the question whether or not the company has been compensated for this depreciation by previous charges for services, the standard of cost recoupment is made to give way to the standard of consumer benefit. This same distinction is revealed by current arguments as to whether or not the rate base (or, alternatively, the nominal rate of return) should contain a write-up for price inflation. If the write-up is designed to protect the corporate stockholders against the "confiscation" of a part of their original investment, measured in dollars of current purchasing power, it implies a cost-recoupment principle of a fair return. But if it is justified on the legally more orthodox ground that a rate base should reflect the current values of the corporate assets as a source of service to the consumers, it invokes a consumer-benefit or value-of-service principle of rate control. Nor is the choice between these two points of view one of merely academic interest. For on this choice must depend the question what kind of an allowance should be made for price inflation, or even whether any allowance at all is justified in a given situation.

THE VICIOUS-CIRCLE OBJECTION TO A VALUE RATE BASE

For almost half a century, down to the time of the Second World War, the fair value principle of rate regulation, distinguished from any version of a cost principle, was held to control both state and Federal regulation of public utility and railroad rates in a series of Supreme Court decisions setting forth the principle and prescribing rules for its application in specific cases. Data on original costs of construction as well as on current reproduction costs were to be "considered" along with other data germane to an appraisal of property.[3] "But the "fact to be found" was "value" or "fair value" (the terms were used interchangeably) at the time of the rate case, just as in a valuation for tax purposes or in a determination of the proper compensation payable to the owner of property expropriated under the law of eminent domain. Indeed, the justices who wrote the ruling opinions in the rate cases sometimes made cross references to the tax and condemnation cases, thereby implying that the meaning of the term "value" remains unaffected by the purpose of the valuation.

Early in the history of the fair-value doctrine, critics of the Supreme Court's decisions objected that the "value" of public utility properties, though acceptable for purposes of tax assessment or as a measure of compensation for a compulsory taking, cannot qualify as a valid rate base since this value necessarily depends on the earnings that the company will be permitted to derive therefrom—the

[3] "We hold, however, that the basis of all calculations as to the reasonableness of rates to be charged by a corporation maintaining a highway under legislative sanction must be the fair value of the property being used by it for the convenience of the public. And in order to ascertain that value, the original cost of construction, the amount expended in permanent improvements, the amount or market value of its bonds and stocks, the present as compared with the original cost of construction, the probable earning capacity of the property under particular rates prescribed by statute, and the sum required to meet operating expenses, are all matters for consideration, and are to be given such weight as may be just and right in each case. We do not say that there may not be other matters to be regarded in estimating the value of the property. What the company is entitled to ask is a fair return upon the value of that which it employs for the public convenience. On the other hand, what the public is entitled to demand is that no more be exacted from it for the use of a public highway than the services rendered by it are reasonably worth." Smyth v. Ames, 169 U.S. 466, 546–547 (1898).

very question at issue in a rate case.[4] Any attempt to test the fairness of the rates by reference to a valuation of the properties is an attempt to reason in a circle, or, if you like, to put the cart before the horse. "Fair value" should therefore either be renounced as a measure of the rate base, or else given a special definition, which distinguishes its meaning from that assigned to it in other legal appraisals, presumably identifying it with some kind of cost or with some "fair" figure derived from cost. This vicious-circle argument against a fair-value rule was based squarely on the charge of logical fallacy. It did not rely on those "practical" objections to reproduction-cost methods of rate control so frequently voiced by critics of these methods.

This apparently unanswerable objection to a value rate base was repeated on many occasions, not only by economists but by legal commentators in the law reviews and elsewhere. As early as 1923, it was accepted by Justice Brandeis (Justice Holmes concurring) in his concurring opinion in the *Southwestern Bell Telephone* case,[5] which contained the famous phrase that "value is a word of many meanings." But although other justices, including Chief Justice Stone, later supported Justice Brandeis in this conclusion, the validity of the circular-reasoning argument was not overtly conceded by the Court itself until 1944, when Justice Douglas gave it express recognition in an opinion speaking for the Court.[6] One may there-

[4] This criticism was urged most vigorously, and perhaps with the most telling ultimate effect on judicial thinking, by Professor Robert L. Hale of the Columbia University Law School, who restated his position in his book, *Freedom through Law* (New York, 1952), Chap. 15, "The 'Fair Value' Fallacy in Rate-Making."

[5] Southwestern Bell Telephone Co. v. Public Service Commission of Missouri, 262 U.S. 276, 289–312 (1923).

[6] Federal Power Commission v. Hope Natural Gas Co., 320 U.S. 591 (1944). At page 601 Justice Douglas wrote: "Rate-making is indeed but one species of price-fixing. . . . The fixing of prices, like other applications of the police power, may reduce the value of the property which is being regulated. But the fact that the value is reduced does not mean that the regulation is invalid. *Block* v. *Hirsh*, 256 U.S. 135, 155–157 (1921); *Nebbia* v. *New York*, 291 U.S. 502, 523–539 (1934) and cases cited. It does, however, indicate that 'fair value' is the end product of the process of rate-making not the starting point as the Circuit Court of Appeals held. The heart of the matter is that rates cannot be made to depend upon 'fair value' when the value of the going enterprise depends on earnings under whatever rates may be anticipated."

In a footnote to the last quoted sentence, Justice Douglas cited Institutional Investors v. Chicago, M., St. P. & P. R. Co., 318 U.S. 523, 540 (1943), for a statement that the meaning of the word "value" is to be gathered "from the purpose for which a valuation is being made."

None of the dissenting opinions in the *Hope* case took issue with Justice Douglas's

fore raise the question how the Court, while under the influence of its older and more tradition-minded members, could have continued for so many years to deliver opinions and decisions in important rate cases without squarely facing a criticism of its underlying premise so plausible and, if valid, so utterly devastating.

To this question of judicial psychology and legal reasoning no complete answer will be attempted here. But a partial answer lies in the Court's success in sidestepping an embarrassing and difficult question of substantive law by resort to rulings on evidence that go a certain distance, at least, toward a recognition of the special character of a so-called valuation for rate-making purposes. Thus, in the *Minnesota Rate Cases,*[7] Justice (later Chief Justice) Hu ;hes referred to vicious circularity as a reason for rejecting railroad-land valuations based indirectly on estimates of corporate earning power. And for similar reasons, the Court in this and other rate cases gave little if any attention to the market prices of railroad or public utility securities as reflecting the values of the underlying properties, despite a suggestion in *Smyth v. Ames* that these prices ought to be taken into account. In short, while the Court talked about value, and while some of its actual rulings were predicated on value concepts, the data on which it laid most emphasis as evidence or "elements" of value were records of actual costs or estimates of reproduction costs. The costs themselves, moreover, were not given that critical review of their relevance that a skilled appraiser would give to them if his interest were confined to their possible bearing on the value of a property, say, for the purposes of a loan by an insurance company.[8] On the contrary, the cost data seem to have been accepted in their own right, as an index of a value which was fair rather than as evidence of a value which, fair or unfair, actually existed.

In any event, the question why the Supreme Court waited so long

---

conclusion that "value of the property," in its usual sense, is disqualified as a rate base. On statutory rather than constitutional grounds, Justice Reed contended that the Federal Power Commission was under a duty to adhere to "traditional concepts of fair value and earnings." But even these concepts were those of a special rate-making value, the essentials of which, he believed, "had been worked out in fairness to investors and consumers by the time of the enactment of this Act" (the Natural Gas Act of 1938).

[7] 230 U.S. 352 (1913).

[8] My *Valuation of Property* (New York, 1937) discusses in detail the relevance of original costs and of replacement costs in proof of the value of property. See, especially, Chaps. 8 to 10.

before conceding, in language and not just in somewhat hazy rulings on evidence, that "fair value" for rate-making purposes must be given a special meaning in order to avoid the circular-reasoning fallacy, has become of historical interest only. It now seems generally agreed, at least by all experts, that a "fair-value" measure of the rate base is not the same thing as a "fair-value" standard in taxation, in the law of damages, or in most other legal appraisals.[9]

CAN VALUE BE DEFINED SO AS TO AVOID VICIOUS CIRCULARITY?

Impressed with the force of the vicious-circle argument, many writers have gone beyond the point of insisting that rate-making value must be given a special interpretation in order to qualify it as a plausible rate base. They have also asserted that the measure of a rate base is necessarily cost and not a value in *any* accurate sense. Consistently with this view, they have insisted that the familiar debates as to the merits of an actual-cost rate base versus a fair-value rate base should be converted into debates as to alternative standards of costs, including various types of replacement cost and various versions or modifications of actual cost. The whole philosophy of a fair-return standard of rate making, they have declared, is a cost-of-service philosophy. On no account, therefore, should it be compromised or confused by a backdoor introduction of value-of-service criteria of reasonable public utility rates in the guise of value standards of the rate base.

Offered as a sound, general approach to the determination of a

[9] But in litigated valuations, there has been a backwash into the tax and condemnation cases of concepts of "fair value" developed in the early rate cases. In the valuation of railroad properties, e.g., tax assessors have been prone to attach to estimates of reproduction costs, or to rate-case valuations derived from these costs, probative significance in disregard of the far lower commercial values of the properties as inferred from current stock and bond quotations or from capitalized earnings.

In the condemnation of private utility systems under programs of public ownership, a somewhat different situation has prevailed. Even in jurisdictions, such as New York State, which apply a net-investment or actual-cost standard of rate control, condemnation awards have been based largely on the much higher, reproduction-cost appraisals. As a result, the expropriation may become a source of large windfall profits to the investors in the private company. See Lewis Orgel, *Valuation under the Law of Eminent Domain*, 2d ed. (Charlottesville, Va., 1953), Chap. XVII. In order to remove or reduce this source of windfall profits, the New York State Legislature passed a law calling upon the Public Service Commission to certify to the condemnation appraisers the allowable amount of the rate base. I have been told, however, that the appraisers have actually paid but little attention to the certifications.

proper rate base, I believe that the above statement is valid. But the statement must be revised, not only for reasons of scientific accuracy but also for important reasons of practice. What it rightly asserts is that for practical reasons to be noted in the following chapters, "value of the property," in any definitive sense of the term "value," cannot qualify as an acceptable measure of the rate base. What it wrongly implies is that, in the determination of the rate base, cost factors must entirely supersede value factors since the latter factors are alleged to be completely ruled out by the vicious-circle fallacy.

In fact, however, these factors are not thereby completely ruled out—not even under a net-investment principle.[10] Thus, the general (though not uniformly applied) rule that the cost of property must disappear from the rate base whenever the property ceases to be of public service is a value principle: the cost is now ignored, even if it has not been recouped by previous charges for service, because the property has ceased to have any value for its intended use. And thus, the reluctance of a commission to include in the rate base the price paid for public utility plants in a transfer from one company to another company, as distinct from original construction cost, again imports a value concept in its choice of the one cost rather than the other. That is to say, the cost entitled to a fair return is the cost incurred for the public benefit, not the price paid to buy out other financial interests. Finally, the principles for the allowance of depreciation, both as an operating charge and as a deduction from cost new in rate-base determination, become hopelessly arbitrary unless they are related, however roughly, to the downward trend in the *values* of the depreciating fixed assets as they become older and more decrepit. In none of these situations, to be sure, does value supersede cost as a measure of the rate base. But in all of them, value factors either influence the choice of the relevant cost or else influence decisions as to the proper rates of amortization and dates of complete write-off.

The vanity of attempts to apply cost-of-service principles of rate

[10] That is, not ruled out under this principle in its modern interpretation. The only way to rule them out, or nearly out, would be to accept rigorously an "unrequited sacrifice" rule of a fair return—a rule under which a company would be permitted to charge such rates as would ultimately indemnify it for all outlays, whether for capital account or for operating expenses, but which would never permit it to receive more than full indemnity. Such a rule, in my opinion, would be utterly impractical.

control in complete disregard of value factors was recognized by Justice Jackson in his brilliant dissenting opinion in the *Hope Natural Gas* case,[11] in which he denied the applicability of the prudent-investment theory to the production phase, as distinct from the transmission phase, of the natural-gas business.

> The prudent investment theory [he wrote] has relative merits in fixing rates for a utility which creates its service merely by its investment. The amount and quality of service rendered by the usual utility will, at least roughly, be measured by the amount of capital it puts into the enterprise. But it has no rational application where there is no such relationship between investment and capacity to serve. . . . The service one renders to society in the gas business is measured by what he gets out of the ground, not by what he puts into it, and there is little more relation between the investment and the results than in a game of poker.

But if one admits, as I think one must, that the net-investment principle of rate control is workable only if it can be made to preserve some degree of correspondence between the amount of the investment and the value of the corporate assets, how can this admission be reconciled with the vicious-circle objection to a value rate base? The answer is that, for rate-making purposes, the values of the corporate assets must cease to be identified with their value as private property and hence as sources of income (or of cost saving) to the corporation or to its investors. Instead, the relevant values must refer to the potential values of the assets as instruments for the production of service *to the community of consumers*. If these assets were not only utterly essential for the performance of the service but also utterly irreplaceable, their value to the consumers would be set by whatever rates of charge for service the consumers would be willing to pay rather than go without—set, in short, by what the traffic would bear. But if the assets are replaceable, their potential value to the consuming public is limited by their replacement costs.

For practical purposes, we may define this "service value" of public utility assets in another way: as the value that these assets *would have* to the corporation *if* it were obliged to continue the rendition of the service at rates determined without reference to any principle of a fair return. Under this hypothesis, the assets are

[11] Federal Power Commission v. Hope Natural Gas Co., 320 U.S. 591, 628 (1944). Quoted at p. 649.

worth to the company whatever amounts would indemnify it for their loss—a loss limited by the opportunity to replace the old assets with new assets, probably of a superior design.[12] Such a hypothesis may seem meaninglessly artificial. But either it or some similar hypothesis is necessary in order to avoid the vicious-circle difficulty. Otherwise, if a public service commission or an appellate court were to value a public utility plant, for rate-making purposes, either at an exorbitant value on the one hand or at a zero value on the other hand, its very action would tend to make its findings of value come true, given enough potential earning power to support the exorbitant valuation.

Needless to say, the indirect admission of value elements even into a net-investment program of rate control is a far cry from the acceptance of "value of the property" as the actual measure of the rate base. We therefore still face the question whether, under a service-value definition of the kind just suggested, the values of the corporate assets, as distinct from their costs, can qualify as a sound measure. This question, I suggest, requires two answers. The first is that the adoption of a value standard, thus construed, is not ruled out by the mere logic of the vicious-circle objection.[13] But the second is that, in actual practice, the application of such a standard would be hopelessly unfeasible. For it would require the acceptance of a replacement-cost-of-service principle of rate control.[14] Reasons for denying the feasibility of this principle will be stated in the following chapters.

THE CONFUSED MEANING OF "FAIR VALUE" IN ACTUAL RATE REGULATION

The somewhat "difficult" interpretation of "fair value" suggested in the preceding section is the one required by any attempt to observe the fundamental economic distinction between "value" and "cost" [15] in a valuation for rate-making purposes, while avoid-

[12] But the hypothesis must exclude from the valuation of the existing plant and equipment any allowance for a temporary value in excess of replacement cost based on the impossibility of immediate replacement.

[13] It does, however, pose some difficult conceptual problems raised by the question what required rate of output, and hence what required plant capacity, should be assumed for the purpose of the appraisal.

[14] See pp. 226–230, *infra*.

[15] The "fundamental economic distinction" is not that between cost and *market* value, since vendibility is not an essential attribute of value despite the jingle that "The worth of a thing is the price it will bring." Instead, the distinction is

ing the pitfall of the vicious-circle fallacy. But, although this conception of value is implicit in *some* of the Supreme Court's earlier rulings and comments on the measurement of the rate base, it is certainly inconsistent with others. A similar statement applies to the postwar rate cases in those jurisdictions which, for reasons of statute law or legal tradition, have continued to adhere to a fair-value rule. Any attempt to rationalize the findings of "fair value" in these cases along the lines just suggested would involve the attempt to substitute the philosophy of an economic theorist for the philosophy of a legal practitioner. While there is implicit in this latter philosophy the same general distinction between value and cost that has been set forth in the preceding paragraphs, the distinction has been far less sharply drawn, so that it fails to reveal those uncomfortable dilemmas presented by uncompromising alternatives.

What has just been said is stated more simply by the observation that, in actual practice as distinct from textbook theory, the "fair value" standard of rate control is an eclectic or compromise standard, which takes simultaneous account of conflicting measures on the ground that there is "something to be said" for each of them.[16]

---

between the actual or anticipated sacrifice involved in producing or acquiring any sought-for object, and the favorable importance or prizeworthiness of the object from whatever standpoint is accepted by the person making the valuation.

[16] This statement applies more clearly to some cases or to some jurisdictions than to others. At one time the Supreme Court, under the influence of Justice Butler, seems to have come close to the point of identifying "fair value" with estimates of reproduction cost of a substantially identical plant minus deductions for opinion estimates of "observed depreciation": McCardle v. Indianapolis Water Co., 272 U.S. 400 (1926), a position from which it receded in later cases. See my *Valuation of Property*, Chap. 31. In Ohio and perhaps in a few other states, the highest state courts have, in effect, interpreted the governing statutes as making depreciated reproduction cost "the value for the purposes of the case." But the more usual position is that original cost and reproduction cost, along with still other data, are "elements to be considered." The only definite requirement about reproduction cost is that it be "given some weight." See, e.g., Diamond State Telegraph Company, July 23, 1954 (Delaware Supreme Court), 54 *Public Utilities Fortnightly* 393 (Sept. 30, 1954). In New York State, the Court of Appeals has recently interpreted a statute governing transportation and telephone rates (but not electric and gas rates) as requiring a commission to take account of data on replacement cost. But it has denied that such data are controlling. N.Y. Telephone Co. v. Public Service Commission, 309 N.Y. 569 (1956). Several years ago, the Illinois Supreme Court, reversing the Illinois Commerce Commission, held that a public utility is entitled to a "fair-value" rate base which gives consideration to replacement costs. Illinois Bell Telephone Co. v. Illinois Commerce Commission, 414 Ill. 275 (1953). But the Commission's subsequent allowances for "values" in excess of depreciated original costs have varied for reasons not made explicit in the opinions.

But the basis of the compromise is itself usually left indefinite, with the result that even the expert, familiar with the decision by a particular commission in the last rate case, cannot predict with confidence how the conflicting data will be resolved by the same commission in a later case.

Is this eclectic standard of rate control good or bad? Persons who like it praise it for its flexibility and for its refusal to be bound by codified rules that may prove embarrassing at later times.[17] Persons who dislike it condemn it for its confusion, its indeterminacy, and its invitation to endless controversy between corporate or investor interests and consumer interests. For reasons suggested in the next two chapters,[18] my own convictions support the latter position.

[17] Persons who object to *any* fair-return limitations on public utility earnings naturally tend to prefer a "fair-value" rate base as the lesser evil because of its very flexibility. This viewpoint was voiced by the late President Hadley of Yale, an eminent economist, in testimony before the New York Commission on the Revision of the Public Service Commission Laws, *Hearings* (Albany, 1930), II, 728–729. Asked whether, if compelled to choose, he would take a rate-making "valuation" based on original cost, or one based on reproduction cost, he replied: "I would select reproduction cost every time." One reason for this choice, he continued, is that "reproduction cost is not an exact thing. You don't know what it is. . . . Now, on the record of uncertainty—the inevitable uncertainty, it seems to me as [is?] an important value in the reproduction cost because the Commission is left to decide things for itself, and the Commissions of the Western States are in general well-meaning men who will try not to do a thing that will hurt business. I think that is the saving clause in the reproduction system, the fact that it is frankly a makeshift, the fact that the results are almost necessarily guesswork, and therefore the Commission simply uses it as a general guide to prevent grave abuses and decides the cases partly at least by its good business sense as far as the law will permit it." While President Hadley was here replying to a question about the relative merits of original cost and reproduction cost as measures of the rate base, his grounds for preferring the latter apply a fortiori to an undefined and flexible "fair-value" rate base.

[18] I have presented these reasons in much more detail, and by reference to the actual history of attempts to apply the fair-value rule, in a minority report of a commission appointed to investigate the alleged breakdown of regulation in New York State. Cited in footnote 1 of this chapter.

XII

THE RATE BASE: ACTUAL COST WITH OR WITHOUT ADJUSTMENT FOR PRICE-LEVEL CHANGES

As indicated in the preceding chapter, the primary economic issue in rate-base determination concerns the question whether or to what extent the rate base should comport with an investor-sacrifice or actual-cost principle of rate control as against a present-value or replacement-cost principle. We now turn to the merits of this question. The present chapter and the one immediately following will discuss the nature and philosophy of an actual-cost rate base, whereas Chapter XIV will discuss the nature and philosophy of a replacement-cost or present-value rate base.

But before comparing these two different principles of rate control, we must again note the confusion imported into the comparison by the factor of general price inflation, which became serious following the Second World War. This inflation has created a wide gap between the recorded depreciated costs or book values of public utility assets and the current service values of these assets as based on their prevailing replacement costs. But this same inflation has also given rise to the plausible contention that, even under a strict actual-cost or cost-recoupment principle of rate control, either the rate base or else the allowed rate of return should be enhanced in order to make the capital costs incurred in early years commensurate with the money income to be allowed on these costs in later years.

The important point to note here is the vital distinction between these two objections to an unmodified original-cost type of rate base. Only the first objection calls for the rejection of actual-cost principle of rate making in favor of a present-value or reproduction-cost principle. The second objection calls merely for a restatement of original costs in the very interest of the cost-recoupment principle. The practical importance of this distinction can be illustrated by an extreme, but by no means impossible, situation in which the (depreciated) replacement cost of the plant and equipment of a particular company has fallen below (depreciated) original cost even during a period of rising general price levels.[1] Here, the acceptance of a rate base *lower* than original cost would be required under a present-value standard of rate making, whereas the acceptance of a rate base *higher* than original cost would be required under a stabilized-dollar version of an actual-cost standard.

By way of minimizing the confusion resulting from the twofold distinction (a) between an actual-cost and a present-value rate base, and (b) between an unadjusted actual-cost rate base and an adjusted actual-cost rate base, these chapters will first discuss the relative merits of the former alternatives under the simplifying assumption of a stable general price level. But since this assumption is not realistic under conditions prevailing today, proposals to make allowance for price inflation or deflation even under an actual-cost rule of rate making will be considered briefly toward the end of this chapter and will receive extended discussion in Chapter XV, "The Fair Rate of Return."

GENERAL NATURE OF THE ACTUAL-COST STANDARD

In its modern form, the actual-cost or net-investment standard may be defined as one which measures the rate base by a summation of the actual legitimate costs of plant and equipment devoted to the public service (including or plus allowances for interest during construction), with appropriate deductions for accrued deprecia-

[1] Justice Brandeis called attention to this possibility in his concurring opinion in Southwestern Bell Telephone Co. v. Public Service Commission of Missouri, 262 U.S. 276, 289–312 (1923). Referring to the use of reproduction-cost standards of rate making in the 1880s and the early part of the twentieth century, he wrote: "At first reproduction cost was welcomed by commissions as evidence of present value. Perhaps it was because the estimates then indicated values lower than the actual cost of installation; for even after the price level had begun to rise, improved machinery and new devices tended for some years to reduce construction costs." P. 299.

tion and with reasonable allowances for working capital.[2] But this definition is obviously indefinite and is even subject to minor violations in actual practice. In consequence, the amount of the rate base may be in dispute even among parties all of whom would accept an actual-cost principle of rate control.[3]

A full and critical review of all of these disputes would fill a large volume and is beyond the scope of a book so largely limited to the basic criteria of reasonable rates. But at least four problems of application are of special theoretical interest, and they will be discussed briefly if only to give to the type of rate base here under review enough definiteness to permit of a comparison of its economic merits with those of a present-value rate base. These problems concern (a) the choice between "original" cost and subsequent "acquisition" cost, (b) the allowance for interest during construction, (c) the inclusion or exclusion of capital outlays previously charged off as operating expenses under earlier accounting conventions, and—most important of all—(d) the allowances for depreciation, both as an annual operating charge and as a deduction from cost new in the measurement of the rate base.

[2] This definition conforms to general regulatory usage in recent years. The standard has been called by various names, used sometimes interchangeably, sometimes with distinctions. "Original cost," in public-utility accounting, has now become a term of art. It means the cost of an asset when first devoted to the public service rather than the cost to a transferee company. "Historical cost," though once used in special senses, has now become a term for *any* cost which, having already been incurred, has now become a "matter of history." The "prudent-investment" or "net-investment" principle seems now to be used interchangeably with the "actual-cost" principle despite earlier suggested distinctions. "Prudent" imports the requirement that the investment, in order to gain recognition in the rate base, must have been prudently incurred in the light of foresight rather than of hindsight. See Justice Brandeis's comment on this point in his concurring opinion in Southwestern Bell Telephone Co. v. Public Service Commission of Missouri, *supra*, note 1. "Net" means net of deductions either for capital investments already recouped from revenues charged to depreciation or amortization, or else for asset depreciation already sustained—an obvious ambiguity. "Investment" refers to the capital funds contributed by the company to the public service as distinct from the current values of the assets acquired by these funds.

[3] A dispute of this nature was the basis of Justice Reed's dissent from the decision of the Supreme Court in Federal Power Commission v. Hope Natural Gas Co., 320 U.S. 591, 620–624 (1944). While Justice Reed phrased his dissent as one in support of the traditional fair-value standard, he raised no objection to the Federal Power Commission's refusal to recognize replacement costs rather than original costs. Instead, he declined to go along with the Court in sustaining the Commission's rate order despite its refusal to include, as a legitimate item of actual capital cost, the $17,000,000 of well-drilling expenses and other outlays that the company had previously "conservatively" charged to operating expenses under its earlier accounting practices. See pp. 180–183, *infra*.

ORIGINAL CONSTRUCTION COST VERSUS SUBSEQUENT ACQUISITION COST

Let us assume, as we must under the rate-making standard now before us, that the rate base of a company which seeks an increase in its rates is to be set at the depreciated actual costs of its properties (with adequate allowances for working capital), regardless of the question whether or not these costs reflect the present values of the assets. But let us also assume that the properties of the present company, while constructed at a cost of $60,000,000, were later acquired by purchase from the original company in an arms' length transaction for a cash price of $75,000,000—a price paid in view of then anticipated earnings, and despite the existence on the vendor company's books of a depreciation reserve of $10,000,000 at the time of the transfer. Under this assumption, which actual-cost figure should govern the rate base—the $50,000,000 depreciated construction cost, or the $75,000,000 acquisition cost? [4] A mere resort to the definition of "actual cost" will not supply the answer, nor would the substitution of terms such as "historical cost" or "original cost" (in its traditional, nontechnical sense). For the $60,000,000 construction cost and the $75,000,000 acquisition cost are equally actual, equally historical, and equally original (to the one company or to the other).

Problems of this nature—usually complicated, however, by transactions between affiliated interests and by noncash purchases or mergers—were faced on a wide scale by the regulating commissions in the 1930s and 1940s, when, after inexcusable delays, they finally

[4] I have discussed this issue at greater length in testimony offered on behalf of an industrial consumer in a case before the Maryland Public Service Commission involving rates charged by the Consolidated Gas and Electric Company (now the Baltimore Gas & Electric Company), testimony later partly published under the title "Original Cost as a Rate Base," 20 *Accounting Review* 441–447 (1945). There is a considerable literature on the accounting aspects of the subject with incidental reference to rate making. For a discussion of the subject from the standpoint of the accounting profession, see James L. Dohr, "Power Price Fixing," a series of articles in the June, July, and August issues of the *Journal of Accountancy* for 1945. For sharply opposing points of view on rate-making aspects, see Homer Kripke, "A Case Study in the Relationship of Law and Accounting; Uniform Accounts 100.5 and 107," 57 *Harvard Law Review* 433–478 and 693–727 (1944); and William A. Paton, "Accounting Policies of the Federal Power Commission—a Critique," 77 *Journal of Accountancy* 432–460 (1944). Finally, see Federal Power Commission, *Report on the Reclassification and Original Cost of Electric Plant of Public Utilities and Licensees* (Washington, D.C., 1950).

undertook to secure a wholesale reorganization of operating company accounts following the financial and moral breakdown of the great holding-company systems. Led by the Federal Power Commission and by the more aggressive state commissions, the country's regulating agencies have come close to a settlement of the *accounting* disposition of these dual costs. The basic property records are now kept at an "original cost" expressly defined as cost to the person (corporate or natural) first devoting the property to the public service. For the most part, this means actual construction costs. But the acquisition costs to the present accounting company are also to be recorded, and the difference between the two figures (with adjustments for depreciation, the provisions for which are not clear, at least not to me) is to be charged (if acquisition cost is higher) or credited (if lower) to a special Acquisition-Adjustment Account —an account subject to later disposition, usually by fairly rapid amortization in the case of a debit entry.

This accounting disposition of the problem, however, has not been conclusive for rate-making purposes; for the question remains whether the rate base must be governed entirely by what is now called the "original cost." There is the further question whether any excess in acquisition cost, even if included in the rate base, should be subject to standard rules of depreciation or whether it should be subject to special types of amortization. These problems are not easy ones, and only their barest elements will be discussed here. But the elements are important, and they have not always been recognized in the partisan debates on the issue.

Our hypothetical case presents the problem in its simplest form in assuming not only that the $75,000,000 transfer price was paid in cash but also that it represented an arms' length transaction between strangers. We may add the further assumption that, viewed as a business transaction, the price paid for the properties by the present company was not extravagant in the light of the generous earnings that might have been anticipated under the influence of the then prevailing rules and practices of rate regulation. This being the situation, what are the merits of a contention by the present company that, even under an actual-cost rule of rate making, it must be permitted to enjoy a "fair" rate of return on the cost incurred by it rather than on the cost to the vendor company?

Subject to a qualification to be noted presently, I think that this contention is without merit and that the relevant cost datum is the \$50,000,000 depreciated original cost. True, the \$75,000,000 transfer price was also an actual cost—in fact, the only cost actually incurred by "the present accounting company." But this cost does not represent a contribution of capital to the public service. Instead, it represents a mere purchase by the present company of whatever legal interests in the properties were possessed by the vendor. Even under an actual-cost standard of rate control, investors are not compensated for *buying* utility enterprises from their previous owners any more than they are compensated for the prices at which they may have bought public utility securities on the stock market. Instead, they are compensated for devoting capital to the public service. The only capital so devoted was the original \$60,000,000, of which \$10,000,000 has already been recouped from revenues earmarked as allowances for depreciation. The present company's claim is therefore merely a claim to be standing in the vendor company's shoes. This conclusion would be equally valid if the figures were reversed and if the acquisition cost were to fall \$25,000,000 short of the depreciated original cost.[5]

The foregoing conclusion is subject to revision if the transfer of the properties to their present corporate owner was an essential, or at least a desirable, part of a program of integration, justified in the public interest for the purpose of securing operating efficiencies that would offset any unavoidable excess in acquisition costs over original costs. In such a situation, and in view of the failure of our prevailing public utility laws to provide for compulsory mergers, a claim by the present company that its purchase of the acquired properties was, in effect, a devotion of capital to the public service, cannot be dismissed as without merit. On the contrary, the

[5] The unfairness, not to say the absurdity, of a uniform rule permitting a transferee of a utility plant to claim his purchase price as a measure of rate-making investment was noted by Judge Learned Hand, speaking for the Circuit Court of Appeals for the Second Circuit in an accounting case involving the valuation of property acquired by the present company through a merger of two predecessors. If the rate base, he said, were to be set at the price paid by the new purchaser, then "the builder who does not sell is confined for his base to his original cost; he who sells can assure the buyer that he may use as a base whatever he pays in good faith. If the builder can persuade the buyer to pay more than the original cost the difference becomes a part of the base and the public must pay rates computed upon the excess. Surely this is a most undesirable distinction." Niagara Falls Power Co. v. Federal Power Commission, 137 F (2d) 787, 793 (1943).

company may properly receive an opportunity to prove its claim, although difficulties of proof are serious. Proof should be more readily adduced with respect to mergers or acquisitions, the terms of which have first been cleared with the regulating commission after a full public hearing and investigation.

There remains the question whether any excess in acquisition cost over original cost, if found includible in the rate base, should later be subject to depreciation or amortization through cost-recoupment allowances charged to annual operating expenses, or whether any required amortization should be cleared through the income statement "below the line" (of operating deductions), thereby taking place at the expense of the corporate stockholders. At least for purposes of accounting, the latter alternative has been chosen by the Federal Power Commission. As to the correlative treatment of the problem for rate-making purposes, something will be said in the next chapter, on depreciation.

INTEREST DURING CONSTRUCTION

In a rate base measured by actual cost, there would be no occasion for an allowance for interest during construction if capital devoted by the company to the public service were given an opportunity to enjoy a "fair rate of return" from the very moment of its investment.[6] Some commissions, indeed, have granted this opportunity by including directly in the rate base amounts expended on work still under construction and hence not yet rendering public service. But unless these amounts are relatively small compared to the investments in completed plant, their immediate inclusion in the rate base would be a questionable departure from the general principle that the public utility consumers of any given year should pay a return only on the costs of those assets that are performing for them a useful service. In any event, the general practice is to withhold from the rate base major plant construction costs until the plant itself has become part of the "used and useful" property.

[6] This section summarizes the conclusions of a report, dated Aug. 25, 1952, entitled "Interest during Construction" that I submitted to Messrs. Arthur Andersen & Co., Public Accountants. The report was largely a commentary on points raised by the Federal Power Commission in its opinion in the *Northern Natural Gas* case, decided June 10, 1952 (Docket Nos. G-1382, G-1533, G-1607). It was included as part of a brochure by Arthur Andersen & Company entitled "Principles Underlying the Capitalization of Interest During Construction," dated March 1, 1953.

As long as this withholding practice exists, as I think it should, at least in times of rapid plant expansion, there arises a need for some rate-making provision whereby the company may eventually receive an adequate compensation for its advance commitment of capital. The standard provision of this nature, and the one that I believe most satisfactory, is that of a "computed" allowance for interest during construction—an allowance not restricted to the contract interest, if any, that the company may pay on loans designed to finance the construction work. "Interest during construction" is not a happy term for this allowance; but it has become traditional, and no clever phrasemaker has yet offered a convenient substitute.

So far as I know, the propriety of some allowance of this type as a component of the rate base has been conceded by all commissions. But, several years ago, the standard form of its recognition was challenged by the staff members of the Federal Power Commission, who insisted that the allowed rates of "interest" should be limited to the low contract rates payable on borrowed funds instead of being set at rates, such as 5 or 6 per cent, typical of the usual "fair rate of return." The staff even intimated that, on grounds of "accounting principles," the compensation payable to companies for their commitment of capital in advance of any claim to a fair return should take the form of later enhancements in the allowed rate of return instead of taking the form of enhancements in the rate base. The relevant "accounting principles" were found to lie in the strict adherence of orthodox fixed-asset accounting to the principle of cost. While the failure of a company to earn any return on capital embodied in work in progress was conceded to justify more liberal later allowances of opportunities to make a profit, it was held not to constitute a "cost" in an accurate sense, at least not in any sense recognized in financial accounting.

This attempt by the Commission's experts to judge the merits of a practical rule of rate control by an appeal to "general principles of accounting" gave some spokesmen for the utility companies a dose of their own medicine.[7] But to me, at least, it seems quite

[7] I refer particularly to attempts by expert witnesses for public utility companies to invoke "accepted principles of accounting" in support of the recognition of acquisition costs rather than of "aboriginal costs" as components of the rate base. Both sides in rate-case disputes have a tendency to find these "accepted principles" relevant when, but only when, they comport with whatever rules of rate making they favor in the case at bar.

unconvincing, since it has little or no bearing on the really important practical question—whether the provisions for compensation on capital that has been tied up in work under construction should take the form of a rate-base enhancement or of a rate-of-return enhancement. Here, I think standard procedure is wise in accepting the former alternative, and for reasons quite apart from the academic question whether the allowance of interest during construction is regarded as a recordation of an actual cost, or alternatively as the allowance of a credit to the corporation for a temporary denial of opportunity to earn a return on cost.[8]

As to the question whether the allowance for interest during construction should be based on percentage rates approximating contract interest on secure loans, or whether it should be based on a higher rate approximating an accepted "fair rate of return," the answer should depend on the answer to the further question, Which comes closer to reflecting the "time discount" that the investment market places on investments on which the anticipated return is subject to a material delay, as compared to investments that promise immediate income? To the best of my knowledge, no thoroughgoing study of this problem has ever been made. But the well-recognized market preference for early, and hence less uncertain, realization strongly suggests a time discount not lower than that reflected by a standard "fair rate of return."

RECLASSIFICATION PROBLEMS IN STATEMENTS OF ORIGINAL COSTS

One difficult problem in the "interpretation" of a net-investment standard of rate making may be illustrated by the most famous example of its occurrence in the history of American public utility regulation: the example of the *Hope Natural Gas* case of 1944.[9] I mention it here as raising the question whether a capital cost, however legitimately incurred, may properly be excluded from an

[8] In my opinion, the latter interpretation is correct, since a failure to earn any return on the cost of a capital asset is not in itself a cost in a strict sense. As to the proposal to disallow rate-base enhancements in favor of a more liberal "fair rate of return," this would impose upon those persons who must calculate the proper percentage rate the obligation of attempting to do indirectly, and without benefit of adequate accounting data, what the overt allowances for interest during construction do directly and systematically. Anyone who has faced the difficult problem of establishing a fair rate of return or of estimating "cost of capital" can hardly welcome a change in the rules of rate-base determination that would needlessly add to his difficulties.

[9] Federal Power Commission v. Hope Natural Gas Co. 320 U.S. 591 (1944).

actual-cost rate base on the ground that the company has already recouped the outlay from earlier customers, or at least has enjoyed an adequate opportunity to secure recoupment.

The reader may recall that, in the *Hope* case, the Supreme Court, expressly renouncing the fair-value doctrine in *Smyth v. Ames*, upheld the Federal Power Commission in its acceptance of original-cost data to the exclusion of replacement-cost estimates in the measurement of a rate base. And on this particular issue there were no dissents from the majority opinion by Justice Douglas. But in its summation of original costs, the Commission had excluded approximately $17,000,000 of outlays, mostly in the nature of well-drilling expenditures, which the Commission's own accounting rules would recognize as capital outlays but which the company, following the earlier practices of the natural-gas industry, had previously charged off as operating expenses. In reply to the company's argument that, regardless of their accounting disposition, these outlays represented actual costs of properties still used and useful in the public service, the Commission insisted that the costs had already been recouped, under the guise of current operating expenses, by charges for service imposed upon the earlier customers.[10] Hence, a later reclassification of the expenditures as items of capital investment, entitled to future compensation, was deemed improper, as involving the vice of double-counting against consumers.

Justice Douglas's opinion did not pass directly on the merits of this issue, since it found the Commission's rate order to be fair regardless of possible infirmities in the technique of rate-base determination.[11] But two of the dissenting opinions (those by Justices Reed and Jackson) took sharp issue on this point; and both the

[10] In support of its contention that the outlays had already been recouped, the Commission stated that, during the period (1898 to 1923) for which the company sought to reaccount for its plant costs, the average rate of earnings on its average invested capital (capital stock and surplus) was more than 15 per cent. But for reasons stated in pp. 210–212, *infra*, the enjoyment of excess earnings does not in itself prove capital-cost recoupment or return of capital. Justice Jackson's dissenting opinion is good on this point.

[11] Referring to his conclusion that the Commission's rate order was fair when judged by practical standards of corporate financial requirement and of reward for risks assumed, Justice Douglas declared: "In view of this disposition of the controversy we need not stop to inquire whether the failure of the Commission to add the $17,000,000 of well-drilling and other costs to the rate base was consistent with the prudent-investment theory as developed and applied in particular cases." 320 U.S. 591 at 605–606 (1944).

legal and the practical merits of the issue remain unsettled to the present day. What made this issue, as presented in the *Hope* case, so difficult was that it raised the frustrating problem of the fairness of retroactive regulation, applied to a company hitherto subject to no Federal regulation and apparently to only limited state regulation. If the Hope Company's earlier practice of charging well-drilling outlays to operating expenses had been sanctioned by the Federal Power Commission and had been accepted by the Commission as a basis on which to account for capital costs and operating costs in the determination of reasonable rates, its later attempt to restore the outlays to its capital accounts would clearly have been indefensible on the ground of double-counting.[12] But such was not the situation in the *Hope* case, since the company had ceased "expensing" its well-drilling costs in 1923, five years before the passage of the Natural Gas Act. One may therefore seriously doubt whether its superseded accounting practices had actually resulted in the imposition upon earlier customers of higher charges than would have been imposed even if the later, superior form of accounting had been followed right from the beginning.[13] In view of this doubt, one may question the fairness of the Commission's refusal to permit the retroactive reclassification of the well-drilling expenses, especially so since the very imposition of a net-investment standard of rate control on a concededly well-managed company which had long been free from any such strict regulation was

[12] A situation like that just mentioned has applied to reclassification problems raised in other rate cases. Thus, in Illinois Commerce Commission v. Commonwealth Edison Company, April 13, 1933, 15 P.U.R. (N.S.) 404, the Commission disallowed, as part of the "true" cost of plant and equipment, approximately $40,000,000 of overhead costs incurred between 1908 and 1934 which, though otherwise properly included as a part of plant construction costs, had been treated as operating expenses by the company itself. "It would be improper," said the Commission, "to allow the company to capitalize on its books now any items (apart, of course, from mere accounting errors) which were charged to operating expenses and so reported in reports to the Commission for the period from July 1, 1913 to the present." To the best of my knowledge, these disallowed overhead costs have never been restored to the Commonwealth Edison Company's plant accounts, with the result that the book costs of the company's present properties are said to understate their original costs.

[13] But the Federal Power Commission had some basis for an affirmative answer to this question because of the fact that, in a 1921 rate case, seven years prior to the passage of the (Federal) Natural Gas Act, the Public Service Commission of West Virginia had allowed the Hope Company's claim that its well-drilling expenses should be included among its current operating expenses. As of 1923 and thereafter, the same state commission required Hope to charge future expenses of this type to capital account.

itself a rugged—though, in my opinion, a justified—innovation in public policy.

Aside, however, from questions of retroactive fairness, I believe that the Federal Power Commission was warranted in attaching far more weight to the manner in which a public utility company has accounted for its capital outlays than would seem to have been deemed proper by Justice Jackson in his dissenting opinion warning against the tendency "to make a fetish of mere accounting." The very nature of an actual-cost or net-investment standard of rate making is such as to impose the necessity of making somewhat arbitrary, conventional distinctions between operating costs and capital costs.[14] In view of this necessity, rate regulation wisely places a high premium on consistency of accounting practice, for reasons analogous to those which also dictate adherence to reasonable consistency in income-tax accounting.

THE DEDUCTION FOR ACCRUED DEPRECIATION

Up to a time ending, perhaps, in the middle of the 1930s, writers and commissions were not in agreement as to the propriety of *any* deductions for depreciation in the measurement of a rate base derived from actual costs rather than from estimates of current replacement costs. Some writers argued that the deduction from cost new of an allowance for the depreciation ("decline in value") of the corporate assets imports into a cost rate base a value element which has no business there.

Today, this issue is closed as a matter of accepted practice. Actual cost, no less than replacement cost, is now recognized as subject to deductions for depreciation. But the question what constitutes a proper allowance for depreciation, both as an annual operating charge and as a negative component of the rate base, is still one of the most difficult and most controversial problems of rate-base determination under the actual-cost principle. Because

[14] The special importance of these somewhat arbitrary distinctions between capital outlays and operating costs lies in the fact that, even under an actual-cost standard of rate making in its accepted form, public utility companies may not seek restitution from later consumers for deficiencies in revenues secured from earlier consumers; nor, on the other hand, may realized earnings in excess of an established "fair rate of return" be credited to the rate base as a return of capital. See Professor R. L. Hale's significant treatment of this point in his article on "Utility Regulation in the Light of the Hope Natural Gas Case," 44 *Columbia Law Review* 488–530 at 503–520 (1944).

of its involved character, its discussion will be reserved for the next chapter. But provisionally, the deduction for depreciation, under the actual-cost standard, may be defined as a deduction for that part of the capital costs of the fixed assets that has already been, or should have been, recouped from past revenues.

THE TWO MAJOR ADVANTAGES OF AN ACTUAL-COST RATE BASE

In introducing the chapters of this book on "reasonable-return" standards of rate control, Chapter X distinguished five different goals of rate-making policy by reference to which one might judge the relative merits, alike of alternative measures of the rate base and of alternative measures of a fair-percentage rate of return. An exhaustive study might now attempt to apply first the one test and then each of the others to an appraisal of the claims on behalf of an actual-cost rate base versus the claims on behalf of a reproduction-cost or present-value rate base. But this procedure would be needlessly confusing, since it would involve endless detail which would draw attention away from the primary issues involved in the controversy.

Stated briefly, the claimed superiority of an actual-cost standard of rate making lies in the two, closely related, virtues of administrative feasibility and of capital-attracting or credit-maintaining efficiency.[15] Also stated briefly, the claimed superiority of a re-

[15] The great classic in defense of the "prudent-investment" principle of rate making, which remains unrivaled by the later literature despite a number of statements whose validity would now be challenged, is Justice Brandeis's minority opinion (concurring as to results) in Southwestern Bell Telephone Co. v. Public Service Commission, 262 U.S. 276, 290 (1923). "The thing devoted by the investor to the public use," he declared, "is not specific property, tangible or intangible, but capital embarked in the enterprise." The Justice found the legal basis for this position in what may be called a quasi-agency conception of a public utility company. "The investor agrees, by embarking capital in a utility, that its charges to the public shall be reasonable. His company is the substitute for the state in the performance of the public service, *thus becoming a public servant.*" [My italics.] Perhaps because the issue was not raised by the contesting parties, the opinion did not consider directly the question whether even the amount of "the capital embarked in the enterprise," as distinct from the fair value of the corporate assets, should be stated in terms of dollars of currently equivalent purchasing power. But Justice Brandeis minimized the weight of the price-inflation argument for a fair-value rate base by suggesting the likelihood of later price deflations, such as had previously occurred in American history.

The Supreme Court's later renunciation of the fair-value rule in the *Hope* case, while going far toward accepting Justice Brandeis's economic and practical de-production-cost or present-value standard lies in its supposed performance of the function of a competitive price in controlling demand for public utility services so as to prevent wasteful overuse on the one hand and inadequate use on the other hand. Postponing the latter claim for the chapter on the replacement-cost principle, we may now consider the affirmative case in favor of actual cost.

1. SUPERIOR ADMINISTRATIVE FEASIBILITY

No program of rate regulation is self-executing. On the contrary, any such program must be administered; and the responsibility for its effective administration falls on a state or Federal public service commission. Among the most important virtues of an actual-cost rate is that of relative ease of administration in terms of speedier disposition of rate cases, definiteness of decision, and minimum expense to all parties and to the commission.[16] It is a virtue reluctantly conceded even by supporters of a "fair-value" rate base. To be sure, these supporters have rightly noted that even an actual-cost standard is by no means free from serious controversy—a controversy likely to shift from rate-base issues to disputes about the "fair" rate of return. But since the fair-value standard, in its currently accepted version, accepts both actual-cost and replacement-cost data as so-called "elements of value," its acceptance merely adds new controversies. Moreover, as commissions and courts get more experience with the application of actual-cost rules of rate control, there is good reason to hope for a material reduction in sources for disagreement.

What has just been said is not meant to imply that, in a jurisdiction operating under a fair-value rule, where a rate-level case re-

---

fense of the alternative "prudent-investment" rate base, did not follow his proposal to substitute an investment standard for a value standard as a test of constitutionally "confiscatory" rates. Instead, the Court seems to have left open the question what, if any, rate-making restrictions may be imposed upon a legislature or a public service commission by the due-process clause of the Constitution of the United States. On this point see the separate opinion of Justices Black and Murphy. See also a later decision, Market Street Ry. v. Railroad Commission, 324 U.S. 548 (1945), in which the Court declined to upset a California commission rate order which fixed a rate base below actual cost minus conventional depreciation.

[16] This great "practical" advantage of an actual-cost rate base over an appraised-value rate base was developed in much more detail, and with historical examples, in the Minority Report of the New York State Commission on the Revision of the Public Service Commissions Law (1930), a report in which I participated as a member of the Commission. See Chap. IX, note 2.

quires an expensive engineering reappraisal of plant and equipment, the sum total of rate cases held in any one year will necessarily impose upon commissions, and upon the disputants, more trouble and expense than would be imposed under an actual-cost rule. Indeed, the contrary situation is likely to prevail; for the difficulties of an engineering valuation are so great that they discourage commissions and parties from engaging in rate cases, with resulting serious delays in needed rate revisions.

2. GREATER EFFECTIVENESS IN CREDIT MAINTENANCE

In hearings before public service commissions, critics of the actual-cost principle have sometimes asserted that even its defenders can claim nothing in its favor save the "practical" virtue of convenient administration. In fact, however, this assertion is false. The primary advantage claimed for the principle is that of superior efficiency in enabling public utility companies to maintain their credit and to secure needed new capital on terms most favorable to the consuming public. This claim rests on considerations of corporation finance, which demand the maintenance of a sound relationship between corporate earning power and the annual revenue requirements imposed by fixed and quasi-fixed charges on outstanding securities. Since these charges depend on the amounts of outstanding bonds and preferred stock, they are far more closely related to the capital actually invested in the corporate properties than to the current reconstruction costs or present values of the corporate assets.

For reasons already noted in Chapters IV and VI, the basic criticism of an actual-cost principle from the standpoint of economic theory runs to the effect that historical costs are sunk costs and hence that they should be disregarded in price fixing, since they fail to set up a proper barrier by which to discourage excessive consumption of public utility service while permitting all worth-while consumption. This is a weighty criticism, and one that will again receive attention in Chapter XIV. But even if valid, it would not belie the offsetting superiority of an actual-cost standard when judged by the quite different criterion of capital-attracting effectiveness. For, although actual costs will have become sunk costs at the time of a rate case, they are not sunk costs when first incurred, and it is at this time that investors must balance, *ex ante,* the expected future return on any proposed investment against the presently required outlay of capital. In other words, what induces investment is the expectation of a return on (or return of) costs that will have become sunk as soon as the investors have made their commitments. Whatever scheme of rate regulation will put investors in a position to draw this balance between present outlay and anticipated future return with the most confidence is the scheme most likely to permit a well-managed company to maintain sound financial health at a minimum "cost of capital" to the consumers.

So clearly is the security structure of modern public utility companies geared to an assumed earning power based on sunk capital investments that I very much doubt whether a so-called "fair-value" rate base would win that widespread favor which it still enjoys within the industry and among professional investors were it not for the conviction that it will always remain, as it is today, the higher rate base or, at least, that it will never fall below a rate base measured by depreciated actual cost.[17] This conviction rests on the belief that, because of the political power of the labor unions combined with other inflationary influences, the long-run trend of prices will continue upward, periods of business recession constituting at most temporary interruptions or slight reversals of the general trend.

THE DANGER THAT ACTUAL COST WILL BE REJECTED IF HIGHER THAN REPRODUCTION COST

Even some supporters of a reproduction-cost principle of rate control have conceded that the actual-cost standard has the advantage of greater effectiveness in credit maintenance and hence in capital-attracting efficiency. Their preference for reproduction cost lies in other directions. But two objections to an actual-cost standard must be noted in this chapter since they challenge actual-cost advocates on their own ground of credit maintenance. The first objection is that, during a period of price inflation, equity capital

[17] The tendency of public utility spokesmen to support the fair-value principle only when it leads to the establishment of a rate base at least as high as one derived from actual costs was noted by Justice Brandeis in his separate opinion in the *Southwestern Bell Telephone* case, *supra*, note 15.

cannot be secured in sufficient amounts under any system of rate making which fails to protect stockholders against further inflation. The second objection is that the original-cost principle will be publicly rejected whenever, for reasons of price deflation or of technological progress, its maintenance calls for the establishment of rates of charge for service higher than current replacement costs. These two objections are obviously related; but we may treat them separately, the second in the paragraphs immediately following.

This objection runs to the effect that original-cost rate making is a deviation from competitive-price determination popular with the consuming public only as long as the deviation is in their favor. But let reproduction-cost appraisals fall in the future as they fell (or would have fallen had they come into widespread use) during the period of falling prices from the 1870s to the 1890s, and the very persons who now so loudly proclaim the fairness and efficiency of the actual-cost tests will shift their position and join in a demand that public utility rates be set free from the bondage of inflated historical costs.[18] Those rare stalwarts who demand consistency even of themselves will be hopelessly outnumbered by newer experts, by more recently appointed commissioners, and by other persons not bound by embarrassing prior commitments.

More serious, perhaps, than the danger of an official reversion from a well-established actual-cost rate base to a lower, fair-value rate base would be the danger of competition from new, publicly owned utility systems, possessed of plant and equipment both more efficient to operate and less costly to construct. To persons who feel sure that this country will never again undergo a major decline in price levels, this danger may not seem serious. But there re-

[18] Writers on the history of rate regulation have often noted that, when the upward trend in prices tended to make "fair values" higher than reported original costs—a tendency that became marked by the time of the First World War—various spokesmen for consumer interests and for company or investor interests shifted sides so as to maintain the more "practical" type of consistency—support for the lower or the higher rate base, respectively. When railroad rates in Nebraska were at issue in the litigation leading to the Supreme Court's pronouncement in Smyth v. Ames, 169 U.S. 466 (1898), Mr. William Jennings Bryan, as counsel for the state, argued that, if the constitutionality of the rate statutes was to be tested by reference to any fair-return standard, this return should be related to current reproduction costs and not to the much higher corporate book values. These latter values, however, were said to have been grossly inflated through the practice of stock watering.

mains the danger that full recovery of the costs of old plant and equipment may be precluded by revolutionary developments in the technique of production, for example, in the field of atomic energy. The actual experience of the railroads, and the even sadder experience of the electric street railways, are examples of such a danger.

A completely reassuring answer to this objection is not possible in my opinion. Hence, one must put down to the debit of an actual-cost standard the distinct possibility that it may later become inoperative. For at least two reasons, however, I do not find these objections sufficient to condemn the standard. In the first place, even when the fair-value principle is in force, as it now is in Ohio and some other states, it may well later be renounced as unworkable or as contrary to the public interest. In the second place, the danger that an actual-cost rate base may be difficult to sustain in the face of falling prices or of technological progress, can be reduced, even though not avoided, by rapid cost-recoupment in the form of liberal allowances for depreciation. As to any danger that may still remain, it can be and should be allowed for in the concession to public utility companies of "fair rates of return" well in excess of interest on secure loans. More will be said on this point in Chapter XIV, which will stress the fact, well illustrated in the transportation field, that not even a reproduction-cost or "fair-value" principle of rate making can survive more than a limited degree of extraordinary technological obsolescence.

FAILURE OF AN UNMODIFIED ACTUAL-COST STANDARD TO SAFEGUARD STOCKHOLDERS AGAINST PRICE INFLATION

For reasons already suggested in this chapter, and for other reasons to be discussed in Chapter XIV, the "practical" advantages of an actual-cost standard of rate making are so great, and the "theoretical" advantages of a reproduction-cost standard are so dubious, that many writers predicted a general shift from the latter standard to the former following the renunciation of the "fair-value" doctrine by the Supreme Court as a mandatory "law of the land." Up to a point, history down to date has confirmed this prediction; for the majority of states, including some that have adhered verbally to a "fair-value" rule of rate making, have actually applied some version of an original-cost rate base with little or no recognition of current reproduction costs.[19]

Even today, however, the so-called "fair-value" rule of rate making has not yet suffered its oft-anticipated complete demise. Indeed, in several jurisdictions it has been restored to some measure of its earlier vigor by the action of a commission, a court, or a legislature. But one would be naive in assuming that the partial restoration has been based on any conversion of influential political interests to a reproduction-cost or present-value theory of rate making as distinct from an actual-cost theory. Almost certainly, the revival of the fair-value doctrine has been based on another consideration: namely, on the failure of the actual-cost standard, in its traditional version, to make any direct allowance for the serious, continuing price inflation. Rightly or wrongly, many fair-minded people have regarded this failure as grossly unfair to public utility stockholders, on the ground that it measures the stockholder investment in terms of dollars that have lost their former significance.

[19] Attempted classifications of the state and Federal jurisdictions by reference to the accepted measure of the rate base are unsatisfactory because of the prevalence of compromise measures, because not all commissions have adhered to uniform practice, and because commissions, courts, and legislatures have played fast and loose with the terms "value" or "fair value" as standards of rate control. For a partial classification, see a published report by the Federal Power Commission, *State Commission Jurisdiction and Regulation of Electric and Gas Utilities*, June, 1954. A more recent study of rate cases prepared by the public accounting firm of Arthur Andersen & Co. finds that the commissions in 21 states and in the District of Columbia, together with the Federal Power Commission, "have explicitly adopted original cost or investment as the rate base"; that the commissions of 7 states, while subject to statutes imposing fair-value requirements, have actually given "sole or predominant weight to original-cost evidence"; and that in 12 states "fair value as defined in *Smyth* v. *Ames* is still the test of a lawful rate base . . . although for various reasons original cost or investment has been adopted in a number of decisions." *Return Allowed in Public Utility Rate Cases*, Supplement #3 (1959).

In New York State, because of an utterly irrational distinction in the statutory provisions as to the criteria of reasonable rates, gas and electric companies are subject to an actual-cost standard, whereas telephone and transportation companies are subject to a standard which must give an unstated degree of consideration to an undefined value of the property—a "value" presumably somehow derived from estimates of reproduction cost. New York Telephone Co. v. Public Service Commission, 309 N.Y. 569 (1956). In 1957, the Maine legislature, following a court decision imposing a fair-value test, passed a bill providing that the commission, in fixing a rate base, "shall not include current value." On the other hand, the Minnesota legislature passed a law making it mandatory for the commission (which lacks control over gas and electric companies) to take "fair value" into account.

Adoption of a fair-value measure of the rate base has been seen as the most expedient way by which to make amends for this failure. For this purpose "fair value" is not identified with a reproduction-cost rate base, since such a rate base would often confer upon the common stockholders enhancements in the dollar values of their equity securities far in excess of any increases in general price levels or in current "costs of living." But, in line with traditions followed at times (though rejected at other times) by the Supreme Court, "fair value" has been interpreted to mean a rate base which, while derived in large measure from actual costs, gives some "equitable consideration" to estimates of depreciated replacement cost. And in this way, what goes under the name of the "fair-value" rule can really be turned into a liberalized version of the actual-cost rule, the liberalization taking the form of some equitable upward adjustment in the rate base as an offset to price inflation.

The question whether or not any such offset is justified, either in fairness to corporate stockholders or as a necessary concession to the need to attract equity capital, presents an extremely complex problem. Its discussion will be deferred to Chapter XV, on "The Fair Rate of Return," since any desired adjustment for inflation can be made through the percentage rate rather than through enhancements of the rate base. But what should be noted here is that, if such adjustments are to be applied to the rate base, they can be made more effectively through price-index revisions of an actual-cost rate base than through engineering revaluations of the public utility properties. Indeed, resort to reproduction-cost appraisals is an absurdly crude device by which to protect public utility stockholders against the impairment of their capital investment or of their "real income" against price inflation: first, because the appraisals apply to the entire corporate property rather than to the common-equity interest therein; secondly, because the price changes that they record are changes in construction costs, not changes in cost-of-living indices; and thirdly, because their allowances for depreciation are not related directly to the actual depreciation reserves. The import of this last statement will be noted in the following chapter.

XIII

THE RATE BASE: ALLOWANCES FOR DEPRECIATION UNDER AN ACTUAL-COST STANDARD

In an effort to explain, free from confusing complications, the general nature and claimed advantages of an actual-cost or net-investment principle of reasonable rates, the last chapter has paid only casual attention to the most difficult problem of measurement—that of proper allowances for depreciation. We must now discuss this problem before turning, in the next chapter, to the reproduction-cost or present-value theory of rate regulation. But in order to simplify a discussion that must remain somewhat complex in any event, we shall assume in this chapter the existence of stable price levels, with a mere footnote reference to the important question whether, in periods of general price inflation or deflation, offsetting adjustments should be made in the allowances for accruing and accrued depreciation.[1] Such adjustments would be re-

[1] See p. 275, *infra*. For a discussion of price-level adjustments in business-income determination, see Study Group on Business Income, report entitled *Changing Concepts of Business Income* (New York, 1952); James L. Dohr, "The Next Step in Depreciation Accounting," 89 *Journal of Accountancy* 114–119 (1950); the same author's "Limitations on the Usefulness of Price Level Adjustments," 30 *Accounting Review* 198–205 (1955); F. Warren Brooks, "Needed Reform for Utility Tax Depreciation," *Public Utilities Fortnightly*, Sept. 24, 1953, pp. 3–11. But in public utility rate making under the general philosophy of an actual-cost standard, there arises the need to avoid the introduction of price-level adjustments that would make the common stockholders the recipients of capital gains and losses resulting from the inflexibility of interest and preferred-dividend requirements. I believe that the most promising solution of this problem would lie in the application of the price-index adjustments to the entire plant of a public utility company (and not merely to the common-stock equity in this plant), combined, however, with the creation of a consumer-equity account to which would be credited all dollar write-ups in excess of those needed to restate the common-stock capital in dollars

quired under that modified version of an actual-cost rate base, noted at the end of the last chapter, which restates historical costs in dollars of equivalent purchasing power. But in other respects, the relevant concepts and procedure of depreciation accounting would be the same as those to be discussed here, contrary to the requirements of a strictly construed reproduction-cost or present-value standard of rate making.[2]

For convenience in exposition, this chapter is divided into three main sections. The first section discusses the meaning of "depreciation" as the negative term of the rate base—as the amount to be deducted from actual or original cost new in arriving at the depreciated cost or net investment on which a public utility com-

---

of current purchasing power. A corresponding split-up of the depreciation reserve would also be required.

[2] But in "fair-value" jurisdictions, courts and commissions play fast and loose with the value concept, with the result that their treatment of depreciation is often impossible to reconcile with any basic "theory" of reasonable public utility rates. In a letter to me commenting on this situation, one of the country's leading rate experts, with a wide experience as a witness in different jurisdictions, writes: "The subject is handled so loosely, erratically, and variously by different regulatory bodies that a general philosophical discussion of depreciation would have little practical importance." Appraisers, he adds, sometimes start with a cost-new figure above original costs (either estimated reproduction costs or book costs adjusted by price indices) and apply to this basic cost a deduction for depreciation equal to the depreciation reserve enhanced by the per cent by which the basic cost exceeds the book cost.

In so-called "fair-value" states, the most obvious departure from any practice consistent with a present-value or reproduction-cost rate base has lain in the almost uniform acceptance of book costs, unenhanced by price indices, as the basis on which to calculate the annual allowance for depreciation as an operating deduction. See National Association of Railroad and Utilities Commissioners, Report of Committee on Rates of Public Utilities, *1954 Proceedings*, pp. 197–199, noting the general refusal of commissions to allow what has been called "economic depreciation." See also the separately printed 1954 Report of the Association's Committee on Depreciation which, in rejecting the "economic-depreciation" argument, notes that a public utility official, in advancing it, had admitted that, in the event of a decline in prices, his position would need to be "reconsidered." 53 *Public Utilities Fortnightly* 556–557 (April 29, 1954).

More recently, the public utility trade journals have hopefully reported a few rate cases which concede a company's claim for annual depreciation allowances based on higher costs than book costs. See Re Indiana Telephone Co., 16 PUR 3d 490 (1957); Iowa-Illinois Gas and Electric Co. v. City of Fort Dodge, 248 Iowa 1201, 20 PUR 3d 159, 85 N.W. 2d 28 (1957); Texas Railroad Commission v. Houston Natural Gas Corp., 155 Texas 502, 13 PUR 3d 90, 289 S.W. 2d 559 (1956).

In earlier days, while still adhering to the "fair-value" doctrine in *Smyth v. Ames*, the Supreme Court held that a public utility is entitled to a depreciation basis measured by the current value of its property rather than by its original cost. United Railways Co. v. West, 280 U.S. 234, 252–254 (1930). But when it renounced the value doctrine in the *Hope Natural Gas* case, it expressly overruled its previous ruling on depreciation. 320 U.S. 591, 606–607 (1944).

pany is entitled to an opportunity to earn a "fair rate of return." The second section discusses the annual allowance for accruing depreciation as an operating expense or operating deduction. The third section returns to the rate base in order to note exceptions to the general principle set forth in the first section—the principle which identifies "accrued depreciation" with recouped capital outlays. A more logical procedure would be to discuss first the depreciation allowance as an item of operating expense, since the concession of this allowance to the company is what justifies the deduction of accrued depreciation in the determination of the rate base. But I have chosen the present sequence in order to start with the question raised first in the actual controversies about a proper rate base.

MEANING OF DEPRECIATION AS A NEGATIVE TERM OF THE RATE BASE

Except for a company operating with brand new plant and equipment, the actual or original cost of the public utility properties does not constitute the usual measure of the rate base under an "actual-cost" standard. Instead, the accepted measure (with allowances for working capital and for interest during construction) is "cost minus depreciation." An applicable definition of depreciation is therefore called for.

Applied to property, depreciation is generally defined as decline in, or loss of, value; and this definition was formerly accepted by some accounting authorities as relevant to fixed-asset valuations in financial statements, although with the qualification that the term should refer only to those diminutions in value attributable to causes eventually leading to complete retirement—notably, wear and tear, action of the elements, obsolescence, and inadequacy.[3] Appraisal engineers have sometimes quoted the same or similar definitions as if they were also applicable to valuations based on estimated replacement costs. But in actual practice they have often used the term in another sense, as referring to a value differential rather than to a loss in value over any given period of time

[3] See the several definitions quoted by the Committee on Terminology of the American Institute of Accountants in a report discussed below, pp. 201–202. The committee noted the lack of uniformity as to what value-reducing forces should be recognized in depreciation accounting.

—to the difference between the lower value of the old and obsolescent existing property, and the higher cost (and hence, the higher assumed value) of a hypothetical new property.[4]

Both of these related conceptions of depreciation in terms of a value decrement or value deficiency would have their place in the determination of a rate base designed to reflect the present value of public utility properties as distinct from their cost. If this value were to be derived from a record of actual construction cost, the allowance for "depreciation" should reflect any fall in value, not otherwise recognized, that may have taken place since the date of the construction. If, as is more plausible for older properties, the value were to be deduced from an estimate of current replacement cost—and the relevant replacement cost, despite some judicial opinions to the contrary, is the cost of a modern substitute plant —the deduction for "depreciation" should summarize the value inferiority of the existing property, including the inferiority due to liability to earlier retirement. This deduction, moreover, should be subject to a partial offset for any superior quality of the present plant over the hypothetical modern substitute—a superiority often claimed in the early railroad valuations, in the form of allowances for roadbed solidification.

But when we turn to an actual-cost standard of rate control, there arises a serious problem of redefinition. For under this latter standard, actual cost is accepted as a measure of the rate base on its own merits instead of being accepted only in proof of the "real value" of the property. What, then, is the relevance of a negative term in the rate base, called "depreciation," which is so generally treated as a value concept?

Earlier writers sometimes answered this question by borrowing for rate-making purposes an accounting definition once suggested by a British authority, P. D. Leake.[5] Leake defined depreciation as "expired capital outlay"—in other words, as "expired cost"— thereby transferring the word from a value to a cost category. But this definition was a dodge rather than a solution, and the fact that

[4] This distinction is discussed in my *Valuation of Property* (New York, 1937), Chap. 10. Its importance has been stressed by Professors E. L. Grant and Paul T. Norton, Jr., in their treatise *Depreciation* (New York, 1949).

[5] *Depreciation and Wasting Assets*, 3d ed. (London, 1920). The Interstate Commerce Commission once defined depreciation as "lessening in cost value." Valuation Docket No. 2, Texas Midland Railroad, 75 I.C.C. 1 at 125 (1918). But "cost value" is self-contradictory, as would be "weight-height," or "length-width."

it still enjoys some currency among accounting writers who must be aware of its spurious character illustrates the tenacity of convenient though specious phrases. For cost does not "expire." What may be said gradually to expire is the economic significance of the asset as it grows older, in short, its utility or its value. "Expired cost" is therefore mumbo jumbo, and a reversion to the old association of depreciation with loss in value would be a far more sensible alternative.[6]

Partly in view of the above-noted difficulties of reconciling the deduction for depreciation with the acceptance of a nonvalue rate base, some writers of earlier days denied the propriety of any deduction.[7] Original cost, they declared, is original cost—a purely historical datum; and it should not be discounted for depreciation unless the resulting rate base purports to reflect what the properties are now "really worth," contrary to the philosophy of a cost basis of rate control.

Fortunately for the financial soundness of the public utility companies as well as for the efficiency of rate regulation in other respects, this view no longer prevails. Under a properly administered actual-cost or net-investment standard of rate control, the necessity of making allowances for accruing and accrued depreciation as operating charges and as rate-base deductions, respectively, is now generally conceded. No doubt the income-tax laws, which

[6] The current uniform systems of accounts prescribed for public utility enterprises by the state and Federal commissions incorporate, with minor variations, a definition proposed in the 1938 Report of the Special Committee on Depreciation of the National Association of Railroad and Utilities Commissioners: ". . . the loss in service value not restored by current maintenance, incurred in connection with the consumption or prospective retirement of utility plant in the course of service from causes which are known to be in current operation and against which the utility is not protected by insurance." "Service value" is in turn defined as "the difference between the original cost and the net salvage value of the utility plants. . . ." Taken literally, this last-quoted statement would identify "service value" with cost and hence would not make sense. I assume, however, that the statement was intended merely to designate the quantum of property to be affected by the depreciation account.

[7] See my *Valuation of Property* (New York, 1937), Chap. 31, revising a view that I had previously taken in 27 *Columbia Law Review* 113–131 (1927). Some of these early writers denied the very existence of depreciation as applied to a large, well-maintained plant composed of a great number of assets due for retirement at various times. A brief critique of this plant-immortality doctrine appears on pp. 207–209 and 1127–1129 of the book just cited; but the more recent literature appears to deem the doctrine unworthy even of discussion. The combined forces of income taxation and of rapid technological progress have made it clearly indefensible.

have given public utility companies an incentive to take for tax purposes deductions as liberal as the Internal Revenue Service will permit, have done much to change current attitudes, even on the part of the corporate managements.[8]

As yet, the problem of rationalizing the modern practice of depreciation accounting by means of formal definitions that will aid in the further development and improvement of the practice itself has not received a completely satisfying solution. But a promising approach to such a solution has already won widespread acceptance, namely, in the identification of "accrued depreciation" with an amortized cost only indirectly related to decline in value. The significance of this nonvalue definition will now be noted.

ACCRUED DEPRECIATION DEFINED AS AMORTIZED COST AND NOT AS DECLINE IN VALUE

Under an actual-cost philosophy as I construe it, the deduction for depreciation as a negative term of the rate base does not purport to measure the loss in value actually sustained by the depreciable fixed assets since their dates of acquisition. What it represents is the amortized costs of the assets in the sense of that part of the costs which has already been charged, or which should have been charged, to previous periods of operation. "Cost minus depreciation" is therefore a shorthand expression for costs remaining to be amortized by future charges to operation and hence indirectly by future charges against the consumers of public utility service.

This denial that depreciation here means fall in value is not tantamount to a denial of any important relationship between the two concepts. As a matter of sound accounting and sound price fixing, the rates at which capital costs are gradually transformed into a series of charges to operating expense should be based on plausible assumptions as to downward trends in asset values in the course of their useful service lives. More will be said on this point in

[8] Rules of depreciation accounting accepted in income-tax administration are not binding, as a general rule, on companies in the publication of their annual reports or on commissions in their calculations of revenue requirements for rate-making purposes. But the presence of wide and chronic differences is a source of justifiable suspicion that either the one rule or the other may be invalid. This suspicion, in my opinion, justly applies to the distinction currently drawn between the diminishing-charge methods of cost amortization permitted under Section 167 of the 1954 Internal Revenue Code and the less rapid methods of depreciation still adhered to for purposes of public utility accounting and rate control. See pp. 218–222, *infra*.

the second section of this chapter. But even so, what is deducted *sub nomine* "depreciation" is the cost that has been, or should have been, amortized and not the actual decline in value, estimated with benefit of hindsight.

Under a systematic and consistently applied program of rate regulation, this procedure of capital-cost amortization through annual charges to revenue account is by no means one of mere bookkeeping. Instead, it is designed to afford a company an adequate opportunity to recoup from consumers its investments in fixed assets during their estimated useful-service lives. The deduction for depreciation makes the rate base portray the company's net investment in used and useful properties—that part of its gross investment which it is entitled to recoup in the future and on which it is meanwhile entitled to earn a "fair rate" of return.

Were it not for qualifications to be discussed in the third section of this chapter, one might therefore correctly substitute "recouped investment" for "accrued depreciation" as the thing to be deducted from cost new in the measurement of the rate base. Temporarily, these qualifications will be ignored for the sake of simple exposition. Hence, in this section and in the second section, we shall assume that the deduction for depreciation is to be measured by the company's existing depreciation reserve, and that this reserve reflects whatever portions of the cost of existing plant and equipment have already been recovered by the company through charges for service imposed on previous customers.

Not all company spokesmen have conceded the validity of the above-stated rationale of a deduction for depreciation in the measurement of the rate base.[9] Some of them have insisted that any re-

[9] Even today, some public utility representatives oppose revisions in the uniform systems of public utility accounts requiring the showing of depreciation reserves, not on the liabilities side of the balance sheet but as a contra to property on the assets side—a showing that has the support of modern accounting authorities. See William A. Paton and William A. Paton, Jr., *Asset Accounting* (New York, 1952), p. 399. The revision is opposed as lending credence to the deductibility of the reserve for rate-making purposes.

A compromise between the deduction and the nondeduction of the depreciation reserve once gained some favor. Here, the rate base is set at undepreciated cost; but, in the determination of the annual revenue requirements for rate-making purposes, the consumers are credited with a notional rate of interest (typically 3 per cent) on the amount of the depreciation reserve. That part of the gross costs of the properties reflected by the reserve is thus treated as if it had been financed by a consumer loan, instead of being treated as a cost already recouped from the customers. This procedure may still be followed in a few states. But its days are numbered since it rests on arguments of such a flimsy nature.

coupment of investment made possible by the allowance for accruing depreciation as an operating charge is a mere recoupment by the corporation and not by its security holders. This contention is quite true in the normal situation of continuous plant expansion, since the revenues that cover the depreciation charges are regularly reinvested in new plant and equipment instead of being paid back to stockholders or bondholders. But when the reinvestment takes place, the cost of the new property, whether in the nature of a replacement or of an improvement, is charged to plant account and added to the rate base. No net reduction in the rate base takes place.

What has just been said would need modification if a company should find it necessary to set aside all or part of its depreciation-covering revenues in a special replacement-reserve *fund* made up of cash or of highly liquid securities yielding low rates of interest. But reserve funds of this nature are seldom required in public utility finance and, if required, are properly treated as creating special situations. A more important qualification is that called for when a company cannot time its replacement and expansion program so as to reinvest its retained revenues without material delay. As a result, the company may be in temporary possession of redundant liquid assets which it must hold until needed. Here again, some readjustment is called for in fairness to the corporate investors. But the readjustment does not require the nondeduction of the full depreciation reserve. Instead, it can take the form of an allowance in the rate base for "construction working capital" over and above the conventional allowance for operating working capital.

THE ALLOWANCE FOR DEPRECIATION AS AN OPERATING EXPENSE

Our provisional identification of accrued depreciation with recouped capital outlays supplies a partial answer to the basic question raised near the beginning of this chapter—that of the relevance of a deduction for "depreciation" in the measurement of an actual-cost rate base. But it leaves unsolved another problem in depreciation accounting for rate-making purposes: the determination of a proper annual allowance for accruing depreciation as an operating expense. This problem is still highly controversial and will doubtless long remain so because of the absence of any single, "theoretically correct" solution and because the choice of a wise, workable solution must be based on a variety of partly conflicting considerations, the quantitative importance of which defies precise measurement.

In one respect, however, the actual-cost principle greatly simplifies the task of securing a workable solution. For, properly administered, it minimizes the issue of fairness to investors because it permits a marked degree of coordination between the allowances for *current* depreciation as an operating deduction and the build-up of the reserve for *accrued* depreciation. With one important qualification and with appropriate adjustments in the "fair rate of return," *any* method of amortization (say, group straight-line) can be made as fair to investors as any other (say, sinking fund). The important qualification is that the method which contemplates the faster recoupment is likely to be the one offering the greater assurance that the recoupment will actually take place. More will be said on this point in a later paragraph.

THE DEPRECIATION ALLOWANCE VIEWED AS A TEMPORAL COST APPORTIONMENT

On the general character and significance of the annual allowance for accruing depreciation, as distinct from its measurement, there is no longer room for serious dispute in the light of modern accounting theory. The allowance for any given year is but one step in a systematic and gradual transfer of capital costs into a series of charges to current operations during the estimated useful lives of the depreciable fixed assets. In this way, the attempt is made to apportion the costs of the assets among the years during which they perform service instead of charging them in lumps either to the year of acquisition or to the year of retirement. Thus the cost of a depreciable fixed asset is treated as a prepaid expense, to be amortized through a series of subsequent charges against current revenues.

This conception of the nature of depreciation allowances for rate-making purposes coincides with the modern view of depreciation accounting as a tool of business-income determination.[10] In

[10] More than twenty years ago, this position was taken by Professor Perry Mason, in his *Principles of Public-Utility Depreciation* (Chicago, 1937).

a much-quoted report on the latter subject, the Committee on Terminology of the American Institute of Accounting expressed this view as follows:

Depreciation accounting is a system of accounting which aims to distribute the cost or other basic value of tangible capital assets, less salvage (if any), over the estimated useful life of the unit (which may be a group of assets) in a systematic and rational manner. It is a process of allocation, not of valuation. Depreciation for the year is the portion of the total charge under such a system that is allocated to the year. Although the allocation may properly take into account occurrences during the year, it is not intended to be a measurement of the effect of all such occurrences.[11]

The view, accepted by the Committee on Terminology with respect to financial statements and accepted in this book with respect to rate regulation, that depreciation accounting is a form of cost apportionment and *not* a process of annual reappraisal, offers a useful approach to the problem under review.[12] The further requirement that the method of amortization must be systematic rather than haphazard or rather than subject to the convenience of the management, while not wholly free from ambiguity, offers one step toward a solution by its exclusion of a number of inadvisable ways of accounting for retirements or replacements that were once practiced by public utility companies. But while retirement accounting and replacement accounting would be ruled out, there would still remain room for choice among a rich variety of alternatives, each expressible by curves of mathematical simplicity or elegance.

As to the basis of choice among these alternatives, the Com-

[11] American Institute of Accountants, *Accounting Research Bulletin*, No. 22 (New York, 1944), reprinted in the Institute's *Accounting Terminology Bulletin*, No. 1 (New York, 1953). In an illuminating article on "Concepts of Depreciation," Louis Goldberg of the University of Melbourne compares this conception of depreciation with that previously expressed by the Institute of Chartered Accountants in England and Wales. 30 *The Accounting Review* 468–484 (1955).

[12] But in one important respect the report of the Committee on Terminology seems to me subject to serious criticism: namely, in its failure to note that the distinction between a cost allocation and a valuation is not that of an antithesis. Perhaps the Committee took this point for granted. But I have seen the report cited as authority for the statement that, *since* depreciation accounting is a procedure of cost allocation, it *therefore* cannot be designed to reflect the decline in asset values between dates of acquisition and dates of retirement. Such a conclusion is a *non sequitur*. Conceivably, the cost allocation itself may be based on plausible assumptions as to rates of asset-value deterioration.

mittee on Terminology was content to declare that the method of allocation, in addition to being systematic, must also be "rational." But it did not supply any criteria of rationality, although it wisely implied that any of various alternatives might be rational —possibly *equally* rational in the dim light of foreknowledge of future events and probable consequences. Other writers have discussed at length criteria of rationality or reasonableness, either with respect to standard financial accounting or with respect to rate regulation.[13] But the subject is highly involved, and what follows is designed chiefly to illustrate the difficulty of the problem and the need for further analysis in the light of actual experience.

THE APPORTIONMENT OF COSTS ON RELATIVE-BENEFIT PRINCIPLES

Applied under an actual-cost philosophy of rate control, the rationale of the systematic transfer of capital costs originally charged to plant account into a series of smaller charges to operating costs is a corollary of the principle that the costs of supplying public utility services should be borne, as far as feasible, by those customers who derive a benefit from the particular outlays in question. It is for this reason that the burden of reimbursing a company for the acquisition of capital assets is distributed over the periods during which customers will enjoy the use of these

[13] In its application to accounting for unregulated industries, the subject is discussed by all the advanced treatises on accounting. See also E. L. Grant and Paul T. Norton, Jr., *Depreciation* (New York, 1949). These authors, rightly in my opinion, insist on the indeterminate character of depreciation accounting: pp. 39–40, 184, 365.

In rate regulation, the law and practice of depreciation accounting are discussed by the textbooks on public utility economics or rate making. The lead in the modern development was taken by the New York and Wisconsin commissions during the 1930s, following two famous Interstate Commerce Commission Reports, *Depreciation Charges of Telephone Companies* (No. 14700) and *Depreciation Charges of Steam Railroad Companies* (No. 15100), written by Commissioner Eastman, 118 I.C.C. 295–411 (1926). A report on *Depreciation—a Review of Legal and Accounting Problems*, by the staff of the Wisconsin Public Service Commission (Madison, 1933) was followed, in 1943, by a more elaborate *Report of Committee on Depreciation* of the National Association of Railroad and Utilities Commissioners (Washington, D.C., 1943). This notable document appeared several years after some state and Federal commissions had made depreciation accounting mandatory under the uniform systems of accounts, thereby making obsolete the retirement-expense accounting hitherto so widely practiced by the nontelephone utilities with commission sanction. 1936 is often cited as the year of the general change-over to depreciation accounting.

assets. By the time when the assets have ceased to perform a useful service, their costs should already have been fully recovered.[14]

Since a "benefit rationale" is basic to the whole process of depreciation accounting, viewed as an instrument of rate regulation, one naturally looks to the same rationale for guidance as to a choice of a reasonable method or rate of cost amortization. Equal apportionments, one may assume, should be made to years of equal benefit; heavier apportionments to years of heavier benefit. Or rather, the amortization of the costs should take place at such a rate that, in combination with the allowed fair rate of return on those capital costs that are still unamortized, plus the annual allowances for taxes imputable to those assets, it will impose annual burdens on consumers related as closely as feasible to the relative benefits conferred upon them by different years of use. In the determination of these relative benefits, account must be taken of the fact that the use of newer and more modern fixed assets will free consumers from the obligation to defray the heavier current operating expenses imposed by the retention of aged and obsolescent assets—expenses exclusive of depreciation allowances but inclusive of all charges for current maintenance.

This relative-benefit criterion of a sound basis of depreciation accounting for rate-making purposes presents serious conceptual difficulties quite aside from the obvious practical problems of *ex ante* measurement—difficulties that have not often been squarely faced in the literature. Indeed, the difficulties are probably insoluble without resort to convenient fictions or hypotheses, the validity of which must rest on their workability for practical purposes of rate control. One way to envisage the problem is to view it as a search for rates of cost amortization that, together with the allowances for a fair return and for taxes, will be roughly correlated with the annual rent values of plant and equipment—rent values which, in all probability, will gradually decline as the property grows older, contrary to the behavior of Oliver Wendell Holmes's "one-hoss shay."

[14] As Dr. Terborgh points out, the replacement of assets is likely to take place continuously, or in a series of steps, *antedating* the retirement of the assets. This gradual process of "degradation" is characteristic of electric generators, which are used less and less as they grow older, finally being relegated to the status of stand-by equipment. See George Terborgh, *The Bogey of Economic Maturity* (Chicago, 1945), pp. 102–108.

But at best, since the cost apportionments must be made *ex ante,* subject only to a minimum of midstream revisions, any correlation between the resulting annual charges imposed on consumers for capital costs (depreciation plus fair return plus taxes) and the relative benefits derived by consumers from the use of older assets as compared to newer assets, must be extremely rough. Hence, the choice of any given method of depreciation accounting must not be premised on any assumption of a close adherence to a relative-benefit standard.[15]

An alternative rationale of the periodic assignment of depreciation charges may be preferred to the one just suggested. But it amounts to a restatement of the same idea in a different form. Here, the annual allowance for depreciation is an allowance for an assumed, probable rate of decline in the service values of the assets—a decline typical of assets of the character and ages of the assets in question. Being systematic and standardized, this allowance cannot portray the many ups and downs in asset values between dates of birth and dates of death. It therefore represents, not the actual changes in value from year to year but a stylized allowance for a downward trend. But with all of its practical and conceptual deficiencies,[16] this indirect association between annual

[15] The absence of any such close association is implicitly recognized by the acceptance of rules of rate regulation that do not require an increase or decrease in rate levels, year by year, designed to maintain a constant or completely stable rate of return for each year. While the revenues of any one year will *usually* suffice to cover all operating deductions including the current allowance for depreciation, and while the instability of these revenues is therefore usually reflected only in the realized rate of return, or only in the reported net income, the apparent uniformity with which the consumers of a given year pay for the depreciation allowance of the same year is merely a uniformity of an accounting convention—a convention which alone determines what part of revenues shall be deemed to constitute a recoupment *of* capital as distinct from that part which shall be deemed to constitute a return *on* capital. See p. 363, *infra*.

[16] The major theoretical difficulty is that of securing an acceptable definition of "service value" applicable to a regulated utility. "Market value" is clearly disqualified except, arguably, for those very limited classes of public utility assets (such as trucks or office buildings) that continue to enjoy a significant sale value even after their acquisition by the operating company. If the company were free from rate regulation, the value of the asset to its present owner (sometimes called "value to the going concern") would be the relevant value. But, within limits, the value of an asset to a public utility company is whatever value a commission is willing to give it for rate-making purposes! Hence, as suggested in Chap. XI, some conception of value to the consuming public, or of value to a hypothetical company, is required.

Despite conceptual difficulties, a value concept of the type just suggested is not merely useful but utterly indispensable for the rational management of public

charges for accruing depreciation and annual decrements in asset values seems to me to come closer to supplying a plausible "theory of depreciation accounting" than has any suggested alternative.

SINKING-FUND, STRAIGHT-LINE, AND DIMINISHING-CHARGE APPORTIONMENTS

A systematic treatment of the subject of this chapter would now call for a review of alternative methods of depreciation accounting, including group methods as well as unit or item methods, by reference to the somewhat vague criterion of "rationality" suggested in the preceding paragraphs. Such a treatment, however, would go far beyond the scope of this book. The reader is therefore referred to the special treatises on depreciation for an exposition and critical appraisal of the different alternatives. But a brief comment is here in order as to the bearing of the preceding discussion on the relative merits of three different methods of systematic cost amortization referred to as straight-line apportionment, interest procedure, and decreasing-charge methods.[17]

The straight-line method, which is the one now most generally applied by public utilities, amortizes the costs of assets (minus net salvage) through equal annual charges over estimated service lives. Interest procedures, which include the sinking-fund method, begin with gentler charges, which are increased year after year as the asset approaches retirement. Decreasing-charge methods are illus-

---

utility companies. Quite aside from any use to be made of it in systematic depreciation accounting, it is essential to rational decisions as to the timing of retirements and of replacements. A decision that an old or obsolescent asset ought now to be retired is tantamount to a conclusion that, in the light of its limited efficiency, its *present service value* is now less than its net-salvage value. And a rational decision to buy a new asset (whether in the nature of a replacement or of an addition) is tantamount to a conclusion that the asset will be *worth* at least its cost!

[17] William A. Paton and William A. Paton, Jr., *Asset Accounting*, p. 268. The authors add a fourth category: production and revenue methods, which relate the periodic charge for depreciation to rates of service output during each period (say, electric energy output in kilowatt-hours) or to operating revenues. While these two related methods are subject to theoretical objections, their chief limitations lie in the practical difficulties of making them keep step with the principle of (complete) cost amortization during the useful service lives of the fixed assets. See the above-noted (footnote 13) *Report of Committee on Depreciation*, pp. 62–64, 71–77. Despite these difficulties, further study of their possible application in public utility accounting and rate making, perhaps in combination with the more orthodox straight-line accounting, is warranted with special reference to the phenomenon of the business cycle. The study should consider the actual experience of the Canadian Pacific Railway in part-way application.

trated by the declining-balance method and the sum-of-the-years-digits method, both of which now enjoy limited acceptance under the current income-tax laws. They impose larger annual write-offs in earlier years than in later years, whereas the interest-procedure methods impose smaller write-offs in the early years.

The sinking-fund method, and not the straight-line method, is the one which, except for taxes, tends to impose upon consumers equal annual capital charges for the use of a fixed asset. For under this method, when applied to rate regulation, the sum of the charge for return *of* capital (the annual depreciation charge) and the charge for the return *on* capital (the "fair rate of return") tends to remain the same during the life of the asset.[18] While the sinking-fund method formerly enjoyed some favor among commissions, it now has but little standing in comparison with the somewhat simpler and more conservative straight-line apportionments.

Although a straight-line method is designed to impose equal annual depreciation charges during the lives of the assets, it actually makes some allowance for declining annual rent values when used in conjunction with a rate base measured by depreciated cost. For, under this combination, the sum of the diminishing fair return plus the constant depreciation charge grows smaller and smaller as the assets approach retirement. Thus, the failure of straight-line expressly to recognize the interest or time-discount factor in the decline of capital values over periods of time serves as an offset, in the nature of a countervailing fiction, to its apparent assumption that fixed assets will have undiminished annual rent values throughout their service lives.[19] But any precise offset

[18] This statement assumes that the sinking-fund accumulations are based on a rate of interest equal to that of the allowed rate of return. But it ignores the income-tax factor as an additional capital charge, a serious omission in these days of high income-tax rates. On this point Mr. Gordon R. Corey, Financial Vice President of the Commonwealth Edison Company, writes me as follows: "Federal income taxes often have strange effects upon carrying charges. For example, the level-premium annual carrying charges on an investment in property with a 35-year life, which can be depreciated on a declining balance basis for Federal income tax purposes, are approximately 9½%, while such annual carrying charges on an investment in perpetuity, which cannot be so depreciated, are about 1½% higher or 11%. This comparison assumes a 6¼% return, a 52% income tax rate, and 52% debt with a 3½% interest rate on the debt."

[19] The significance of the failure of straight-line to recognize the interest or time-discount factor in asset valuation is illustrated by the valuation of property assumed to have a constant annual rent value, as typified by an annuity certain. Consider, e.g., the valuation of a twenty-year, $1,000 annuity, valued on the basis of a 5 per cent interest rate compounded annually. When first purchased, one year

would be purely coincidental, since there is no reason to infer a causal or statistical relationship between the phenomenon of time discount and the phenomenon of deteriorating annual rent values.

At the present time, income-tax considerations are leading many public utilities to adopt some permitted version of diminishing-charge depreciation in their determination of taxable income. This practice will be discussed in the third section of this chapter. But some authorities have insisted that, quite aside from tax aspects, straight-line depreciation accounting fails to give adequate recognition to the typically sharp downward trend in annual rent values during the earlier and rosier years of useful service life—a trend so marked that it is not adequately counterbalanced by the interest-factor offset noted above.[20] Applied to public utility rate control, this contention would imply that diminishing-charge methods of cost amortization should be accepted in the determination of the annual allowance for accruing depreciation and hence in the future build-up of the depreciation reserve. The validity of this contention cannot be proved or disproved by mere "depreciation theory"; [21] but it derives practical support from an argument, presently to be noted, in favor of rates of amortization which minimize the more serious danger of going too slow by running the less serious danger of going too fast.

before the first annuity payment, the annuity will be worth $12,462.21. In ten years, when half of its life has expired, the value will fall to $7,721.73, about 62 per cent of the initial value. Yet straight-line amortization would reduce the value by 50 per cent, to $6,231.10.

[20] This states the defense of diminishing-charge methods of depreciation accounting in its most plausible form. But the more frequent form of this defense completely ignores the fact that even straight-line methods give some recognition to declining-rent values through their countervailing failure to allow for the time-discount factor. The denial by some accounting authorities that this latter factor has any bearing on the proper allowance for fixed-asset depreciation, as an item of operating expense, may perhaps be explained by the refusal of standard financial accounting to recognize interest requirements as a production cost.

[21] One of the country's distinguished authorities on depreciation draws the following conclusions from theoretical and empirical evidence: "(1) It seems quite clear that the straight-line writeoff is a gravely retarded method of depreciation for productive equipment. Any realistic allocation procedure should get rid of at least one half of the initial value over the first third of the service life and at least two-thirds over the first half. (2) The straight-line method is apparently less retarded for buildings and structures than for equipment, how much less it is impossible, on the basis of the available evidence, to say." George Terborgh, *Realistic Depreciation Policy* (Chicago, 1954), p. 47. For an illuminating discussion of the most plausible assumptions as to the probable future rates of electric power plant depreciation, see Joseph Lerner, "Capital Costs of Private vs. Public Power for A.E.C.—a Review," 33 *Land Economics* 113–126 (1957).

"PRACTICAL" CRITERIA OF COST APPORTIONMENT

Aware of the vague and unsatisfying nature of any attempt to rationalize the choice of a method of depreciation accounting by reference to some basic "theory" of temporal cost apportionment, such as a relative-benefit theory, one may be tempted to conclude that, within wide limits, any method of amortization is as good as any other. This conclusion, however, does not follow. For the relative merits of alternative methods are subject to appraisal in the light of criteria of usefulness other than those suggested by any assumed ideal standard of fairness among the consumers of different periods. The importance properly attached to other criteria is all the greater in view of the fact that most public utility consumers are fairly steady customers and hence that *any* chosen method of cost amortization, consistently followed, will to a material degree offset in a later period of time deficiencies or excesses of an earlier period.

Aside from the marked advantages of methods of depreciation accounting that are relatively simple, understandable, and uniform throughout the country, one may properly give weight to the criterion of conservatism from the standpoint of corporation finance. This criterion suggests that, as between two proposed methods of cost amortization, one of which undertakes faster write-offs than the other during the early years of useful service lives, any reasonable doubt may well be resolved in favor of the former unless, in consequence, the resulting temporarily higher rate levels will be a serious deterrent to the development of a demand for utility services commensurate with plant capacity.[22] Accept-

[22] But the qualification just noted may be important. In a brilliant review of the development of modern depreciation theory in America, Mr. George O. May suggests that the development of railroad transportation would have been retarded had the railroads been induced or compelled to practice depreciation accounting from the beginning, with resulting higher early freight and passenger tariffs. *Financial Accounting* (New York, 1943), Chap. 9. A possible reply is that, while early transport development would probably have been slower, it might also have been more rational and more conducive to later technological progress. But a confident appraisal of this historical issue is rendered impossible by the virtual certainty that the adoption of systematic depreciation accounting could not have taken place except in company with many other changes in the direction of "commercial law and order"—changes in financial practice, in business ethics, and in rules of rate regulation of a kind not associated with pioneer days.

From the standpoint of price economics, the danger of excessively rapid capital-cost amortization is that the resulting rates of charge for service may eventually

ance by consumers of a slightly higher current rate level may be a price well worth paying for lower rate levels in the long run.

There are at least two reasons in support of this canon of conservatism. The first is the more obvious one that a public utility company may be better protected in its opportunity to enjoy a complete recoupment of its capital investments, combined with a fair rate of return on the unrecouped portions, if the attempted rate of recoupment is fairly rapid during the earlier years of the service lives of its assets. If recoupment is delayed until obsolescence has proceeded beyond a somewhat ill-defined limit, the traffic may no longer bear, nor public opinion tolerate, rates of charge for service high enough to continue the procedure thereafter. The experience of the railroads and of the street railways strikingly illustrates this danger.

The second reason in favor of liberal depreciation allowances during early years of service life concerns the effect of the allowances on the replacement policies of the management. As guardians of the investor interests, a management will naturally be reluctant to replace obsolete equipment with new equipment, even if otherwise desirable, if the accompanying retirement of the old assets will result in a reduction in the "valuation" of the company's properties for purposes of rate regulation. Hence, a management may be impelled to delay the replacement, against the dictates of sound economics, until the cost of the old asset has been completely amortized. The danger of this drag on progressive technology will be reduced if the assets are written down to minimum book values safely ahead of the dates on which they ought to be retired or relegated to stand-by service.[23]

---

become too low. That is, if these rates, when freed from the burden of fixed charges, should fall below marginal or incremental costs of production (including the costs of necessary increments of plant and equipment), wasteful overconsumption would probably result. See p. 302, *infra*. One may also note the Keynesian argument that rapid depreciation accounting, like rapid public debt amortization, constitutes a "disinvestment" adding to what Lord Keynes thought to be the danger of chronic unemployment. J. M. Keynes, *The General Theory of Employment, Interest, and Money* (New York, 1936), p. 95. Compare the opposing views of Dr. George Terborgh, *The Bogey of Economic Maturity* (Chicago, 1945).

[23] This position would have been challenged by the late Mr. Samuel Ferguson, the distinguished Chairman of the Hartford Electric Light Company, who expressed the fear that straight-line depreciation accounting, combined with the deduction of the resulting depreciation reserve in the measurement of the rate base, would tempt managements to make unjustified, premature replacements. See his "Further Comments on the NARUC Report," 21 *Journal of Land & Public*

One of the many practical virtues of an actual-cost standard of rate control, as distinct from a fair-value standard, lies in its compatibility with a policy of regulation which resolves reasonable doubts about proper annual charges for accruing depreciation in favor of the more liberal charge. Under a fair-value rule, strictly construed, any excess in the depreciation reserve over "actual depreciation," estimated in the light of hindsight, will result in a windfall benefit to the corporate investors even though the reserve reflects funds supplied by the consumers.[24] But under an actual-cost rule, strictly construed, any reserve built up by customer contributions is deductible however far it may exceed the "actual depreciation." The excess in the reserve becomes, in a sense, a "consumer equity," comparable to the consumer equity in publicly owned plants originally financed by bond issues that have already been paid off, in whole or in part, long before the plant has become obsolete.

QUALIFICATIONS OF THE STATEMENT THAT "ACCRUED DEPRECIATION" REFLECTS RECOUPED INVESTMENT

In the first section of this chapter, "depreciation" as the negative term of the rate base was first defined as amortized cost and then associated with recouped investment. But the reader was warned that the association is not one of complete identity; and it now remains to discuss the reasons for the warning.

The costs of fixed assets can be said to have been "amortized"

Utility Economics 181–184 (May, 1945). But, as Emery Troxel notes, Ferguson's position lacks plausibility or factual support. *Economics of Public Utilities* (New York, 1947), p. 346.

The point of view expressed in the text is similar to that often voiced with respect to depreciation accounting for unregulated business enterprise. That is to say, inadequate rates of cost amortization have been held responsible for undue delays in the making of economically desirable replacements. But with unregulated business, the asserted reluctance of management to replace an old asset prior to its accounting amortization is generally assumed to be irrational; whereas, with public utilities, the corresponding reluctance of management to retire an asset in a manner adverse to the rate base is not irrational in the same sense.

[24] For the leading case illustrative of this result, see Board of Commissioners v. New York Telephone Co., 271 U.S. 23, 31–32 (1926), in which Justice Butler, speaking for the Supreme Court at a time when it was still adhering to the "fair-value" doctrine, declared: "By paying bills for service they [the customers] do not acquire any interest, legal or equitable, in the property used for their convenience or in the funds of the Company."

when they have been gradually written off the plant accounts (or credited to the depreciation reserve as a contra to the plant accounts) through a series of charges to current operations. But the costs thus amortized cannot be said to have been "recouped" unless they have been matched by the receipt of revenues from the sale of the services. What part of these revenues shall be considered cost recoupment and what part income or return on investment cannot be answered with precision except in terms of accounting conventions.[25] Under these conventions, all revenues in excess of those equal to current operating expenses, exclusive of depreciation, are considered *available* for recoupment. But not even these revenues are held to constitute a return of capital beyond the point at which they cover the current depreciation charges. Any additional revenues are treated as income—as a return *on* capital rather than as a return *of* capital.

Accepting the above-noted conventional definition of "recouped investment," one may say that the depreciation reserve of a public utility company reflects capital already recouped if, but only if, it has been built up through charges to previous periods of operation, in all of which periods the company has enjoyed revenues at least equal to its operating expenses including the depreciation charges. In other words, in no year must the company have suffered an operating deficit.

But suppose that a company has suffered such a deficit, as have many of the railroads in years of poorer revenues. In that event, and under the actual-cost principle, should its entire depreciation reserve be deducted from cost new in the measurement of the rate base? Or should the deduction be limited to the cost that has already been recouped?

Partly, no doubt, because the Interstate Commerce Commission has not purported to apply a strict, actual-cost standard of rate control, this question has not often been raised in the literature. But, in my opinion, the entire depreciation reserve should be deducted.

[25] The conventional nature of the distinction between the return *of* and the return *on* capital is emphasized by the refusal of the Committee on Terminology of the American Institute of Accountants to identify depreciation accounting with periodic asset revaluations. This refusal throws doubt on the meaning of a traditional concept of property income as whatever portion of the gross yield of property can be withdrawn without "impairing the capital." As my accountant colleague Professor James L. Dohr reminds me, "impairment of capital" is itself a convention.

This conclusion is in harmony with the general principle, applied alike under a fair-value rule and under an actual-cost rule, that past deficiencies in corporate revenues do not give rise to valid claims for offsetting recoveries in later years. In refusing to grant such recoveries, rate regulation should and does recognize the resulting necessity of giving to companies the somewhat risky opportunity to enjoy higher rates of return than might otherwise be required in order to attract capital. But any resulting higher "costs of capital" are well worth paying for as a means of avoiding the evils of capitalized deficits.[26]

THE DEPRECIATION RESERVE AS A MEASURE OF "ACCRUED DEPRECIATION"

Under a systematical and well-administered program of actual-cost rate regulation, a public utility company's depreciation reserve, as presented in its balance sheet, should reflect capital costs already amortized through previous charges to operation and hence already recouped unless the company has been unfortunate enough to have suffered operating deficits. But this is an ideal situation, and the ideal has not always been realized in actual practice. One reason for its nonrealization is that some commissions, including that of New York State, have not been given the legal power to approve or disapprove rates of depreciation for income-statement purposes, while other commissions possessing this power have lacked the funds or the will to exercise it.

In these situations, the depreciation reserve per company accounts may not be a reliable measure of the costs that the company has already recouped, or has had the opportunity to recoup, from previous consumers. Its place should then be taken by a hypothetical or pro-forma reserve, sometimes called "the reserve requirement." This reserve requirement reflects the reserve that the company would have built up had it embodied in its financial statements methods of depreciation accounting acceptable for rate-making purposes. Not all commissions, however, have made clear what methods they deem acceptable; and in this event they are under at least moral pressure to accept the method (if any) consistently and carefully applied by the company in question, so long as it falls within the limits of recognized good practice.

[26] Compare pp. 92 and 167, *supra*.

TREATMENT OF PROPERTY RETIRED PREMATURELY BY VIRTUE OF EXTRAORDINARY OBSOLESCENCE

In modern regulation, the allowances for "depreciation" both as operating expenses and as deductible reserves are designed to cover so-called functional depreciation including obsolescence and not merely physical deterioration or wear and tear. Hence the allowances must be based on estimates or plausible assumptions as to the effect of obsolescence on useful-life expectancies. But neither a corporate management nor a commission can hope accurately to predict, years in advance of the event, the dates as of which old properties may need to be retired for reasons of "extraordinary obsolescence."

To a material extent, the harm arising from this lack of prophetic vision can be minimized by midstream re-estimates of remaining-life expectancies and hence by reasonable step-ups or step-downs in annual depreciation charges, so that the dates of complete amortization can be made to correspond fairly well to the dates of the actual retirement. This procedure is much more readily employed under "group methods" of depreciation accounting, whereby the premature retirements of some assets are offset by the longevity of other assets in the same group.

But there occasionally arise extreme cases of unexpected obsolescence, in which a company faces the necessity, or at least the economic desirability, of retiring expensive portions of its entire plant and equipment years before it has received a fair opportunity to recover its investment therein under a routine procedure of depreciation accounting. Striking examples of this necessity have arisen in recent years in the gas industry, in which the distribution companies have abandoned their old manufactured-gas plants in favor of the purchase of the much cheaper natural gas from the newer pipeline companies.

Under a strictly construed present-value theory of rate making, the fact that a company may have failed to recover its outlay in outmoded plant should not give it even a shadow of a claim to a recovery of this outlay from future consumers. But under an actual-cost or net-investment principle, the problem illustrated by the premature retirement of the manufactured-gas plant presents a dilemma. On the one hand, the cost principle suggests that a company should receive an opportunity to recover from later customers compensation for all capital outlays for which it has not yet received full compensation from earlier customers. Yet, on the other hand, the same cost principle has usually been held to entitle a company to compensation only for such capital outlays as reflect the costs of property still "used and useful in the public service."

Faced with this dilemma, commissions have tended—wisely, in my opinion—to prefer the former alternative to the latter. Their precise rulings on this issue have not been uniform. But the New York Commission, which accepts a net-investment standard of rate control, has sanctioned the amortization as an operating expense, over a "reasonable" period of time, of capital losses incurred by gas-distribution companies in the premature retirement of manufactured-gas plants accompanying conversion to natural gas.[27] I do not understand, however, that it has also approved the inclusion of these losses, prior to their amortization, in the rate base under the guise of an "intangible asset." This compromise seems illogical; but it may be defended as a workable intermediate position between the complete denial and the full-scale recognition of the claim of a public utility company to post-mortem compensation for capital costs on which it has not received a reasonable opportunity to secure full compensation prior to retirement.

THE PROBLEM OF RETROACTIVE APPLICATION

The acceptability of a company's depreciation reserve as the proper measure of the rate-base deduction for depreciation has been an especially acute issue when the reserve was built up, in whole or in part, before the company had come under the jurisdiction of the regulating commission, or else before the adoption by this commission of its present rules of depreciation accounting. This latter situation was one of general experience in American rate regulation. For in the 1920s and the early 1930s, most commissions, influenced by shortsighted arguments of spokesmen for the gas and electric companies,[28] had approved or acquiesced in a

[27] See the New York Commission's Cases 15613 (Feb. 26, 1952); 15604 (Feb. 19, 1952); 16189 et al. (June 29, 1953). All of these cases concerned change-over to natural gas by the metropolitan gas companies.

[28] As an outstanding exception, the American Telephone & Telegraph Company has long practiced accrued-depreciation accounting on a straight-line basis and has made notable contributions to the development of this accounting through the application of actuarial techniques. But its well-deserved credit for this

modified version of retirement-expense accounting, which rested content with a relatively small "retirement reserve"—a reserve sufficient only to avoid the serious instability in annual reported operating expenses that would result from the immediate charge of all retirements to current operation. In the mid-1930s, the commissions prescribed uniform systems of public utility accounts requiring the practice of accrued-depreciation accounting of the character described in the second section of this chapter. This change left those companies that had practiced retirement-expense accounting with reserves grossly inadequate by the newer standards. There arose a serious problem of retroactive application.

To the best of my knowledge, no writer has yet attempted a country-wide survey of the subsequent rate cases with the object of determining to what extent the burden of the required shift to depreciation accounting may have fallen on the corporate stockholders or on the consumers. Indeed, such a study would necessarily be inconclusive, since so many factors other than rulings on depreciation have determined commissions' rate-making policies.[29] But there is little reason to doubt that, in most jurisdictions, the burden has been shared by both of these interested groups. Fortunately, the problem has grown less and less serious with the passing years because of the gradual build-up of reserves equal, or nearly equal, to modern reserve requirements.[30]

Although most problems of retroactive adjustment have been

---

accomplishment was sadly marred by its insistence, for a period of years, that the reserves built up, in effect, by charges to its consumers were chronically far in excess of "actual depreciation" and hence should not be deducted in the determination of the "fair value" of its properties for rate-making purposes. See New York Telephone Co. v. Prendergast, 36 F (2d) 54 (1929). The inconsistency of this position was sharply noted by Chief Justice Hughes, speaking for the Supreme Court in Lindheimer v. Illinois Bell Telephone Co., 292 U.S. 151 (1934).

[29] One cannot properly assume that, during the period in which the companies practiced retirement-expense accounting, the consumers must have received the benefits of any temporarily lower operating deductions. Regulation in these years was very deficient in most areas and was seriously crippled by the unrealistic rulings of the appellate courts in rate-making valuations under the prevailing doctrine in *Smyth v. Ames*.

[30] In one opinion, the citation of which escapes me, the Federal Power Commission intimated that it would not deduct more than the actual reserve unless it should be *grossly* inadequate. In the Consolidated Edison Rate Case of 1952, 96 PUR NS 194, the New York Public Service Commission found the electric-property depreciation reserves grossly deficient and deducted the higher reserve requirement in the determination of the rate base. But the Commission stressed the fact that, under its governing statute, it had not received any authority to prescribe the company's depreciation accounts.

concerned with inadequate depreciation reserves, the opposite situation has also arisen on occasion. Indeed, it arose in the famous *Hope Natural Gas* case.[31] Before the Hope Company had come under the jurisdiction of the Federal Power Commission, it had enjoyed complete or partial freedom from rate regulation, with the result that the book values in its accounts exercised little or no control over its rates of charge for natural gas. Under a highly "conservative" accounting practice—a practice that may possibly have been due to its control by the Standard Oil Company of New Jersey—it had built up a depletion and depreciation reserve that the Commission found grossly excessive in the light of modern "reserve requirements." Declaring that the reserve, however excessive, nevertheless represented contributions made by past consumers, Commissioner Scott favored its complete deduction from cost new in the measurement of the rate base. But the Commission disagreed and deducted only the "reserve requirement" of $22,328,016, which fell short of the actual reserve by about $18,000,000. Taken in combination with other rulings, adverse to the company, the Commission's action on this issue seems to me to have been wise. But neither this action nor the alternative desired by Commissioner Scott can be rationalized by reference to any grand "theory" of retroactive equities in rate regulation taking place under a change in the rules of the game.[32]

THE DEPRECIATION OR AMORTIZATION OF ACQUISITION-ADJUSTMENT COSTS

Since this chapter is concerned with the depreciation problem under an actual-cost principle of rate making, it has taken for granted the acceptance of actual cost as a "depreciation base"—as the capital cost to be amortized through systematic annual charges to operations. But, as already noted in Chapter XII, "actual cost" is ambiguous; and one source of ambiguity arises with respect to property which, having been constructed to the order of the original operating company at a cost of, say, $10,000,000, has later

[31] 3 F.P.C. 150 (1942). In the case on appeal, the Supreme Court did not take occasion to pass on the merits of this issue. Federal Power Commission v. Hope Natural Gas Co., 320 U.S. 591 (1944).

[32] See Robert L. Hale, "Utility Regulation in the Light of the Hope Natural Gas Case," 44 *Columbia Law Review* 488, 503–520 (1944); Henry F. Lippit II, "Net Investment Rate Making—the Deduction for Depreciation," 62 *Harvard Law Review* 1155–1179 (1949).

been acquired by "the present accounting company" for, say, $12,000,000. Which cost is relevant as a measure of the rate base is a question already discussed in this earlier chapter, which concludes that the appropriate answer depends on the reasons for and consequences of the transfer. We must now *assume* an answer to this primary question, since the answer governs any rational treatment of depreciation, at least for purposes of rate control.

Let us first assume a decision that the only relevant cost for rate-making purposes is the original cost of $10,000,000. Under this assumption, $10,000,000 should constitute the depreciation base; and the accounting disposition of the $2,000,000 excess should have no recognized effect in a rate case. Even if the excess cost is gradually amortized instead of being charged in a lump against some kind of a surplus account, the annual amortization charge should be against income, "below the line," and not against operation.

If, on the other hand, the $2,000,000 excess acquisition price (or any part thereof) has been held to be a proper component of the rate base, as reflecting capital devoted to the public service, it should then receive corresponding treatment in the manner in which it should be depreciated or (in other words) amortized. But how rapidly it should be amortized is a difficult question to answer with confidence unless the excess purchase price can be intelligently distributed to the various plant accounts, tangible and intangible. If this is not feasible, an arbitrary rate, such as characterizes accounting practice with respect to some intangibles, may be chosen. But in any event, the amortization should be treated as an operating charge for rate-making purposes—a conclusion which militates against a speed of amortization seriously burdensome to present consumers.

It is my understanding that, for purposes of accounting, the Federal Power Commission has uniformly required the amortization of acquisition-adjustment-account debits over a "reasonable period of time" (typically 10 to 15 years)—not, however, by charges to an operating expense account.[33] But it is also my understanding

[33] Federal Power Commission, *Report on the Reclassification and Original Cost of Electric Plant of Public Utilities and Licensees* (Washington, D.C., 1950). See also the articles by Messrs. Clader, Colbert, Dohr, Kohler, May, and Paton, cited in footnote 33 of my own article, "Contributions of the Federal Power Commission to the Establishment of the Prudent Investment Doctrine of Rate-Making," 14 *George Washington Law Review* 136–151 (1945).

that the Commission has not yet been compelled to decide a rate case in which acquisition-account allowances have been at issue. Pending a decision on rate-making aspects, the Commission's disposition of the accounting question seems to me defensible only on the assumption that it is necessarily of a provisional character.

ALLOWANCES FOR INCOME TAXES UNDER DIMINISHING-CHARGE DEPRECIATION ACCOUNTING

In the public utility field, the most recent important controversy about depreciation has concerned the accounting and rate-making effects of the provisions of Section 167 of the Internal Revenue Code of 1954, permitting business corporations, in calculating taxable income, to use diminishing-charge procedures of depreciation accounting: specifically, a declining-balance method and a sum-of-the-years-digits method. These liberalized tax-accounting allowances were supported in the Congressional committee hearings partly on the ground that they would stimulate business to expand its capital investments, and partly on the ground that they come closer than straight-line depreciation accounting to a reflection of the rates at which most fixed assets actually depreciate in value from the dates of acquisition to the dates of retirement.[34] But many public utility companies have chosen to stress the first point and to ignore the second. That is to say, they have fairly generally decided to take advantage of the diminishing-charge deductions for tax purposes, while resting content with straight-line depreciation procedures for their financial statements and, presumably, for rate-making purposes.[35] As a result, and since they are in an era of heavy plant expansion rather than in an era of stable equilibrium between

[34] Although these two grounds have often been cited as supplying two separate arguments in favor of the accelerated-depreciation provisions of the tax code, they can also be considered as complementary. That is to say, the provisions may have been incorporated in the belief that business corporations would be stimulated to undertake more rapid capital expansion if permitted to claim for tax purposes rates of depreciation equal to the rates at which their assets actually depreciate.

[35] I am informed that many of those industrial companies which use diminishing-charge methods of depreciation for Federal income-tax purposes use the same methods for their own financial statements. Failure of the public utilities to follow suit might be due to refusal by commissions to let them do so. But no serious attempts at persuasion on the part of the companies have yet come to my attention.

acquisitions and retirements, their current Federal income taxes are reduced by the accelerated rate of tax depreciation, whereas their annual allowances for depreciation as reported to the public service commissions remain unaffected.

By way of making accounting adjustments for this discrepancy between their income reports for tax purposes and their income reports for regulatory purposes, many public utilities have sought leave from the various public service commissions to include, as operating charges, the higher income taxes to which they would be subject were they to report taxable income on a straight-line basis. The excess in these "normalized" taxes over current tax liabilities is to be carried to a special deferred-tax account, against which to charge any later, offsetting enhancements in income taxes. This accounting procedure has already been sanctioned by the Federal Power Commission and by perhaps a majority of those state commissions that have yet issued definite rulings on the subject.[36]

But the really important issue is concerned with the rate-making aspects of this accounting problem, and here each of three major alternatives (along with some rather question-begging compromises) has derived support from some commissions. The first position, accepted in Pennsylvania, California, and elsewhere, is that a public utility company which elects to pay income taxes on a diminishing-charge basis of depreciation accounting may receive no allowance for any taxes beyond those for which it is actually liable in a given year. The second position is that a rate-making allowance shall be made for "normalized taxes" as an operating deduction but that no offsetting deduction shall be made in the measurement of the rate base, since the account for deferred taxes is deemed to constitute a restricted surplus and not a reserve representing amortized capital costs. The third position is that (both for rate-making and for accounting purposes) "normalized" taxes shall be accepted as operating deductions but that any excess in such tax allowances over actual taxes shall be credited to a special reserve account, the

[36] The periodical *Public Utilities Fortnightly* undertakes to keep the reader up to date on commission treatment of accelerated depreciation for accounting purposes as well as for rate-making purposes. See, e.g., Owen Ely, "Deferred Taxes—How the Issue Is Working Out in Rate Cases," 62 *Public Utilities Fortnightly* 104–109 (July 17, 1958). See also an address by F. M. Beatty, Partner, Arthur Andersen & Co., before the New York Society of Security Analysts, Oct. 16, 1957: "Current Regulatory Status of Deferred Tax Accounting for Utilities."

amount of this reserve being deducted from cost new in arriving at the rate base just as is the ordinary depreciation reserve.[37]

The second alternative is the one most popular with the public utility industries since, from their point of view, it has the charm of imposing upon the consumers the obligation to pay deferred-tax allowances which, instead of being transmitted forthwith to the United States Treasury, are to be treated as capital investments entitled indefinitely to the enjoyment of a "fair rate of return" for the benefit of the corporate stockholders.[38] In this respect it has the same charm as that once possessed by the practice under which some public utilities would demand straight-line allowances for accruing depreciation while insisting on the deduction of nothing but a minimum "observed depreciation" in the measurement of the rate base. Support for this position of the industry has been forthcoming from the Federal Power Commission [39] and from a few state commissions. But I have never seen a plausible defense

[37] This arrangement was proposed by the company and approved by the commission in the last Commonwealth Edison Company general rate case before the Illinois Commerce Commission: Docket 44391, June 18, 1958, 24 PUR 3d 209. In another case, the same commission, while not deducting the credit to deferred taxes from the rate base, undertook to reach the same or similar results by an adjustment in its rate-of-return allowance, a far less effective alternative. Re Peoples Gas Light and Coke Co., 27 PUR 3d 209 at 227 (1959). The Wisconsin Commission has held that amounts equal to the allowances for deferred taxes must be added to the depreciation reserve—a position defended by its distinguished Chief of Accounts and Finance, A. R. Colbert.

[38] Noting that his company was one of the first public utilities to use the sum-of-the-years-digits method of depreciation for tax purposes, President Walker L. Cisler of the Detroit Edison Company said: "Consequently we have received and will continue to receive through tax deferrals the maximum possible amounts of interest-free capital. At the present time, through rapid amortization and accelerated depreciation, we have invested about $20 millions of interest-free capital, and this figure will assume somewhat larger proportions." Brochure containing address of Nov. 7, 1956, to the New York Society of Security Analysts, Inc., New York City, p. 7.

[39] The Federal Power Commission's position was first clearly set forth in Matter of Amere Gas Utilities Co., 16 F.P.C. 880 (Aug. 29, 1956), where Commissioner Connole wrote a vigorous dissent. The Commission had indirect legal support for this position in a decision of the Circuit Court of Appeals for the District of Columbia: City of Detroit v. Federal Power Commission, 230 F. (2d) 810 (1955), although that case dealt with accelerated amortization under Sec. 168 of the Tax Code, not with liberalized depreciation under Sec. 167. Apart from its assumption that it was under Congressional mandate to give corporate stockholders the direct benefit of the tax provisions, the Federal Power Commission itself seems to have entertained doubts on the merits of the issue. For, in a ruling of July 2, 1956, relative to accounting for Federal income tax accruals under liberalized depreciation methods, after citing the above-noted court decision as "completely controlling," it added that "the extraordinary ability and willingness" of natural gas companies to attract capital might lead it to call Congressional attention to the

of this claim to the enjoyment of a profit on funds not contributed by the corporate investors. The defense usually offered is that plant expansion financed by these funds enhances management costs and increases the risk factor. But management costs are covered in the allowances for operating expenses, not in the rate of return. And the risk factor (which may even be reduced, not increased, if the company is permitted to accrue a so-called deferred-tax reserve, as it will under Alternative Number 3) is properly taken into account in the allowance of a fair rate of return on capital contributed by the investors. Hence, there is no need to concede to stockholders a return on capital contributed, in effect, either by the taxpayers or by the consumers.[40]

As I see it, the only reasonable controversy as to the choice among the three above-noted alternatives is that between the view that, for rate-making purposes, companies should receive no allowances for taxes other than for actual current taxes, and the view that, if they practice liberalized-depreciation accounting for purposes of income taxation, they should receive an annual allowance for deferred taxes combined with a deduction of the resulting deferred-tax reserve from what would otherwise be the rate base. Here I am convinced that the weight of the argument lies with the latter

---

question whether the incentive provided by Sec. 167 of the Internal Revenue Code is necessary or desirable for the natural-gas industry.

Whatever interpretation may be placed on "intent of Congress," one might suppose that state commissions would not feel bound by this intent in developing rules for the regulation of intrastate utility rates. Yet the utility companies have pressed the Congressional-intent argument on these commissions—not without success in some cases. Ironically, the accounting firm of Price, Waterhouse & Company is reported to have urged the Federal Power Commission to broaden its accounting treatment of deferred taxes so that the views of state regulatory commissions shall not be superseded by Federal rules of accounting! 60 *Public Utilities Fortnightly* 872 (Nov. 21, 1957).

[40] As supporting the adverse position just expressed, see City of Alton v. Commerce Commission, 19 Ill. 2d 76, 165 N.E. 2d 513 (1960), overruling a Commission rate-making order which conceded to a water company an operating deduction for deferred taxes without making an offsetting deduction in the determination of the rate base. In a very able opinion speaking for the Court, Justice Schaefer said: "The Commission's order therefore can not stand. Funds generated by accruing deferred tax expenses, and any facilities financed out of these funds, must be excluded from the rate base." 165 N.E. 2d 524. In this case the Court had at first adhered to the "flow-through" principle in denying to the company any rate-making allowance for taxes in excess of current tax liabilities. But it changed its ruling following a petition for rehearing filed by Arthur Andersen & Co., public accountants, as Amicus Curiae: Brief by Max Swiren, Esq. The final ruling permitted the Illinois Commerce Commission to apply what I have called Alternative Number 3 to the problem under inquiry.

position,[41] and this for three reasons: first, that this position is in harmony with the modern tendency to regard straight-line depreciation as erring on the side of a retarded allowance for cost recoupment rather than an excessive allowance as was once often thought to be the case; secondly, that the very practice of taking rapid depreciation for tax purposes tends to reduce more rapidly the actual values of the depreciating assets—namely, their tax-saving values; and thirdly, that unless utility companies are permitted to set up reserves against so-called deferred taxes, thereby protecting themselves against the possible repeal of the diminishing-charge provisions of the present tax law, they are likely to exercise what has been held to be their option to ignore these provisions in favor of the orthodox straight-line tax accounting—an option adverse to the long-run interests of their consumers. Already, indeed, utility companies in Pennsylvania and elsewhere have exercised this option.[42]

Instead of ending this somewhat involved chapter with a summary of its major conclusions, I return to the primary question raised at the beginning—the question as to the relevance of any

[41] The main argument for a commission's refusal to make any deferred-tax allowance in a rate case—for the so-called flow-through principle—is that, as long as the tax law remains unchanged and as long as additions to depreciable corporate assets exceed retirements, the "tax deferment" will be continuous and hence will amount, in effect, to a permanent tax saving. With qualifications this contention is correct in that a reduction in current taxes below what these taxes would be under straight-line accounting will not later be offset by an increase in these taxes beyond what they would be under straight-line. Robert Eisner, "Accelerated Amortization, Growth, and Net Profits," 60 *Quarterly Journal of Economics* 533–544 (1952); Sidney Davidson, "Accelerated Depreciation and the Allocation of Income Taxes," 33 *The Accounting Review* 173–180 (1958). But under flow-through, the major benefit of the tax reduction would go to the earlier consumers, in the years in which the tax payments have been reduced, instead of being apportioned among consumers more nearly in proportion to their relative responsibility for payments for services resulting in eventual tax liabilities. As an argument against the accrual of a tax-deferral reserve, the permanent-deferral theory is suspiciously similar to the now discredited "plant immortality" theory of depreciation, mentioned early in this chapter, which was once adduced by the utility industry as an argument against the deductibility of accrued depreciation from cost new in the determination of the rate base.

[42] But see Professor Ben W. Lewis's contention that commissions should have the power to deny to companies, for rate-making purposes, any income-tax allowances in excess of the taxes to which they would be subject if they were to choose accelerated depreciation in determining their tax liabilities. "The Duty of a Public Utility to Reduce Its Income Tax Liability by Using Accelerated Depreciation," 35 *Land Economics* 104–114 (1959).

deduction for "depreciation" in the measurement of a rate base under an actual-cost standard as distinct from a present-value standard. These different standards require different rules for the treatment of depreciation and even require different definitions of "depreciation" as the amount to be deducted from cost new in the determination of the rate base. But it does not follow, as has sometimes been stated or implied, that the allowance for depreciation under an actual-cost standard "has nothing to do" with the decline in the values of the depreciable assets. For even an actual-cost rate base, while not measured by the value of the property at the time of a rate case, should be indirectly related to this value and, especially, should be kept from exceeding the value to a serious extent. The rules for depreciation accounting or capital-cost amortization summarized in this chapter are designed to serve this practical purpose. Hence, the annual allowances for depreciation as an operating deduction are properly designed to reflect the general downward trend in the service values of the depreciable fixed assets during their "irresistible march to the junk heap." To this extent, as in other respects noted in Chapter XI, an actual-cost standard of rate control is not a standard completely devoid of value concepts.

XIV

THE RATE BASE: REPLACEMENT COST AS AN ALTERNATIVE STANDARD

Bearing in mind the fact that, even when modified by price-index adjustments designed to restate past costs in terms of present dollars, the actual-cost basis of rate control is not a value basis, we now turn to the fundamentally different replacement-cost [1] or present-value principle. Let it be noted at the outset, however, that, when strictly interpreted, this principle refers to a paper theory rather than to any actually accepted, operative rule of rate making. Even so, the principle is worth reviewing if only to call attention to the superficial character of its resemblance to that ill-defined measure of reasonable utility rates which goes under the title of the "fair-value" standard.

Chapter XII, on the actual-cost or net-investment standard, first undertook to define this standard in its modern form before considering its claims for acceptance as a preferred basis of rate regulation. In turning to the alternative of replacement cost, the present chapter will follow, in part, a reverse procedure. First will come a brief exposition of the economic argument in favor of replacement-cost rate making; then a discussion of the relevant definition of replacement cost. This seemingly illogical sequence is dictated by the need to consider first the economic philosophy of the replace-

[1] In this chapter, following a widespread practice, I use "replacement cost" and "reproduction cost" as mere synonyms. Those writers who have drawn a distinction have not been in agreement as to what the distinction should be. But there has been a tendency to associate "reproduction cost" with the cost of replacing the existing plant with a substantially identical new plant. An alternative distinction, though one seldom drawn in rate-making literature, would be that between the cost of replacing the *present* output of service and the cost of producing *additional* units of service (reproduction cost).

ment-cost principle in order to find any rational grounds for a choice among alternative definitions of the principle, especially for a choice between the legally orthodox tradition in favor of the reproduction cost of a substantially identical plant and the economists' preference for cost of replacement by means of a modern, substitute plant. But the economic philosophy can be stated very briefly here, since it has already been discussed in Chapters IV and VI.

ECONOMIC ARGUMENT FOR A REPLACEMENT-COST STANDARD

Viewed, then, from the standpoint of price theory, the case for the replacement-cost principle rests primarily on its supposed effectiveness in securing a proper allocation of the country's limited resources of labor, raw materials, and capital as between the production of public utility services and the production of other goods and services. The attainment of such an allocation is one of the major functions of prices in general, including public utility rates —the third of the four functions distinguished in Chapter III. According to the argument now under review, if public utility rates were set at levels higher than would be required by a replacement-cost standard, they would unduly restrict consumption of public utility services, since they would deter consumers from taking the full amounts of service for which they are ready to pay the costs of rendition as measured by current costs of production. On the other hand, if these rates were set below current replacement costs in compliance with the "false" standard of a return on actual cost, they would tend to stimulate a demand for excessive, wasteful consumption. The consumption would be wasteful because the increased output required in order to satisfy the high demand could be produced only at a cost in excess of the price charged for the service.

In thus serving to secure a well-balanced demand for utility services, replacement-cost pricing is supposed to be a better standard than actual-cost pricing, since it is said to reflect costs that are still escapable rather than outlays which, having already been made, cannot be avoided or minimized by any restrictions on future consumption. Even if, for reasons set forth in Chapter VI, this type of price control does not completely adopt the ideal norm of a competitive price, its supporters would claim that it comes as close to this norm as does any feasible yardstick of monopoly pricing.

Thirty years or more ago, the case for the replacement-cost principle, as just set forth in a very sketchy and inadequate manner, was developed with great skill, and with particular reference to railroad rates, by Professor Harry Gunnison Brown of the University of Missouri.[2] Similar views have been expressed by later writers,[3] but they have lacked both the incisiveness and the firmness of conviction that make Brown's earlier analysis a classic in the history of rate regulation.

REPLACEMENT COST OF SERVICE AS THE RELEVANT COST

Persons familiar with this economic defense of the replacement-cost principle, as expounded by Brown and others, will need no reminder that the version of the principle on which it relies is quite different from the one formerly accepted by the Supreme Court under its interpretation of the "fair-value" rule and still apparently accepted in states that retain this rule even after its repudiation as a constitutionally imposed "law of the land." According to the orthodox legal doctrine, estimates of reproduction or replacement cost are germane to a rate case only as evidence or "elements" of the fair value of the properties devoted to the public use. But because of a widely shared confusion as to the precise relevance of a replacement-cost estimate as a possible index of the actual value of replaceable property—a confusion especially prom-

[2] See, e.g., his *Transportation Rates and Their Regulation* (New York, 1916); "Railroad Valuation and Rate Regulation," 33 *Journal of Political Economy* 505–530 (1935) and 34 *ibid.* 500–508 (1936); "Economic Basis and Limits of Public Utility Regulation," 53 *American Bar Association Reports* 717–737 (1928). For a practicing lawyer's defense of reproduction-cost rate-making valuations similar to that of Professor Brown's, see Frederic G. Dorety, "The Function of Reproduction Cost in Public Utility Valuation and Rate Making," 37 *Harvard Law Review* 173–200 (1923).

[3] Expressed usually, however, in support of the contention that public utility rates should be derived from a rate base which makes allowance for a major inflation in general price levels rather than in support of a strict replacement-cost-of-service basis of rate control. See, e.g., William A. Paton and Howard C. Greer, "Utility Rates Must Recognize Dollar Depreciation," 51 *Public Utilities Fortnightly* 333–356 at 335 (1953). "The inescapable fact," declare these writers, "is that utility rates in general are so unjustifiably low that they stimulate more consumption than can be provided by available facilities, retarding the expansion of the services avidly sought by consumers, and hampering the full enjoyment of what could be supplied if charges were more nearly in line with today's price level."

inent in valuations for different legal purposes including tax and condemnation cases as well as rate cases—reproduction cost has usually been taken to mean the hypothetical cost of a substantially identical new plant: [4] one of the same capacity and location, the same size and number of generating units (with electric utilities), the same motive power (with transportation utilities), the same type of telephone switchboards (with telephone utilities), etc., etc. The resulting "valuation" of the property is therefore an economically meaningless application of up-to-date prices to out-of-date properties. Rates derived from such a rate base do not reflect even a serious attempt to follow the general principles of competitive pricing in a dynamic economy. Nor do they avoid the objection that the costs which they reflect are a form of sunk costs even though restated in terms of current unit prices.

Defenders of the principle of replacement-cost rate control would doubtless reply that the criticisms just advanced apply, not to the principle itself but rather to its fallacious interpretation by tribunals and by engineering experts for the public utility companies.[5] But critics of the whole philosophy of replacement-cost

[4] This was the position of Justice Butler, speaking for the Court in McCardle v. Indianapolis Water Company, 272 U.S. 400, 417 (1926). See my *Valuation of Property* (New York, 1937), pp. 1124, 1184. For a different position, see the concurring opinion of Justice Brandeis in the Southwestern Bell Telephone case, 262 U.S. 276, 312 (1923), and the dissenting opinion of Justice Stone in St. Louis & O'Fallon Ry. Co. v. U.S., 279 U.S. 461, 548–553 (1929). In his dissenting opinion in the *Hope Natural Gas* case, Justice Reed attached the following footnote to his reference to reproduction cost as a relevant factor in the determination of fair value: "'Reproduction cost' has been variously defined, but for rate-making purposes the most useful sense seems to be, the minimum amount necessary to create at the time of the inquiry a modern plant capable of rendering equivalent service. See I Bonbright, *Valuation of Property* (1937) 152. Reproduction cost as the cost of building a replica of an obsolescent plant is not of real significance." 320 U.S. 591, 622 (1944). Even in the early cases, commissions and courts declined to assume replacements of an altogether incredible type. For example, in its railroad appraisals under the Valuation Act of 1913, the Interstate Commerce Commission declined to assume the replacement of ties with outdated kinds of wood such as bois d'arc. I. L. Sharfman, *The Interstate Commerce Commission*, Vol. III A (New York, 1935), pp. 161 *et seq.*

In more recent public utility valuations on a replacement-cost basis, the appraisal engineers have offset, to a limited and indeterminate extent, the fallacy of identical-reproduction-cost appraisals by allowances for obsolescence far more liberal than those conceded in earlier years. But offsets of this kind cannot be reliable in the absence of estimates both of the construction costs and of the operating costs of modern, substitute plants.

[5] In 1942, one of these defenders, Mr. Joseph Jeming, described an actual attempt by a public utility company to apply the reproduction-cost-of-service principle in

regulation, whose views I have supported for many years, may fairly raise the question why a fallacy so elementary from the standpoint of appraisal theory can have persisted to this day among rate-making tribunals, almost sixty years after the Supreme Court, in its dictum in *Smyth v. Ames,* had referred to current costs of construction as one of the important "elements of fair value." In part, at least, the answer must lie in a recognition by practical-minded judges, commissioners, and experts, that estimates of the cost that would be incurred in replacing the service by means of a new type of plant if the existing plant were to disappear into thin air are altogether too speculative and too litigious for purposes of feasible administration. When one considers, moreover, that the replacement might well take the form of a change in the very nature of the service, and not just of a change in the design and location of the plant—a change, say, from street cars to buses, from branch-line railroad service to common-carrier truck service, from house-heating by gas to heating by electrically operated heat pumps, etc.—one need not be surprised that the replacement-cost-of-service theory of rate making remains today a paper theory. What goes under the name of a replacement-cost or reproduction-cost principle is a principle (if such it can be called) which, in its concessions to the real or supposed needs of feasible administration, loses the very charm that has endeared it to some of the academic economists.

---

a case concerned with the rates for manufactured gas "in a large metropolitan community"—I assume, Brooklyn. "Unfortunately for the complete development of the proposed principle of rate-making, the case was settled before any cross-examination in respect of these studies was conducted and before any rebuttal testimony was presented." "An Actual Application of the Reproduction-Cost-of-Service Principle in Rate Making," 18 *Journal of Land & Public Utility Economics* 188–203 (1942). See also Mr. Jeming's earlier article on the same subject in Vol. 17 of the same journal (1941), pp. 138–150.

It is by no mere coincidence that the company which proposed to make use of the replacement-cost-of-service principle was a manufactured-gas company, since technological progress in that area has been relatively slow. In electric power production, e.g., postwar improvements in efficiency accompanying the use of larger generating units are said to have kept pace with increases in the prices of labor and raw materials. See Mr. Philip Sporn's announcement of his system's decision to buy two 450 megawatt generators: American Gas & Electric Company news release of April 27, 1956. In the telephone field, compare the statement of Mr. George C. McConnaughey, Chairman of the Federal Communication Commission, that "by 1965 about 50 per cent of expenditures for new construction is expected to be in electronic categories." 58 *Public Utilities Fortnightly* 563 (1956).

IDENTITY BETWEEN A REPLACEMENT-COST-OF-SERVICE STANDARD AND A PRESENT-VALUE STANDARD

Before discussing further the limitations of a replacement-cost-of-service principle of rate making, we may note that, in theory, the principle is the same as that of a correctly interpreted present-value basis of rate making. The present service value of the existing plant of a public utility enterprise can be measured by the estimated cost of a modern, substitute plant minus an allowance, traditionally called "depreciation," for the value inferiority of the old plant including any inferiority due to higher operating costs. If the rate base were set at a value thus measured, the deduction for depreciaion would offset the higher operating expenses of the existing plant.[6] In actual practice, however, public utility properties are seldom valued in this manner for purposes of rate control. Instead, the valuation is based on a compromise between depreciated actual costs and engineering estimates of depreciated identical reproduction costs. Whether or not the resulting permitted rates of charge for service are higher or lower than those rates which would be called for under a replacement-cost-of-service test must be a matter of almost pure coincidence.

EVEN IF RATES WERE BASED ON REPLACEMENT COSTS OF SERVICE, THEY WOULD STILL NOT CONSTITUTE "OPTIMUM" PRICES

Despite the strong probability that a replacement-cost-of-service principle will never become a widely accepted operative rule of rate making, a serious attempt to go as far in the direction of this principle as practical difficulties permit might still be worth making if the principle itself could qualify, in theory, as a principle of optimum resource allocation. In fact, however, it cannot so qualify, since it identifies reasonable rates with rates sufficient, in the aggregate, to cover total costs of service replacement including a capital-attracting rate of return on the hypothetical investment in a new plant. With a public utility or railroad system still operating at a scale at which further enhancements in rates of output can

[6] See Maurice R. Scharff's discussion of the theory of replacement-cost appraisals in *Technical Valuation*, Jan., 1948, and in 114 *Transactions of the American Society of Civil Engineers* 907–925 (1949); also the articles by Joseph Jeming cited in the preceding footnote.

take place with less than a proportionate increase in operating and capital costs (conditions of decreasing unit costs), such rates will exceed the incremental or marginal costs of the service. Yet, under the economists' theory of optimum pricing, the important relationship between prices and costs is an equality, under equilibrium conditions, between prices and *marginal* costs. Hence, if optimum resource allocation were to be accepted as the primary objective of rate-making policy, as the replacement-cost advocates insist, what would be required is not a mere transfer from an actual-cost standard to a replacement-cost standard, but rather a transfer from *any* standard of total cost to a standard of incremental cost. This transfer could be accomplished only by the aid of tax-financed subsidies as long as incremental costs are less than average total costs.[7]

In view of this theoretical deficiency of a replacement-cost principle of rate control, any claim that it may have for acceptance on grounds of resource allocation must be stated merely in terms of *relative* advantage. That is to say, the claim must rest on a contention that the discrepancies between public utility rates and incremental costs of service, though unavoidable under *any* attempt to make rates as a whole cover costs as a whole, will be reduced if the latter costs are measured in terms of current costs of replacement rather than in terms of costs already actually incurred.

The merits of such a claim cannot be inferred from any a priori theory of cost behavior or cost analysis. Instead, it would have to be put to an empirical test, the validity of which would be limited to the particular type of public utility under review. But if we accept provisionally the usual American assumption that most public utility enterprises, including most railroads, are even today operating under conditions permitting the enjoyment of further economies of scale, and if we also assume that current replacement costs of service would be higher than historical costs, the acceptance of a replacement-cost principle might be a step in the wrong direction.

[7] See Chap. XX. In a debate on the subject with Professor Brown in 1927 (see footnote 11, *infra*), I argued that the logic of his position would call for support of marginal-cost pricing rather than for support of his proposal to make railroad rates cover total replacement costs. Brown's reply, if I understand it correctly, was that marginal-cost pricing for railroad services is financially impractical but that replacement-cost pricing comes as close to the ideal of competitive pricing as is feasible under private ownership.

THE RESOURCE-ALLOCATION ARGUMENT FOR REPLACEMENT COST IGNORES THE TAX FACTOR

The claim on behalf of a replacement-cost standard, to the effect that it will secure a better allocation of resources than can the rival standard of actual cost, is further weakened by a factor seldom taken into account in debates about the relative merits of the two types of rate base, namely, the tax factor. Under our prevailing systems of state, local, and Federal taxation, a material fraction of the revenues received by the railroads and by the private utility companies from the sale of services must be paid out in taxes. For example, in the year 1956, the total taxes reported by the country's larger private electric power companies in their income accounts amounted to more than 21 per cent of gross revenues.[8] In addition, special excise taxes have been imposed at times upon the purchase of electric service, telephone service, and passenger transportation tickets—taxes imposed directly on the consumer but collected and transmitted to the governments by the utility companies.

Viewed from the standpoints of the consumers and of the companies that render utility service, the payment of these taxes is a part of the necessary costs of service. Viewed, however, from the standpoint of the nation or of the community, this payment cannot be relied upon to reflect, even in a rough and ready way, the *social* costs of service rendition—the burdens that the community as a whole could save by refraining from supplying the service. Yet the resource-allocation defense of replacement-cost rate control involves the assumption that the expenses incurred by the private producer of the service reflect social costs.[9]

In the 1920s, when Professor Harry Gunnison Brown wrote his trenchant criticisms of an original-cost standard of rate making,

[8] Federal Power Commission, *Statistics of Electric Utilities in the United States, 1956, Classes A and B, Privately Owned Companies*, p. xv.

[9] It might be argued that the inclusion of taxes in the prices charged for utility services will bring *relative* rates and prices into harmony with *relative* social costs. But this argument would require the dubious assumption of a close comparability between the tax components of public utility rates and the tax components of the prices of nonutility products. The prospective importance of atomic power, to be produced with atomic fuel supplied by the Atomic Energy Commission at prices not closely related to production costs, and to be aided by expensive government-financed research and by government liability of insurance, will throw even more doubt on the reliability of relative enterpriser costs as a measure of relative social costs.

he paid special attention to the danger that this standard would result in too high rates in a period of declining reproduction costs. For many years, however, replacement-cost valuations have been associated with high rather than low valuations, and the critics of the actual-cost principle have expressed the fear that the resulting rates would be too low and hence would encourage wasteful consumption. But if this danger were really believed to exist despite the heavy prevailing levels of utility taxation, the remedy would not need to be sought for by the adoption of a replacement-cost principle of rate making. Instead, it could be secured far more easily and expeditiously by an increase in utility taxes!

WHENEVER ECONOMIC CONDITIONS DISQUALIFY AN ACTUAL-COST BASIS, THEY WILL PROBABLY ALSO DISQUALIFY A REPLACEMENT-COST BASIS

The rate-making philosophy presented by Brown and other economists on behalf of a replacement-cost principle of rate making, though failing in my opinion to support this principle, is much more effective in revealing both practical and theoretical limitations of the actual-cost principle. Under certain conditions these limitations, inherent in *any* sunk-cost standard of rate making, become fatal or, at least, too serious for toleration. But the point that I would now stress is that, in the recent experience of American rate regulation, when these conditions have prevailed, the remedy has seldom if ever been found to lie in a resort to the replacement-cost standard. Instead, the sought-for remedy has taken the form of a departure from, or even of the complete abandonment of, *any* fair-return standard of rate making. Three episodes in the history of rate regulation will illustrate this point, although others will occur to the reader.[10] The first concerns the regulation of railroad rates; the second concerns the regulation of natural-gas rates; the third concerns local-transit fares.

In December, 1927, when railroad valuation by the Interstate

[10] The Supreme Court's review of the California Commission's unorthodox rate-making valuation of the Market Street Railway in San Francisco suggests another case in point. Here the Court sustained the rate order against the charge of unconstitutionality despite the Commission's refusal to concede a return either on actual cost or on replacement cost, its much lower valuation having apparently been based on the price that the City of San Francisco had bid for the property. Market Street Railway Co. v. Railroad Commission, 324 U.S. 548 (1945). The facts were unusual and the decision is of uncertain import as a legal precedent.

Commerce Commission was an extremely live issue, I presented a paper before the American Economic Association defending an "actual-cost" or "prudent-investment" measure of the rate base in preference to a rate base derived from estimated replacement costs.[11] (The railroad carriers themselves were staunchly supporting replacement-cost appraisals.) This paper was followed by several commentaries from other economists, including a dissent by Professor Brown, who came to the defense of replacement cost on grounds summarized in the preceding paragraphs.

The subsequent history of the Interstate Commerce Commission valuations would make a long story, and there is no need to relate it here. But what is important to note is that this history has belied both Brown's assumptions and mine. For in actual practice, rate regulation by primary reference to the standard of a normal "fair rate of return" on *any* rate base has proved unfeasible. Competitive conditions, combined with the early shortsightedness of the railroads in failing to amortize capital costs by reference to anticipated obsolescence, and aggravated by governmental subsidies to road, air, and water carriers, have prevented the railroads from earning a so-called fair return even on depreciated historical costs and have prevented the Commission from making a serious attempt to regulate rates on any such standard of a fair return. In short, the actual experience of the railroads has revealed the limitations of an actual-cost basis of rate control when applied to a highly competitive industry. But this same experience has also revealed the limitations of a replacement-cost standard. Indeed, railroad valuations, if based on current reproduction costs, would be even higher than those based on net investment, with the result that any attempt to apply such a standard would be even more clearly out of the question.[12] This situation has created a rate-making prob-

[11] "Railroad Valuation with Special Reference to the O'Fallon Decision," 18 *American Economic Review, Supplement* 182–205 (1928). A revision of this paper was published under the title "The Economic Merits of Original Cost and Reproduction Cost," 41 *Harvard Law Review* 593–622 (1928).

[12] That is to say, out of the question under any standard of replacement cost or reproduction cost that has ever been applied or that has any likelihood of being applied in the future. Under a "paper theory" of replacement cost, as presented in some of the economic treatises, "reasonable" railroad rates might be derived from estimates of the cost of replacing the service by whatever forms of transport, including possible new railroads of modern design and location, would justifiably be adopted if the nation's transport system were to be planned *de novo.* The acceptance of these rates would then give to the existing railroad properties values

lem as yet unsolved. But the most widely discussed solutions involve the complete or partial abandonment of any fair-return test of reasonable railroad rates in favor of rates designed to meet the actual competition of rival forms of transport without too serious distortions of the railroad rate structure.[13]

For a second illustration of the failure of a replacement-cost standard to cure the deficiencies of an actual-cost standard, one may turn to the regulation of natural-gas prices by the Federal Power Commission. In the *Hope Natural Gas* case, the Commission had included in the pipeline company's rate base its actual net investment in gas-producing properties. Although the Supreme Court accepted this disposition of the problem of allowing for the costs of gas at the wellhead, Justice Jackson wrote a brilliant dissent.

> The prudent investment theory [he wrote] has relative merits in fixing rates for a utility which creates its services merely by its investment. . . . But it has no rational application where there is no such relationship between investment and capacity to serve.[14]

Yet this conclusion did not lead the Justice to support a reversion to a replacement-cost or "fair-value" basis of price control. Instead, it led him to reject *any* rate base as a sound determinant of the commodity component of the price of gas. The Commission, he said, should be left free

based on capitalized estimated earnings–values on which the railroad companies would be earning what might be regarded, in a backhanded way, as a "fair rate of return." Compare an editorial in the March 8, 1954, issue of *Railway Age*, entitled "Suppose Railroading Were a New Industry."

[13] See Chap. IX, pp. 139–142, *supra;* also former Interstate Commerce Commissioner Clyde B. Aitchison's article, "Fair-Return-on-Value-Theory in Rate Making Loses Force," 25 *I.C.C. Practioners' Journal* 11–15 (Oct., 1957). Mr. Aitchison refers to the Interstate Commerce Commission's rate decision in *Ex Parte 206* as having completely overthrown the "fair return on value" test of reasonable rates. Under competitive conditions now facing the railroads, he declares, this test "has proved to be an illusion."

In an earlier draft of this chapter, I incautiously wrote that, today, not even any railroad management would claim that railroad rates could feasibly be fixed at a level that would yield a conventional "fair rate of return" on a reproduction cost appraisal. I was wrong! In 1956, following a holding of the New York Court of Appeals that the statutes governing telephone and transportation rates require the state commission to take into consideration estimates of reproduction costs, the New York Central Railroad, in seeking leave to increase intrastate passenger fares on its Hudson, Harlem, and Putnam Divisions, was reported to have set as its goal a 6 per cent return on the "replacement value" of properties that had acually cost only a fraction of this value to construct. Washington *Star*, Oct. 9, 1956.

[14] 320 U.S. 591 at 649 (1944).

to fix the price of gas in the field as one would fix maximum prices of oil or milk or coal, or any other commodity. Such a price is not calculated to produce a fair return on the synthetic value of a rate base of any individual producer, and would not undertake to assure a fair return to any producer. The emphasis would shift from the producer to the product, which would be regulated with an eye to average or typical producing conditions in the field.

This is not the place to discuss the merits of Justice Jackson's dissent in the *Hope Natural Gas* case, or the merits of the still highly controversial and difficult problem of natural-gas price regulation. His position is cited here merely as another example of the failure of a replacement-cost rate base to meet the practical needs of price control whenever the economic situation is such as to suggest the rejection of an actual-cost standard.

For a third and final illustration we may turn to municipally owned transportation systems. The New York subway system will serve as a concrete example. At the current 15¢ fare the system is now running an operating deficit, and the present management is doing its best to retard the necessity of a further increase in fares by resort, partly, to cost-cutting devices resulting in impaired service outside of rush hours. Any attempt to make the system earn interest on the subway-incurred debt, or to earn "constructive" interest on original construction costs, would be generally conceded to be out of the question. In short, if one includes capital costs as a part of the costs of service, the system must be run at a heavy loss. But by the same token, any attempt to go beyond this point and to secure a return on the high hypothetical replacement costs of subway construction would be even more clearly hopeless. Yet no sane man proposes to shut the subways down, and few would contend that they *ought* to be shut down unless the subway riders are prepared to cover operating expenses plus a fair return or fair interest rate on the current "fair value" of the property.[15] Again we come to the point of this discussion: that the serious limitations of an actual-cost standard of public utility pricing are unlikely to be overcome by a transfer to a replacement-cost standard.

[15] The New York subway situation illustrates the theoretical limitations of the assumption, implicit in replacement-cost rate theory, that public utility services are not worth rendering unless they can be made to yield revenues to cover total operating expenses plus a fair return. See Chaps. VII and XX.

It is perhaps by no mere accident that one looks in vain in the more recent economic journals for any elaborate defense of the replacement-cost theory of rate making such as was forthcoming in the 1920s and 1930s from the skillful pen of Harry Gunnison Brown. Brief statements leaning toward such a defense may still be found; [16] and the public utility trade journals such as *Public Utilities Fortnightly* still carry articles in support of a "fair-value" basis of rate control not clearly distinguished from a replacement-cost basis. But the emphasis is now on the failure of an unmodified actual-cost rate base to give stockholders protection against inflation. The more theoretical, optimum-resource-allocation argument is either left out or else remains very much in the background.

This current loss of interest in replacement cost on the part of the academic economists has several possible explanations. Among the more probable ones, however, has been the emphasis attached by modern "welfare economists" to pricing based on marginal costs, rather than on total costs or average costs, as a basis of sound resource allocation. Even today, to be sure, those economists who would abandon total-cost pricing in favor of marginal-cost pricing are in a decided minority, if one may judge by a count of published articles on the subject. But their arguments have served to take the bloom off the replacement-cost standard as an asserted means of avoiding the defects of a sunk-cost basis of pricing. More promising ways of reducing these defects lie in the manipulation of utility taxation to this end and in the appropriate design of the rate structure.

Among the actual or proposed measures of the rate base discussed

[16] Of the recent statements in support of replacement-cost public utility pricing that have come to my attention, the most interesting ones are from British economists who have urged that the nationalized utility and railroad industries, instead of being under Parliamentary mandate to charge rates designed to cover fixed-charge requirements on debt incurred in the acquisition of properties from their former private owners, should be instructed to cover "notional" interest and depreciation charges on plants valued on a replacement-cost basis. See W. Arthur Lewis, *Overhead Costs* (London, 1949), Chap. 2; the same author's chapter on "The Price Policy of Public Corporations," in W. A. Robson, ed., *Problems of Nationalized Industry* (New York, 1952), Chap. 10; A. M. Henderson, "Prices and Profits in State Enterprise," 16 *Review of Economic Studies* 13–24 (1948–1949). D. J. Bolton, the leading British expert on electric tariffs, concedes a theoretical case in favor of replacement-cost determination of total revenue requirements for the British Electricity Authority but feels that the resulting excess in revenues over contractual interest requirements would become the target for opposing demands from labor and from consumers. *Electrical Engineering Economics*, Vol. 2: *Costs and Tariffs in Electricity Supply*, 2d ed. (London, 1951), pp. 42–45.

in the last three chapters, the two that seem to me to deserve highest rating are the two versions of the actual-cost standard set forth in Chapter XII: the orthodox form of a net-investment rate base, and a modified form in which the common-stock-equity component of the rate base is restated by the application of a price-index corrective—preferably, by a so-called cost-of-living index. Under favorable conditions, either probably has a better chance of successful operation than has a rate base derived in whole or in part from estimates of reproduction costs. Under unfavorable conditions, no rate base can be expected to yield satisfactory results, since reasonable rates cannot then be identified with rates designed to yield a "fair" rate of profits under *any* definite standard of fairness.

In 1926 Dr. John Bauer, then as now a leading exponent of the prudent-investment principle, closed an article in reply to Professor Brown with a paragraph that brilliantly summarizes the different philosophies underlying, respectively, the use of actual cost and the use of replacement cost as a measure of the rate base. With Dr. Bauer's permission, I quote his statement in concluding these chapters on alternative measures of the rate base.

> Brown has his eyes fixed constantly upon possible economic waste—and quite properly. But he seems to think only or chiefly in terms of possible discrimination between users, or in retarded or overstimulated use of service—all conjectural and improbable consequences—while I see the waste chiefly in the periods of speculation and financial stagnation, in the cross-purposes between public bodies and the companies, the terrific direct costs of regulation, and the failure to achieve economies of operation and better service which would be available through effective regulation and co-operation with the companies, but are lost through the incessant conflict and deadlock over rates.[17]

[17] John Bauer, "Rate Base for Effective and Non-Speculative Railroad and Utility Regulation," 34 *Journal of Political Economy* 479–513 (1926). While I would not go as far as Bauer here went in minimizing the resource-allocation objection to actual-cost pricing, I would emphasize even more than he did the failure of the replacement-cost standard to supply a remedy. Compare Emery Troxel, "Valuation of Public Utility Property—A Problem in Efficient Resource Use and Efficient Regulation," 23 *Journal of Business of the University of Chicago* 1–21 (1950).

XV

THE FAIR RATE OF RETURN

As noted in Chapter X, the measurement of the rate base is merely the first step in the calculation of a fair return on the cost or "value" of the corporate property. The second step is the allowance of a "fair" or "reasonable" annual rate of return on this rate base. Up to recent years, the percentage rate was usually arrived at in a much more casual or conventional manner than was the rate base, largely because of the absence of that specious aura of expertise and objectivity that has surrounded engineering valuations for rate-making purposes. Since the Second World War, however, rate-of-return specialists have supplied a similar aura to their subject, with the result that, today, the most critical and controversial part of a general rate case is likely to be given over to the conflicting testimony and exhibits of the finance witnesses for the different parties and, sometimes, for the commission or its staff.[1] The conflict is by no means limited to disagreements on questions of fact or prophecy. It also extends to express or implied disagreements as to the very meaning of a "fair" rate of return or as to the relative weights that should be given to multiple standards of fairness.

The present chapter will review briefly the basic controversies about the meaning and measurement of a fair rate of return.[2] Only

[1] "I have been engaged in a lot of these rate cases, a great many of them, and I know that the most difficulty in all rate cases is the setting of the rate of return." Leslie M. Jones, Vice President and General Counsel, Illinois Bell Telephone Co., at a hearing of City Council Committee on Utilities, Chicago, March 4, 1957, transcript, p. 121.

[2] The general treatises on public utility economics or regulation have chapters on the subject. For an excellent monograph on the earlier cases, see Nelson Lee Smith, *The Fair Rate of Return in Public Utility Regulation* (New York, 1932). Ellsworth Nichols, *Ruling Principles of Utility Regulation: Rate of Return* (Washington, D.C., 1955) is a compendium of relevant statements by courts and com-

incidentally, however, will it refer to pronouncements on the subject by courts and commissions. Generally speaking, these pronouncements have been couched in language too broad to serve as clear-cut precedents. Commissions, particularly, have usually carefully refrained from accompanying their "findings" as to a fair rate of return in any given case with a disclosure of the methods and presuppositions on the basis of which they have arrived at these findings. In consequence, a rate-of-return witness who himself takes a definite position on some disputed issue of principle —say, on the choice between a historical-cost and a current-cost measure of the cost of debt capital—may later read the commission's rate order and written opinion without being able to discover whether or not his own views on the issue have carried conviction to the tribunal to which they have been addressed. From the standpoint of one who hopes for progress in the development of more rational systems of rate regulation this is a frustrating experience.

In one respect the Supreme Court's renunciation of the fair-value rule of rate making as the law of the land has added to the complexity of the subject. Since this rule purported to set a single, country-wide standard for the measurement of the rate base, it was also taken to carry with it a uniform set of standards for the determination of a fair rate of return. But today the situation is different. For no longer is there a legal requirement of rate-base uniformity throughout the United States, with the result that the question arises whether a rate of return that would be fair if applied to, and under the philosophy of, a net-investment rate base would also be fair if applied to, and under the philosophy of, a higher or lower fair-value rate base. This question is still subject to sharp dispute; and the dispute is all the more confusing because, at least in most states, a so-called "fair-value" rate base is a compromise rate base, thereby lending itself to the application of ambiguous standards of a fair rate of return.

---

missions. Numerous articles on rate of return have appeared in the periodical *Public Utilities Fortnightly*. See its index references to "Return." The quarterly *Land Economics* (formerly *Journal of Land Economics*) has articles on rate of return by Professors Walter A. Morton (May, 1952), E. W. Clemens (Feb., 1954), Lionel W. Thatcher (May, 1954), Fred P. Morrissey (Aug., 1955), and myself (Feb., 1951). See also Paul L. Howell, "The Rate of Return in Air Transport," 24 *Law and Contemporary Problems* 677–701 (Autumn, 1959).

For reasons already suggested in the preceding chapters, this chapter will attempt to clarify the relationship between rate base and rate of return by a threefold division of subject matter. The first section will discuss the determination of a fair rate of return when applied to an actual-cost rate base under assumed conditions of stable general price levels. The second section will remove the assumption of price-level stability in raising the question whether, even under an actual-cost philosophy of rate regulation, allowance should be made for price inflation through adjustments in the rate of return if not in the rate base. Finally, the third section will discuss the question what measures of a fair rate of return are consistent with the acceptance of a fair-value standard of rate making.

THE FAIR RATE OF RETURN ON AN ACTUAL-COST RATE BASE: ASSUMPTION OF STABLE GENERAL PRICE LEVELS

"COST OF CAPITAL" AS THE BASIC STANDARD OF A FAIR RATE OF RETURN

The acceptance of an actual-cost measure of the rate base by no means completely predetermines, as a matter of "logical consistency," the applicable criteria of a fair annual rate of return. On the contrary, a variety of criteria would be applicable; and each of them might be made both fair and workable if applied with skill and with reasonable consistency over a long period of years. Among the possible alternatives would be any of several schemes for differential rates of return on invested capital designed to reward high managerial efficiency and to penalize inefficiency.

Nevertheless, the basic principle of rate making implicit in an actual-cost measure of the rate base goes a considerable distance toward establishing the relevant tests of a fair rate of return. This principle is that of service at cost. Here, in contrast to any supposed norm of competitive pricing, "reasonable" rates of charge for public utility services are held to be rates sufficient, but no more than clearly sufficient, to cover the total costs actually and prudently incurred by a company in supplying these services. Even the fair-value rule of rate making, as interpreted by courts and commissions, applies this principle in its allowance for operating expenses. But the actual-cost rule attempts to extend the principle, as far as feasible, to the allowed return over and above the annual operating deductions. That is to say, this allowance is itself designed to cover a part of the total costs of service, namely, those costs incurred by the company in securing the necessary capital. Thus the twofold rule that a public utility may charge rates designed to cover its operating costs *plus* a fair return is converted into the apparently simpler rule that the rates of charge shall cover the company's total costs *including* its costs of capital.

This close association—I won't say complete identification—between a fair rate of return and "cost of capital" under an actual-cost standard of rate regulation has the same general claim for acceptance as has an actual-cost measure of the rate base: namely, the claim of compliance with the requirement of credit maintenance and of corporate ability to attract capital on terms favorable to the company and hence to its consumers. For a company that cannot meet its costs of capital, including its fixed charges plus "reasonable" dividend requirements, cannot long continue to supply adequate public utility service to a growing community —not, at least, without violation of express or implied commitments that it has already made in order to secure capital for the construction of its existing plant. In an extreme case, to be sure, failure to cover existing costs of capital may ultimately be cured, from the standpoint of the community, by a drastic financial reorganization under the National Bankruptcy Act or otherwise. But the cure is costly, prolonged, and painful.

With this introduction to the basic concept of a fair rate of return implicit in an actual-cost philosophy of rate regulation, we may now turn to problems of application. It will soon be apparent that the most formidable problem here concerns, not just the technique of measurement but the very meaning of "cost of capital" when applied to that portion of a company's invested capital represented by its common-stock equity.

Derivation of over-all "cost of capital" by the cost-of-capital formula. Even those public service commissioners and financial experts who derive their conclusions as to a fair rate of return in large measure from their estimate of "cost of capital" seldom completely identify their conclusions with their estimates.[3] More often than not, the estimated cost is offered as a *minimum* standard of adequacy, subject to a "judgment-made" additional allowance for possible error, for real but immeasurable items of cost, for "regulatory lag," and even perhaps as a reward for efficient operation. More will be said on this point later. Meanwhile we may sketch very briefly the technique of estimating the so-called cost of capital as it has been developed by financial witnesses and by the expert staffs of a number of regulating commissions including, notably, the Federal Power Commission. This technique involves the use of the "cost-of-capital" formula.

Under this formula, a utility company's total invested capital is expressed as 100 per cent and is divided into percentages representing, respectively (a) the capital secured by the issuance of funded or long-term debt, (b) the capital secured by the issuance of preferred stock, and (c) the capital represented by the common stock.[4] In harmony with familiar principles of corporation finance and with generally accepted principles of rate regulation, any capital investment reflected by retained earnings (earned surplus) is treated as a component of the common stock.

This threefold division of the corporate capital by reference to

[3] This discussion largely follows the more elaborate treatment that I gave the subject as a rate-of-return witness for the Commonwealth Edison Company in 1953 and again in 1957 (Illinois Commerce Commission Cases 41130 and 44391).

[4] Unless a utility company has heavy investments in nonutility assets, this invested capital should approximate an actual-cost or net-investment rate base except for investment in plant under construction, for which adjustments are usually made by the temporary exclusion of the investment from the rate base combined with the later introduction into the rate base of an allowance for "interest during construction" after the plant has gone into service. But if the company has large investments in other business enterprises, there arises the problem of deriving a cost of capital for the utility properties from an estimate of cost of capital to the corporation as a whole. A precise determination of this former, hypothetical cost is impossible; and the best practical solution may lie in a presumption of equality between total cost of capital and cost of utility capital. The obvious deficiencies of this solution suggest one reason why American public utility companies are wisely restricted in their ownership of nonutility assets.

With "composite" public utilities, such as those supplying electricity, gas, and steam heating, the usual, though not the universal, rule (Maryland has supplied one notable exception) has been that each major class of service must stand on its own feet financially. That is, deficient earnings from one class of service may not be made good by the allowance of an excess rate of return to another class. But the determination of differential "fair rates of return" for the different departments of a combined utility corporation, based on estimated risk differentials, presents problems which would be hopeless except for resort to largely arbitrary solutions.

its major sources permits the analyst to compute separately the cost of the debt capital, the cost of the preferred-stock capital, and a cost or cost equivalent of the common-equity capital. The separation is very helpful, since the costs of the two classes of senior capital can usually be computed with a close approach to accuracy, leaving only the so-called cost of the equity component subject to major differences of opinion.

Let us assume, for example, that the company in question has a capital structure composed of 3½ per cent bonds to the extent of 50 per cent, of 4½ per cent preferred stock to the extent of 15 per cent, and of common stock (par value plus earned and capital surpluses) to the extent of 35 per cent. Let us also assume that an allowed return of 10 per cent on the common-stock capital is deemed adequate. The over-all cost of capital will then be computed as follows, in per cent:

	Per Cent of Total Capital	*Annual Cost*	*Return on Total Capital*
Funded debt	50%	3½%	1.750%
Preferred stock	15	4½	0.675
Common-stock equity	35	10	3.500
Over-all computed cost of capital			5.925

The assumed capital structure. In the application of the cost-of-capital formula the first problem concerns the choice of the assumed security structure—an important choice since the computed over-all cost may be much lower when the debt ratio is high than when the common-stock equity is conservatively thick. Having in mind this differential, some analysts base their estimates on what they themselves accept as a "typical" or an "ideal" security structure without regard to the actual capitalization of the company under review. Other analysts prefer to base their calculations on the capital structure as it now stands or as it is expected to stand in the not-distant future.

In my opinion, the latter alternative is preferable in the usual case, since it does not put a company under the temptation to "trade on a thin equity" as a means of enhancing the earnings on its common stock. Indeed, the use of a hypothetical or "typical" capitalization substitutes an estimate of what the capital cost *would be* under nonexisting conditions for what it *actually is or will soon be* under prevailing conditions. But if the existing security structure is clearly unsound or is extravagantly conservative, the rule must be modified in the public interest. *Actual* cost of capital may then be disqualified in favor of *legitimate* cost.

It has sometimes been argued that, aside from income-tax differentials, over-all cost of capital cannot be materially affected by a corporation's security structure, since a difference in structure means merely a difference in the distribution of risks that are constant for the investment taken as a whole. This argument, however, has been belied by market experience, which indicates that, up to some ill-defined limit, utility companies can secure capital on more favorable terms by substantial senior-security financing.[5] It does not follow, however, that the public interest is served by resort to debt ratios high enough to minimize the cost of capital. Lower ratios may well be worth their higher costs by reducing the risks of financial adversities which would have a serious impact on the quality and expansion of the supply of public service. But the question where the line should be drawn between needlessly low and dangerously high debt ratios is subject to major differences of opinion. Thus, in postwar rate cases, the operating subsidiaries of the American Telephone and Telegraph Company have defended the objective of a one-third debt ratio for the consolidated system, whereas various noncompany witnesses (including myself on occasion) have argued that a somewhat higher debt ratio would be justified.

Determination of the cost of senior capital. Having determined the appropriate capital structure, such as the one used above for purposes of illustration, the analyst usually finds no difficulty in computing the cost of senior capital with fair precision. Actual

[5] This, at least, is the conclusion which I reached, several years ago, on the basis of a number of comparative studies of the security yields of various electric or gas-and-electric companies with different capital structures. I take it that most security analysts hold the same conviction. But this position has been sharply challenged by Professors Franco Modigliani and M. H. Miller, partly on theoretical grounds, partly by reference to earlier statistical studies: "The Cost of Capital, Corporation Finance and the Theory of Investment," 48 *American Economic Review* 261–297 (1958). See also comments on this article by J. R. Rose and David Durand, and a reply by the authors: "The Cost of Capital and the Theory of Investment," 49 *American Economic Review* 638–669 (1959). The subject calls for a thorough re-examination.

fixed charges on the debt plus actual dividend requirements on the preferred stock represent the annual cost in terms of dollars. The dollar figures are then converted into percentages of each type of capital as measured by the net contribution which the company has received from the issuance of the bonds and of the preferred stock.

In rate cases, some rate-of-return witnesses have contended that the allowance for interest costs and for preferred-dividend requirements should be based, not on charges actually imposed by securities now outstanding, but rather on the hypothetical cost of doing senior financing under current conditions of the bond and stock market. As will be noted in a later section of this chapter, this position comports with the logic of a reproduction-cost theory of rate control; and it is arguably applicable even to a "fair-value" rate base which gives material, though not controlling, weight to reproduction-cost appraisals.

But in a calculation based on an actual-cost standard of reasonable utility rates, the objective is to determine, not what the senior capital would cost if it had to be secured *de novo*, but rather what it really does or will soon cost in view of the fact that much of it has been secured at an earlier date and under market conditions differing from those prevailing today. For rate-making purposes, this actual or "experienced" cost is significant as indicating what the present company will need in order to meet its interest and preferred-dividend requirements. Hence, the estimated *current* costs of new bond money and of new preferred-stock money are directly relevant only as evidence of the probable costs of new senior issues that the company must contemplate in the near future for purposes of refunding or of new financing.

When the cost of debt capital is taken to mean actual cost rather than hypothetical current cost, company witnesses have argued that this cost should include an allowance designed to amortize, over the lives of the outstanding bonds, any unamortized debt discounts, call premiums, and finance expenses on refunded bonds that have been called prior to maturity for the purpose of interest savings. For income-tax purposes, these prematurity refunding costs are treated as an immediate loss rather than as a deferred charge; and accounting-minded witnesses have urged similar treatment for rate-making purposes. But the other position is defended on the ground that the losses, net of tax savings, should be borne by later consumers and not by the stockholders, since these consumers will be the primary beneficiaries of the refunding action.

Consistently applied, either of these alternative rules of rate making would be tenable and fair. For, if any losses from refunding operations are to fall on the stockholders, in the form of a resulting erosion of corporation surplus, the allowed fair rate of return can be made high enough to compensate stockholders for the risk of exposure to such losses in the future. But there is a practical ground for preferring the other alternative: namely, that a management may well hesitate to call high-yield bonds if the immediate financial loss must fall on the stockholders while the reduction in annual interest charges must be passed on to the consumers.

The problem of estimating the so-called cost of the common-stock capital. The really critical problem in the determination of the over-all cost of capital is that of estimating the cost of the common-stock component, or rather that of estimating a capital-attracting allowed rate of return which can be said to reflect cost in a very loose sense of that word. Here, the primary difficulty lies in the very nature of the common stock of ordinary business corporations, including most American public utility corporations; namely, in the absence of any express or implied commitment as to the rates of dividend. In this absence, the annual cost actually incurred by a company in floating stock issues, whether by rights offerings to old stockholders or by public offerings, is simply indeterminate. Dividend payments are contingent on earnings; yet the allowable amount of earnings is the very objective of inquiry in a rate case. There thus arises a vicious-circle difficulty somewhat similar to that which precludes the acceptance of the commercial or market value of a utility property as a measure of the rate base.

This difficulty could be avoided if American rate regulation were willing to follow the older British practice under which even the common stocks of public utility corporations would be subject to a fixed maximum annual rate of dividends, such as $4 per share or 10 per cent of par value.[6] Each new issue of stock could then

[6] Robert H. Whitten, *Regulation of Public Service Companies in Great Britain*, Appendix F, Vol. 1, Annual Report of Public Service Commission of New York, First District, 1913. For proposed application to American regulation, see Arthur P. Becker, "Fixed Dividends for All Public Utility Stock," 21 *Journal of Land &*

be sold at the highest price that it would command in the current market, and the resulting total annual "dividend requirements" could then be accepted as reflecting the company's costs of equity capital. In prosperous times, to be sure, the company should be given an opportunity to earn revenues in excess of these requirements in order to protect the stability of its regular dividends and also, perhaps, in order to raise new capital. But all excess earnings would then be credited to a dividend-stabilization account or to a special proprietorship account representing "consumer equity." Wisely in my opinion, however, American practice has been adverse both to such a rigid limitation and such a quasi guaranty of distributable public utility earnings. Hence, a public service commission, in its allowance of a reasonable rate of return, lacks any predetermined measure either of required earnings or of required dividends on the common stock.

In a number of rate cases, witnesses on rate of return have attempted to estimate what they have called the "historical cost" of a company's common-stock capital by retrospective studies of the market conditions prevailing at the times of the original flotations.[7] In this manner they have hoped to determine approximately what rates of earnings or rates of dividends the subscribers to the stock must have anticipated as a basis for their willingness to pay the subscription prices. These anticipated earnings or dividends would

---

Public Utility Economics 243–249 (1945). Dr. John Bauer has advanced a somewhat similar proposal in *Transforming Public Utility Regulation* (New York, 1950), Chap. X, "Establishing Definite Rates of Return." The really serious objection to all of these plans for a definitive rate of return lies in the danger of a resulting loss of managerial incentive. See pp. 262–265, *infra*.

[7] The application of a historical-cost rather than a current-cost standard even to the common-stock capital derives support from Justice Brandeis's dissenting opinion (concurring as to result) in Southwestern Bell Telephone Co. v. Public Service Commission of Missouri, 262 U.S. 276 at 304–308 (1923). There the Justice insisted that a rule which would fix the return for capital raised in the past by the rate of return which happens to prevail today "opens the door to great hardships." The hardship objection, I may add, is particularly applicable to the initial investments in relatively young public utility enterprises which, while already enjoying good credit as "seasoned" enterprises, were of a speculative, promotional character in their early stages. Today, the natural-gas pipeline companies supply examples in point.

In earlier years, a very crude way of giving recognition to the seasoning factor was through the allowance for "going value" as a component of the rate base. This allowance is now fortunately out of fashion. But its absence provides one justification for the commission practice of conceding "fair rates of return" sufficiently liberal to permit the shares of well-established utility companies to attain market values substantially in excess of book values.

then be accepted as a measure of the company's current earnings requirements. But any attempt to surmise investor anticipations in different periods of past history is obviously wildly speculative; and so far as I am aware, the procedure is not in good repute with public service commissions.

Substitution of estimates of current cost of common-stock capital for attempts to determine experienced cost. Aware of the absence of any specific dividend requirement or earnings requirement actually imposed upon a public utility company in raising common-stock capital, many analysts, including the present writer, have substituted a so-called current-cost test for an actual-cost test. This they have done by defining the current cost as the rate of return on new common-stock capital, the anticipation of which would be required in order to attract this kind of capital to the company in question under prevailing market conditions.[8] Thus, if a public utility could issue new common stock only on terms which would give subscribers thereto an expectation of earnings on subscription price on the order of 10 per cent per annum, a 10 per cent rate would be taken to represent the current "costs of equity capital." This rate would then be applied to the common-equity component of the existing capital in deriving the company's "over-all cost of capital."

The practical defense of this current-cost test of the cost of common-stock capital, when combined with an actual-cost test of the cost of senior capital, is that a company which succeeds in establishing an earning power sufficient to cover both of these costs, year by year, will be able to meet its actual interest and preferred-dividend

[8] The term "capital-attracting rate of return" contains an ambiguity which I do not recall seeing discussed in the literature. When the analyst finds that the anticipation, say, of a 10 per cent return would attract new equity capital, does he mean that the anticipation of a 10 per cent perpetuity would attract capital? Or does he mean that the investor will be attracted by the expectation of a 10 per cent return for the not-distant future, perhaps for the next few years, to be followed by whatever increase or decrease he thinks may result from later attempts of rate regulation to keep abreast of changes in the prevailing costs of capital? I suggest that he cannot mean the former, since he has no way of finding out what investors would pay for a nonexisting type of public utility security. He must, therefore, mean that investors are content to accept the expectancy of a 10 per cent return in view of their own surmises as to whether this return will go up or go down later on. In most periods, perhaps the most plausible assumption is that, under regulation subject to a flexible rate of return, investors accept the chance of a later increase in the allowed rate of return as an offset to the chance of a later decrease.

requirements while maintaining a dividend record on its common stock sufficiently attractive to permit it to issue more of this stock, from time to time, at prices, net of flotation costs, not less than the per-share book value of the old stock. In this way it can float required amounts of bonds and preferred stock while keeping its capital structure in balance by the issuance of new common stock at prices high enough to avoid "impairing the integrity" of the investments of the old stockholders.

Critics of this procedure, of whom there are many among spokesmen for the public utility investors, have made the point that it inconsistently combines actual-cost with current-cost tests of a fair rate of return.[9] Their contention is correct. But the inconsistency of the combination may well be deemed a virtue rather than a vice, on the ground that it is a necessary means of dealing with the "inconsistency" of American corporation finance in combining fixed-charge securities with common stock entitled to no specified rate of dividends.

In one important sense, moreover, the transfer from an actual-cost to a current-cost measure of the cost of common-stock capital represents only a modification, rather than a violation, of the actual-cost principle of rate regulation. For under the current-cost measure, the market prices of the utility stocks are designed to be kept in a reasonable relationship to their book values and hence to the actual net costs, per share, of the corporate assets. "Reasonable relationship," of course, is question begging, and more will be said on this point in later paragraphs. But, generally speaking, the book values (with allowances for the probable need to underprice new common-stock offerings) should set a floor to the market values in periods of normal business conditions.

For the reasons just suggested, the charge of "inconsistency" does not constitute a valid objection to the determination of a fair over-all rate of return which combines experienced cost of senior capital with estimated current cost of common-stock capital. But other objections cannot be brushed aside so readily; and the one that has greatest weight lies in the extreme difficulty of making a reliable estimate of "current cost" in the sense of a current capital-attracting rate of return. Indeed, the only practical solution of this

[9] See Leonard A. O'Connor, "Some Critical Thoughts on Cost of Capital," 62 *Public Utilities Fortnightly* 93–97 at 96 (July 17, 1958).

difficulty lies in the acceptance of a *liberal* estimate, or else (what really amounts to the same thing) in the allowance of a "fair rate of return" in excess of the rate derived by formula. Otherwise regulation under an actual-cost standard of rate making would seriously threaten the ability of utility companies to keep abreast of their vast requirements for new capital.

Earnings-price ratios as evidence of current cost of equity capital. In estimating what anticipated or allowed rate of earnings on new common-stock capital will or would attract this new capital, the analyst has at his command two relevant types of statistical data: first, a several-years' record of earnings-price ratios both for the stock of the very company under review and for the stocks of comparable utility companies; secondly, a similar record of dividend yields in terms of ratios of annual cash dividends to average annual market prices.

Until very recently, at least, some analysts placed greater reliance on dividend yields as an indication of a capital-attracting rate of earnings than they placed on the earnings-price ratios. This preference was based on almost conclusive evidence that the investment market was then placing much higher values on currently distributed earnings than on reported retained earnings. Having this evidence in mind, the analyst might therefore first use the dividend-yield data to estimate what rate of dividends would need to be paid on the stock of the company in question in order to make the stock worth at least its book value plus an allowance for underpricing. He would then decide what rate of earnings on the common-stock capital would be reasonably called for in order adequately to support the required rate of dividends.

But this procedure was seriously roundabout, and it has now lost most of its claim for acceptance save, perhaps, as a check method. For the more recent studies indicate that the market is now valuing public utility common stocks by reference to total per-share earnings, with no clear, consistent tendency to "discount" retained earnings.[10] Earnings-price ratios therefore now deserve more at-

[10] See Fred P. Morrissey, "Current Aspects of the Cost of Capital to Utilities," 62 *Public Utilities Fortnightly* 217–227 (Aug. 14, 1958). The newer attitude toward retained earnings is probably due in part to the growing importance attached by stockholders to the personal income-tax factor. It is doubtless also partly due to the postwar experience of a fairly steady increase in earnings and dividends per share resulting chiefly from the reinvestment of realized earnings. The earlier experience of the gas and electric companies, with their heavy erosions of surplus

tention than current dividend yields as evidence of capital-attracting rates of return.

These earnings-price ratios express the relationship between the reported annual earnings per share and the quoted prices of the shares on the stock market. Usually, though not invariably, they are stated in concurrent form, in that the earnings of a given year will be applied to an average of the prices of the stock for that same year. Needless to say, they do not necessarily reflect the anticipated rates of earnings, and hence the prospect of a future flow of dividends, on the basis of which the market price of the stock is supposedly established by the actions of buyers and sellers. Nevertheless, especially when averaged over a period of several years, they are counted on to give some indication of a capital-attracting anticipated rate of earnings. For example, if the analyst finds that the stocks of a group of comparable utility companies have been selling, fairly consistently, at prices ranging around thirteen times reported concurrent earnings (equivalent to an earnings-price ratio of 7.7 per cent), he has a basis for a *tentative* inference that the stock of the company under review will also sell for about thirteen times its earnings after its earning power has been more or less well "established," following the rate case. If this inference is correct, the stock will sell in the neighborhood of its book value if its post-rate-case earnings come to 7.7 per cent of this book value. But there must be a step-up in the allowed rate of earnings to provide for the underpricing and stock-flotation expense involved in the issuance of additional capital stock. A 10 per cent discount for these last-named items is not infrequently held to be reasonable. This discount would bring the estimated minimum "cost of common-stock capital" to 8.55 per cent (7.7 per cent divided by 0.90).

A variant form of this earnings-price technique is sometimes deemed more reliable, especially when applied in a period of rapid expansion of utility capital. Here, the reported earnings per share are related, not to the price quotations on the stock market but rather to the prices, net to the company after underwriting

---

and enforced capital write-downs coming as an aftermath of holding-company control, had given investors good reason to doubt whether reported retained earnings would be forerunners of increased rates of dividend. The recent initiation by the Commonwealth Edison Company of the practice of "capitalizing" a large share of its retained earnings by the issuance of annual stock dividends will give new interest to the study of the effect of regular stock dividends (as distinct from stock splits) on market prices.

fees and other flotation expenses, at which utilities have actually offered or issued new stock through pre-emptive rights or through underwritten public offerings. These "earnings-net-proceeds" ratios have the advantage of reflecting in themselves any underpricing and cost of financing actually incurred by the utility companies. They must be used with caution, however, in order to guard against the possibility that some new stock issues may have been deliberately underpriced to the old stockholders beyond the dictates of successful financing.

Useful to the analyst as are both of the above-noted types of earnings-price ratios (the earnings-market-price ratio and the earnings-net-proceeds ratio), they have a serious limitation when offered as evidence of a capital-attracting rate of return on equity capital. This limitation is due to the fact that securities are rationally bought and sold, not directly on the basis of earnings already realized but rather on the basis of *anticipated* earnings. It follows that a current or recent earnings-price ratio of, say, 5 per cent, by no means proves a readiness on the part of investors to accept such a slim rate of return on somewhat speculative common-stock commitments. Contrariwise, a current earnings-price ratio of 15 per cent by no means proves that investors count on the continuance of such a handsome rate of return as their price for the contribution of equity capital. Hence, the analyst must consider, to the best of his ability, the extent to which investors may be "discounting" anticipated increases and decreases in corporate earnings per share.

The difficulty faced by the analyst in estimating what allowed rate of return on a company's common-stock capital would bring the market value of the shares into accord with their book values has been greatly enhanced in recent years by the high premiums over book values commanded by many of the public utility stocks in the postwar market. "Growth stocks" like those of Florida Power & Light Company, Florida Power Corporation, and Houston Lighting and Power Company (fast-growing companies operating in jurisdictions notably liberal toward utility investors) have been selling at from two times to more than three times their book values. Indeed, the common stocks of the electric utility companies in general have at times sold at premiums averaging 60 per cent or more over the book values that purport to represent net capital investment per share. These relatively high prices, which have resulted in current dividend yields as low as 2 per cent in extreme cases, and in earnings-price ratios as low as 5 per cent, almost certainly reflect investor anticipation of a continued upward trend in earnings and dividends.[11]

The problem here under review can be restated as follows. It is the problem of deriving the current cost of common-stock capital to any given company from the earnings-price ratios of the stocks of otherwise comparable companies that themselves may be earning, and expected to earn in the future, rates of return well in excess of their own costs of capital.[12] In such a situation, the shares will command premiums over book value based on the anticipation of corporate ability to maintain excess earnings, not just on existing capital but on a capital to be enhanced partly by a reinvestment of earnings and partly by the issuance of new stock at more than book value.[13] Needless to say, the current earnings-price ratios of stocks of this character do not represent "cost of capital" in any significant sense of that term. Indeed, if the allowed rates

[11] In addition to the discussions of growth factor in the references cited in footnote 2, see Charles Tatham's chapters on public utility stock valuation in Benjamin Graham and David L. Dodd, *Security Analysis*, 3d ed. (New York, 1951), supplemented by his article on "The Growth Factor in Electric Utility Earnings," in the March, 1952 issue of *The Analysts Journal*.

[12] This particular problem might be avoided by deriving the cost of equity capital from a record of earnings-price ratios and dividend yields of the stock of the very company in question, rather than from a set of comparable stocks, as long as this stock has been selling at prices not grossly in excess of book values. But the analyst then runs into another problem: that of determining to what extent the market prices of the stock of the company at bar may be based on an anticipation of an increase or decrease in earning power resulting from commission action.

[13] A study of twenty of the largest electric utility operating companies in the United States indicates that, for the period from 1948 to 1956, average earnings per share increased at a rate equivalent to a constant annual increase of 4.6 per cent per annum (6.6 per cent for the shorter period, 1952–1956). The corresponding average increase in dividends per share was at the rate of 4.4 per cent (5.0 per cent for 1952–1956). Both computations were made after adjustments for stock splits and stock dividends. Testimony of James C. Bonbright, Commonwealth Edison Rate Case of 1957, Case 44391, Illinois Commerce Commission, Edison Exhibits 13.25 and 13.26. In my accompanying testimony I surmised that 60 per cent or more of the growth in per-share earnings was attributable directly to the factor of retained earnings (which had averaged about 30 per cent of earnings on the common stock for the companies under review), and that 40 per cent or less was due directly to increases in net capital investment per share resulting from stock offerings between the beginning and the end of the period. These offerings had yielded net proceeds per share averaging about 133 per cent of the pre-offering book values of the old stock. The twenty selected companies did not include growth-stock companies like those of Florida and Texas.

of earnings on such stocks were to be permanently limited, by commission action or otherwise, to a rate of return on their book values equal only to their present earnings-price ratios, their market values could be expected to fall well below their book values.

So far as I am aware, no adequate solution of the problem just set forth has ever been offered. What makes the problem doubly difficult is that *any* constant, realized rate of return on public utility equity capital will tend to be converted into a continuously increasing rate of earnings and of dividends per share, given the maintenance by the company of a constant dividend-payout ratio (ratio of current dividends to current earnings) of less than 100 per cent, and absent any erosion of corporate surplus. Hence, the deficiency of current earnings-price ratios as measures of current costs of equity capital lies, not in the mere probability that these ratios reflect an investor expectation of subsequent increases in earnings and dividends per share, but rather in the probability that they reflect an anticipation of a rate of increase faster than that within the power of a company whose earnings are limited to its cost of capital.

I conclude this discussion of the difficulty of determining the cost of equity capital for any given company by expressing the conviction that the only such cost that can be determined with confidence is a *minimum* or *partial cost.* That is to say, the analyst, by a study of earnings-price ratios and of other market data, may be able to reach a credible conclusion that the cost of common-stock capital comes to *at least* some specified per cent; but the extent of the probable deficiency is necessarily a matter of surmise. Hence, if the minimum estimated cost is to be used in the determination of a computed "overall cost of capital," the resulting computation should be subject to a material, "judgment-reached" enhancement in order to give reasonable assurance of full-cost coverage.

Should the allowed rate of return be designed to prevent the market prices of public utility equities from rising to substantial premiums above book values? A rigorous and literal application of a cost-of-capital measure of a fair rate of return in the above-outlined sense of this measure would mean an attempt by a commission to regulate rates of charge so as to maintain the market prices of utility equities on a par with their book values or rate-base values plus some stipulated allowance for necessary underpricing. Yet a mere reference to any such attempt should suffice to suggest its absurdity. In the first place, commissions cannot forecast, except within wide limits, the effect of their rate orders on the market appraisals of the stocks of the companies subject to these orders. But in the second place, whatever the initial market appraisals may be, they are sure to change not only with the changing prospects of earnings but with the changing outlook of a notoriously volatile stock market. In short, market prices are beyond the control, though not beyond the influence, of rate regulation. Moreover, even if a commission did possess the power of control, any attempt to exercise it in the manner just suggested would result in harmful, uneconomic shifts in public utility rate levels.

This situation is recognized even by supporters of a cost-of-capital standard of a fair rate of return, who undertake to meet the difficulty in two ways. First, "current" cost of equity capital is rarely identified with spot cost. Instead, it is taken to mean a normal or average capital-attracting rate of return characteristic of the market of the past several years and typical of the market anticipated in the not distant future. Secondly, the estimated "over-all cost of capital" resulting from the application of this normalized estimate of the current cost of the common-stock capital is seldom accepted as a full measure of a fair rate of return. On the contrary, as already noted, the computed over-all cost is usually taken as a minimum allowance, subject to a "reasonable" upward adjustment for good measure.

It follows that the common stocks of public utilities which actually succeed in earning a "fair rate of return" as derived by a cost-of-capital technique can be expected to command substantial premiums over their book values or rate-base values except in periods of a seriously depressed stock market—premiums well in excess of any customary allowance for the necessary underpricing of new stock offerings. And the question arises whether the prevalence of these excess premiums is persuasive evidence of a corporate earning power higher than enough to give adequate assurance of continued corporate ability to attract the desired amounts of new capital on terms that do not impair the integrity of the existing capital.

In my opinion, the answer to this question is in the negative.

Regulation is simply powerless to assure the purchasers of public utility equities that future corporate earnings will suffice to maintain market prices on a par with book values or with any other dollar figure. Lacking this power, regulation wisely concedes to the public utility industries opportunities for corporate earnings liberal enough to bring to substantial market premiums the stocks of those well-managed companies that actually succeed in realizing these earnings fairly continuously. But while the allowance of a rate of return, during periods of business prosperity, liberal enough to let utility equities command substantial premiums over their book values seems to me to be called for in the interest of long-run corporate ability to meet capital requirements, the question what constitutes a proper *degree* of liberality has not yet received a convincing answer. Indeed, I doubt whether a conclusive answer can ever be found under such an indefinite standard of a fair rate of return as that of a flexible rate designed to rise and fall with changes in the anticipated rates of income necessary to induce new investments of equity capital.

THE ALTERNATIVE TEST OF A RATE OF RETURN EQUAL TO THAT EARNED ON INVESTMENTS IN OTHER ENTERPRISES OF CORRESPONDING RISKS

As outlined in the preceding paragraphs, a cost-of-capital criterion of a fair rate of return is a criterion of credit maintenance and capital-attracting adequacy. The so-called cost-of-capital formula is simply a useful aid in the estimate of rates of return, the systematic allowance of which over extended periods of time is designed to enable a well-managed, soundly financed company to secure needed new capital on terms that are fair to the existing investors.

But the criterion of credit maintenance or capital-attracting sufficiency is by no means the only one that has enjoyed support from courts, commissions, and expert witnesses. One alternative or complementary standard, particularly, has received at least widespread *verbal* endorsement from the appellate courts, following its repeated approval by the Supreme Court: the so-called "comparable-earnings" test. This test was first clearly set forth in the *Bluefield* case of 1923 [14] as applicable to a fair-value rate base. But

[14] Bluefield Waterworks & Improvement Co. v. Public Service Commission, 262 U.S. 679, 692–695 (1923).

it has been repeated in later Supreme Court opinions, even in the majority opinion in the *Hope* case [15] in which Justice Douglas apparently deemed it also applicable to a net-investment rate base. Since the *Bluefield* case is often cited as the leading judicial pronouncement on rate of return, the pertinent part of the opinion in this case, written by Justice Butler, may be worth quoting:

> A public utility is entitled to such rates as will permit it to earn a return on the value of the property which it employs for the convenience of the public equal to that generally being made at the same time and in the same general part of the country on investments in other business undertakings which are attended by corresponding risks and uncertainties; but it has no constitutional right to profits such as are realized or anticipated in highly profitable enterprises or speculative ventures. The return should be reasonably sufficient to assure confidence in the financial soundness of the utility, and should be adequate, under efficient and economical management, to maintain and support its credit and enable it to raise the money necessary for the proper discharge of its public duties. A rate of return may be reasonable at one time and become too high or too low by changes affecting opportunities for investment, the money market, and business conditions generally.

Here, as in the *Hope* case, are suggested not just one standard of a fair rate of return but two. In the first place, the rate must be equal to that currently earned on "investments" in other equally risky business enterprises. But, in the second place, it must also suffice to maintain the credit and the capital-attracting ability of the very company whose case is at bar. And the question arises what should be done in the likely event that the rate indicated by the one test is higher or lower than the rate indicated by the other. A severely literal construction of the *Bluefield* opinion would seem to require the acceptance of whichever rate of return happens to be higher in any given case. But this interpretation would run so contrary to common sense that it has not won acceptance.

Faced with this problem of judicial interpretation, my own preferred interpretation has been that the courts have not intended to set up two conflicting standards of reasonable utility rates. Instead, the credit-maintenance or capital-attraction standard is primary, while the comparable-risk standard is secondary and ancillary. That is to say, the fair rate of return is a rate, the allowance of which will permit the company in question to support its credit

[15] Federal Power Commission v. Hope Natural Gas Co., 320 U.S. 591, at 603 (1944).

and to raise required supplies of new equity capital on terms fair to the old investors; but this rate is necessarily related to the rates of return that investors, while still free to commit their capital on the competitive market, could expect to secure on investments in enterprises of comparable reputed risk. For the most part, such investments must take the form of the purchase of common-stock equities at prevailing prices on the stock market. What rates of return may be earned by the comparable corporations on their own invested capital, or on the appraised values of their assets, is of no direct concern.

The above-suggested interpretation of the comparable-earnings test seems necessary in order to reconcile it with the other test in the *Bluefield* case—the credit-maintenance test. But one must admit that this interpretation is somewhat strained, and I am not at all sure that it reflects the actual thinking of the judges who have cited the *Bluefield* opinion as a legal precedent. An alternative interpretation, consistent with a present-value standard of reasonable rates rather than with an actual-cost standard, is that regulated enterprises should be permitted to earn on the current values of their corporate assets, as based on replacement-cost appraisals, rates of return similar to the rates actually being earned by unregulated enterprises on the values of their assets, similarly appraised. But even this interpretation is a mere attempt by a layman to spell out a criterion which the Supreme Court itself has never undertaken to rid of its ambiguities.

Quite aside, however, from these ambiguities, the comparable-earnings test can hardly qualify as more than an ancillary measure of a fair rate of return if only because it would impose upon the financial analyst the impossible task of finding a group of unregulated business enterprises, investments in which or by which are comparable to utility investments in their esteemed safety and in the cyclical behavior of their earnings. This does not mean that investments in all industrial companies are necessarily more risky than most public utility investments. Indeed, some of the country's great industrial concerns may be so well entrenched in prestige, organization, and command over their market that their greater vulnerability, if it exists, to competition is more than offset by their freedom from restrictive regulation. But the analyst lacks the means of balancing these offsetting factors.

One type of comparable-earnings test has a real, though limited, usefulness. The rate of return proposed as fair for any given company may be compared to the rates actually earned by similar companies in the same jurisdiction or in other jurisdictions. But this kind of a comparison, while suggestive, is not conclusive in the absence of adequate reasons to assume that the realized earnings of the other companies fall within the range of a fair rate of return.

THE FAIR RATE OF RETURN IN A PERIOD OF BUSINESS DEPRESSION

Most of the detailed development of rate-of-return analysis has taken place since the end of the last war and during a period of high prosperity interrupted only by short recessions that have not seriously affected public utility earnings. But this prosperity may not last forever; and the question may arise, as it did in the severe and prolonged depression of the 1930s, what rates of return will then be deemed "fair," or even whether the whole conception of a "fair" rate of return must give way to the exigencies of an economic emergency.[16] Indeed, this question is pertinent at the present time and despite the lack of imminence of a severe depression. For the adequacy of allowed rates of profit during prosperity must depend in part on anticipated rates of profit or of deficit in depressions.

So far as concerns the more or less regular ebbs and flows of business activity associated with an ordinary business cycle, the generally accepted principle of rate-of-return determination is

[16] See E. M. Bernstein, *Public Utility Rate Making and the Price Level* (Chapel Hill, N.C., 1937); Ralph C. Epstein and John D. Sumner, "Effect of the Depression upon Earnings and Prices of Regulated and Nonregulated Industries," 26 *American Economic Review, Supplement* 36–45 (March, 1936); Ben W. Lewis, "State Regulation in Depression and War," 36 *American Economic Review, Proceedings* 384–404 (May, 1946); David E. Lilienthal, "Regulation of Public Utilities during the Depression," 66 *Harvard Law Review* 745–775 (March, 1933); J. D. Sumner, "Public Utility Prices and the Business Cycle: A Study in the Theory of Price Rigidity," 21 *Review of Economic Statistics* 97–109 (1939); J. D. Sumner, "Public Utility Rate Making in Depression: A Comment," 52 *Quarterly Journal of Economics* 713–717 (1937); Donald H. Wallace, chapters on the structure of prices in regulated industries, Part IV of Monograph No. 32, *Economic Standards of Government Price Control*, Temporary National Economic Committee (Washington, D.C., 1941); National Bureau of Economic Research, *Regularization of Business Investment* (Princeton, 1954), especially section by Edward W. Morehouse on "Regularization of Business Investment in the Electric Utility Industry," pp. 213–281.

fairly clear, although the practice gives rise to difficulties. With rare exceptions, public utility tariffs are not raised and lowered year by year with the object of maintaining a constant or stable rate of return on capital investment. Instead, they are set for the indefinite future, subject to revision possibly in a year or less but more probably only after a period of several years. When a rate case arises, a commission may set rates designed to yield a "fair rate of return" on the average for the next several years and not necessarily for the next twelve-month period. Minor variations above and below this figure, unless they are expected to persist or to grow wider with the passage of time, will be allowed to reverse themselves in later years.

But the occurrence of a major business depression, such as the catastrophic depression of the 1930s, presents a quite different problem, and one that simply cannot be solved in advance by the present fixation of rates designed to yield any predetermined fair average rate of return, counting in the substandard returns (or even the deficits) that may possibly be realized during the next depression, if and when it takes place. At least under the prevailing types of rate regulation, a commission, during a period of prosperity, is pretty well limited to two kinds of action designed to take account of depression possibilities of a severe nature: first, insistence on security structures sufficiently conservative in their use of debt capital to permit public utility companies to remain solvent during periods of substandard earning power; and secondly, approval, during prosperity, of rates of earnings on equity capital which will attract stock subscription from an investment market on warning that even utility companies cannot expect to go through major depressions with unimpaired earnings and undiminished rates of dividend.

Unfortunately, however, the investment market must rest content with a warning of a very indefinite character, since the right of a public utility to an opportunity to earn a conventional "fair rate of return" in the event of another severe business depression is subject to much uncertainty, and since no commission is in a position to commit itself in advance of the event. On the whole, the utility companies were treated very liberally by courts and by most commissions during the 1930s, with the result that their rates of charge were either maintained about at predepression levels or else reduced only to a minor extent—in some cases voluntarily. But only rarely did the utilities find it feasible or permissible to raise their rates in order to make up for loss of output; and any such violation of good economic sense is not likely to be countenanced in another depression.

Among economists in the 1930s, the view was widely held that recovery from the business depression was being seriously interfered with by the distortion in price relationship resulting from the severe decline in the prices of flexible-price commodities including, notably, farm products, unaccompanied by any comparable decline in the "administered" prices of manufactured commodities and in the charges for public utility and railroad services. This view influenced one of the country's leading commissions, that of Wisconsin, to institute a rate case in which it ordered the Wisconsin Telephone Company to reduce its rates on the order of 10 per cent—a moderate reduction but one that could hardly have been supported by a fair-return rule of the orthodox type. But the order was annulled by the highest state court.[17]

In more recent years, business-cycle experts have become skeptical of proposals to combat a depression by enforced reductions of "sticky" prices, and attention has been turned to other alternatives including the possibility of using the combined machinery of regulation and taxation to encourage private utilities to maintain their construction and equipment budgets even when their existing plants are partly idle because of a temporary drop in the demand for service.[18] Unless a reversion to advocacy of price flex-

[17] Wisconsin Telephone Co. v. Wisconsin Public Service Commission, 232 Wis. 274, 287 NW 122 (1939), affirming lower court reversal of commission order in Re Wisconsin Telephone Co., 13 PUR NS 224 (1936). In this case before the commission, a group of economists, including myself, supported a rate-reduction order. For other rulings on rate of return during the depression of the 1930s, see Ellsworth Nichols, *Ruling Principles of Utility Regulation: Rate of Return* (Washington, D.C., 1955), pp. 141–153.

[18] See a recent symposium of economists on *Policies to Combat Depression*, a report of the National Bureau of Economic Research, New York (Princeton, 1956). But economic fashions may change again! As to attempts to encourage advance construction of plant and equipment by private utility companies during a business depression, see Edward W. Morehouse's paper on "Regularization of Business Investment in the Electric Utility Industry," cited at the end of footnote 16, *supra*. Mr. Morehouse was not optimistic about the feasibility of this encouragement. Compare Michael Gort's study of the extent of flexibility in electric utility capital budgeting: "The Planning of Investment: A Study of Capital Budgeting in the Electric Power Industry," 24 *Journal of Business* 79–95 and 181–202 (1951). During the long depression of the 1930s, the great capital investments in electric power were those made by, or under subventions from, the Federal government, notably, the hydroelectric developments and the rural electrification program.

ibility wins general support from economists, the likelihood of widespread public utility rate reductions in even a severe depression will be remote.

Today, the long period of sustained prosperity in America, interrupted by no recession worse than that of 1957–1958, has turned the interest of economists away from problems of depression economics. No doubt the widely held hope that severe depressions are a thing of the past helps to explain this lapse of interest. But optimistic hope is far from certainty; and modern public utility rate theory suffers from its lack of progress in suggesting answers to the question what to do in anticipation of, and in the event of, a major business depression. The problem is by no means confined to that of the general *level* of public utility rates. It extends to the determination of rational rate *structures* during periods of heavy idleness of labor and of plant capacity, when the relative money costs of supplying utility services are unreliable measures of relative social costs.

POSSIBLE ALLOWANCES FOR THE EFFICIENCY FACTOR

Elsewhere in this volume, attention has been called to what is perhaps the most serious of all of the objections to a cost-of-service standard of reasonable public utility rates—to the objection that, as long as rates are fixed so as to assure even a company under mediocre management that it can cover its costs, including a "fair rate of return," and as long as any higher return is denied even to a company under exceptionally able management, there will be lacking under regulated private ownership a stimulus for efficiency comparable to the stimulus of actual competition. American experience with regulated private ownership hardly justifies the unqualified indictment that some writers have made against it on this score. But a plausible case, at least, could be made for the thesis that what has saved regulation from being a critical influence in the direction of mediocrity and tardy technological progress has been its very "deficiencies" in the form of regulatory lags and in the form of acquiescence by commissions in fairly prolonged periods of theoretically "excessive" earnings on the part of companies whose public repute and whose comparative rates of charge for service have not made them vulnerable to popular attack.

But while a situation of this kind may be tolerable, it suggests the wisdom of more systematic and deliberate efforts on the part of regulating agencies to distinguish, somewhat as competition is supposed to do, in favor of companies under superior management and against companies under substandard management.[19] The distinction might take the form of an express and publicly recognized differential rate of return—a differential, for example, under which otherwise comparable companies might be allowed a 6 per cent rate of return under standard management, a 5 per cent rate under substandard management, and a 7 per cent rate of return under top-grade management. Objection might be raised to the substandard rate of return, on the ground that it would make bad matters worse.[20] But one might hope that the restriction of a company, by virtue of a commission finding of inferior management, to a minimum rate of return measured, say, by a "barebones" estimate of cost of capital, would become so intolerable to the corporate stockholders that they would soon enforce a change of management.

Proposals along these lines have often been made[21] but have

[19] See Dr. Charles S. Morgan's monograph on the subject entitled *Regulation and the Management of Public Utilities* (Boston, 1923). The paucity of later studies does no credit to the modern theory and practice of public utility regulation.

There is ground for the conviction that the opportunity of a well-managed utility to earn a return *liberally* adequate to attract capital is in the public interest as encouraging rapid technological progress and long-run policies of operation. See a suggestive, but self-serving, study designed to support this point of view, printed under the auspices of the American Telephone and Telegraph Company: *Profit, Performance and Progress—A Study of Regulated and Non-Regulated Industry* (May, 1959).

[20] See the opinion of Commissioner Eddy in a recent rate-increase application by the Rochester Telephone Company: Case 18837, New York State Public Service Commission, July 29, 1958, 24 PUR 3d 262. Here, the relatively high rates and asserted inferior service of the company were attributed, in part, to delay in conversion to dial service. Yet the resulting unhappy service situation, said Commissioner Eddy, "can only be corrected by the expenditure of very large sums of money in the future—money which must be obtained in the marketplace in competition with other utilities." With some exceptions the Commission, on Mr. Eddy's recommendations, permitted the company to file the rate increases which it had requested. Commissioner Blach, dissenting, would have reduced the requested revenue increase on the ground that "The record in the proceeding shows that the Company's service in at least three of its exchange areas is of poor quality."

In some jurisdictions a commission may lack the legal power to deny a rate increase upon grounds of poor service. See a comment on this point by Commissioner Cyrus J. Colter of the Illinois Commerce Commission, reported in 62 *Public Utilities Fortnightly* 410–412 (Sept. 11, 1958). Compare Ellsworth Nichols, *Ruling Principles of Rate Regulation: Rate of Return* (Washington, D.C., 1955), Chap. 20, "Efficiency of Operation and Management."

[21] The best-known schemes designed to stimulate managerial efficiency have been the various versions or modifications of the former British Sliding-Scale-Dividend

never as yet made headway for a number of reasons, two of which deserve notice here. The objection most frequently raised is that, with most public utilities, there is an almost complete divorce between stockholders and managers. Hence, if any system of bonuses and penalties is to be desired, it should be applied directly to the executive officers and not to the shareholders, who probably have no control whatever over financial or operating policy.

This is a plausible argument. But it is not completely convincing, since it oversimplifies the motivation of modern corporate management. A responsible management, while concerned to render good public service "for its own sake," properly regards itself as representing the interests of its security holders and, particularly, of its stockholders. Success in maintaining a stable dividend record in the face of business recessions and in developing a net income that will permit a gradual growth in dividend payments as the business expands and as technology improves is almost certainly a prized goal of many public utility executives. The psychology of loyalty and of approval applies even within the executive offices of the giant corporations.

But a second objection to specific efficiency differentials in rate-of-return allowances is far more serious. It lies in the absence, at the present stage of public utility regulation, of adequate objective tests of relative efficiency in the performance of public services. Crude comparisons of the rates charged for service by different companies, in different areas, are altogether unreliable because of differences in operating conditions, in local costs of labor and fuel, in taxes, etc. Refined comparisons, which undertake to allow for these many differences, are more promising. But up to the present time, the technique of their development has not reached the stage of high trustworthiness.

Until more headway in this development has been made, proposals for the systematic use of efficiency differentials in rate-of-return allowances would probably be premature. Meanwhile, however, commissions, in cases requiring a determination of a fair rate of return, would be well advised to encourage the submission, both

schemes, under which the allowed rate of dividends or rate of return varies inversely with changes in a company's rate levels. See Irvin Bussing, *Public Utility Regulation and the So-Called Sliding Scale* (New York, 1936). Aside from other deficiencies, these schemes have been brought to grief by the forces of price inflation.

by the utility companies and by opposition parties, of relevant evidence on the performance record of the company under review as compared both to the performance of other utility enterprises, public and private, and to the performance of the same company in earlier years.[22]

Even if rate regulation will not find it feasible to make systematic use of differential rates of return as incentives to managerial efficiency, it should at least take pains to avoid rules of rate making that positively penalize stockholders for efficient or otherwise desirable action by the management.[23] An example of such a rule would be one which requires the removal from the rate base of the unamortized costs of assets prematurely retired because of obsolescence, but which permits the retention of these costs in the rate base as long as the assets are not actually replaced by superior substitutes. Another example is that of a rule which allows recovery of the actual interest charges on high-interest-bearing callable bonds as long as these bonds remain outstanding, but which compels the company, if it chooses to call the bonds, to charge off the call premium and the unamortized debt discount as a loss rather than as a financial expense to be recovered in later years. Similar examples of rules or practices that create a head-on conflict between managerial action in the consumer interest and action in the stockholder interest will occur to anyone familiar with American rate regulation.

[22] The feasibility of attempts to devise and apply standards of relative efficiency in the operation (though doubtfully so in the construction) of different electrical utilities was affirmed by Professor Robert T. Livingston of the Columbia University School of Engineering in testimony before the New York State Commission on Revision of Public Service Commissions Law, 1930, *Hearings*, Vol. III, pp. 2414–2423. The Fourth Annual Report of the Power Authority of the State of New York (year 1934) presented an engineering study comparing the costs of distribution of electricity in various municipalities, some served by public plants, some by private companies. Significant comparisons between rates charged by publicly and privately owned utility systems are made difficult by the relative freedom of the former systems, and of the interest on securities issued to finance these systems, from the burden of taxation. See Twentieth Century Fund, *Electric Power and Government Policy* (New York, 1948), esp. Chap. 7. But the attempted solution of difficult problems is the task of modern public utility regulation.

[23] Even under public ownership there is urgent need for devices of accounting control designed to reveal uneconomic practices that might otherwise appear to make a showing of good performance. As the late Professor A. M. Henderson remarked in a study of price policies for the British nationalized utilities, "No pricing policy can do much to ensure economical operation of state enterprises but at least it can avoid doing positive harm." "Prices and Profits in State Enterprise," 16 *Review of Economic Studies* 13–24 at 16 (1948–1949).

THE FAIR RATE OF RETURN ON AN ACTUAL-COST RATE BASE IN A PERIOD OF PRICE INFLATION

All of the rate-of-return problems discussed so far in this chapter would be present in a dynamic economy even in a period of stable price levels. But most periods of American history have not been of this character. Instead, the period since the end of the 1930s has been one of more or less continuous inflation. Even if we measure this inflation by the *relatively* stable Bureau of Labor Statistics Index of Consumer prices, which comes as close as any well-known index to reflecting changes in "the cost of living," the dollar has lost about half the purchasing power that it possessed shortly before the Second World War. Of even more significance for the problem at hand is the expectation, widely held among economists and businessmen, that the long-run trend of prices will continue upward, interrupted perhaps by short-run plateaus or even minor dips during times of business recession. The combination of this recent history and this present anticipation of rising prices has had a decided influence on accepted and proposed measures of a fair return: an influence in the direction of more liberal allowances, measured in terms of dollars, than would otherwise probably be conceded. But in many jurisdictions the influence has been indirect and of uncertain import.

A strictly applied, present-value basis of rate control would make full allowance in the rate base for any definitely established increases in the current construction costs of public utility properties. More will be said on this point in the third section of the present chapter. But what we are now concerned with is a different question: namely, whether or to what extent the common stockholders in public utility companies are entitled to the protection of their "real" income against a fall in "the value of the dollar" even under the general philosophy of an actual-cost principle of rate control. Such protection could be given in either of two ways: (a) through an upward adjustment in the rate base of the type discussed briefly at the end of Chapter XII, or (b) through an index-number enhancement of the nominal, percentage rate of return. But the choice between these alternatives, while not unimportant, presents only a secondary question, on which I shall comment briefly in a later paragraph. The primary issue is that between those persons who contend, and those persons who deny, that the "fair return" to public utility companies is a return designed to compensate the corporate stockholders for any deterioration (or any *major* deterioration) in the purchasing power of the dollar that has occurred between the time of the original capital investment and the time of a rate case.

Needless to say, the affirmative side of this argument has been pressed with most force by spokesmen for stockholder interests and by public utility counsel and witnesses in rate cases. But it has also received considerable support from disinterested sources. As early as 1945, the late John E. Benton, speaking as counsel for the National Association of Railroad and Utilities Commissioners, declared that, in a severe inflation, an offsetting increase in the rate base or in the rate of return "would be required not only in justice to investors but in obedience to economic law. . . . Investors [he added] simply will not buy utility securities, if they find that progressive inflation operates to destroy progressively their right to receive just and reasonable compensation for the service their properties render." [24]

The above-cited remark by Mr. Benton illustrates the twofold character of the claim for inflation adjustments as presented by company witnesses in rate cases: considerations of "fairness" or "equity" as between consumers and investors; and considerations of financial necessity or expediency. To be sure, these claims are not independent of each other, since the question whether existing investors have a fair claim to indemnity for *past* inflation may be held to depend on the question whether uncommitted prospective investors are willing to supply new capital without receiving any protection against *future* inflation. Nevertheless, the answer to the one issue does not necessarily carry with it the answer to the other. Hence, the two questions may best be reviewed separately: first the question of capital-attracting necessity, then the question of fairness.

[24] From excerpts of an address of Dec. 4, 1945, quoted in a brochure on *The Impact of Inflation on Utility Earning Requirement* by W. J. Herrman, Rate Advisor, The Commonwealth & Southern Corporation (New York), Sept. 1, 1948.

THE PROBLEM VIEWED FROM THE STANDPOINT OF FINANCIAL NECESSITY OR EXPEDIENCY

Viewed from the standpoint of financial expediency, the argument for inflation adjustments in the dollar returns allowed to public utility companies—an argument which should carry with it the admitted desirability of downward adjustments in the event of price deflation—rests on the contention that, unless the utility companies receive assurance of such adjustments in the future, they will not be able to raise adequate amounts of new capital, at least not on reasonable terms, in an investment market convinced of the probability of further inflation and of the unlikelihood of later deflation.

The validity of this contention cannot be proved or disproved by a priori reasoning based on assumptions as to the actions of "rational" investors. But, in an article on the subject published in 1951,[25] I noted the failure of company witnesses and of other advocates of rate-making allowances for price inflation to adduce any convincing empirical evidence of the necessity of such adjustments as a means of preventing the sources of public utility capital from drying up. Since the end of the Second World War, the public utility industries have raised huge amounts of new capital for plant expansion and improvements. Yet, in the face of rising price levels, they have secured most of this capital by the issuance of bonds, debentures, and nonparticipating preferred stocks—securities without any offsets for price inflation. Moreover, up to within the past few years, the bond financing and refinancing was done at interest rates made extremely low by the interest-pegging policies of the United States Treasury.

To be sure, the demonstrated ability of the utility industries to do heavy bond financing during postwar years does not prove their ability to market adequate amounts of common stock without inflation protection, since the market for the bonds has come so largely from financial institutions, notably life-insurance companies and pension funds, with long-term liabilities fixed in dollars. But during this same period the utilities have also raised large amounts of

[25] "Public Utility Rate Control in a Period of Price Inflation," 27 *Land Economics* 16–23 (Feb., 1951).

common-equity capital.[26] And success in equity financing has not been limited to companies operating in "fair-value" states or in states (if any) that make express allowances for price inflation in their calculations of a fair rate of return.

When my 1951 article was written, the operating companies of the American Telephone and Telegraph Company system (The Bell System) were taking the lead in presenting to commissions applications for rate increases based, in part, on claims for inflation adjustments. These claims continue to be pressed with much vigor and with great skill in current rate proceedings. Yet the great success of the parent company in raising unparalleled amounts of new equity capital since the end of the Second World War might be cited as a refutation of any such claim when couched in terms of asserted financial necessity. For the annual rate of dividends was maintained at $9 during every year between 1921 and 1959, with neither a decrease to take account of the price deflation of the 1930s nor an increase to take account of the subsequent inflation. While the stockholders have also received opportunities for supplementary gains through the occasional issuance of "rights," these supplements have not constituted full inflation offsets.[27]

The above-noted postwar history of public utility finance still warrants the conclusion of my 1951 study to the effect "that the alleged necessity of safeguarding utility stockholders against price inflation as a means of attracting new equity capital has not yet been supported by convincing evidence." But even aside from the issue of fairness to investors, the bearing of this negative conclusion on the question whether some such safeguard is desirable is not as clear as one might take it to be. For even in those jurisdictions that

[26] Professor Fred P. Morrissey writes that "the utility industry has accounted for approximately 50 per cent of new equity issues in the post-war period—a trend likely to continue." "Current Aspects of the Cost of Capital to Utilities," 62 *Public Utilities Fortnightly* 217–227 at 217 (Aug. 14, 1958). In an earlier article in the same journal he presented tables on utility bond and stock issues in recent years. "Dividend Policy and Reduction of Tax Liability," 59 *ibid* 15–22 (Jan. 3, 1958). As to future needs for common-stock capital he said (p. 18): "If the predicted expansion is achieved, it is likely that the utilities will be raising at least $600–$800 million annually through stock issues under present methods."

[27] I am informed that, for the period 1946–1957, the subscription rights to convertible debentures and to stock offered to stockholders of the American Telephone and Telegraph Company have had market values equal to an average annual value of about $1.80 per share. This includes the more than $6 value of the 1956 rights to acquire new stock at par ($100).

have not conceded any express inflation adjustments either in the rate base or in the allowed rate of return, most commissions have permitted the public utilities to earn returns sufficiently liberal to support a fairly continuous upward trend in the rates of dividends per share. This favorable trend has been due primarily to the combined effect of a material investment of earnings plus the issuance of new common stock at premiums over book value. In consequence, the owners of many utility stocks, even in original-cost jurisdictions, have not suffered a postwar impairment of their "real" income or a confiscation of their "real" capital, as have the holders of fixed-charge securities or of preferred stocks. Instead, both their annual dividends and the market values of their equities have more or less kept pace with their rising costs of living.[28]

In view of this financial history, the question really at issue is that of a choice between a scheme of rate regulation expressly designed to make the dollar earnings on stock equities rise and fall with changes in general price levels, and the more familiar regulatory practice which makes no direct provision for inflation but which undertakes to compensate stockholders for running the *risks* of inflation by the concession of what might otherwise be needlessly high "fair rates of return." On this difficult question I reserve further comment pending a discussion of the fairness issue.

THE FAIRNESS ISSUE

We come now to the question whether public utility companies are entitled to inflation adjustments in their return allowances on grounds of fairness as between investors and consumers, even assuming that no such adjustments need be conceded for the sake of capital attraction and hence in the long-run interests of the consumers themselves. On this question my own position has not been free from doubt, in view of the fact that the holders of fixed-income investments, including investors in savings banks, U.S. Savings Bonds, and corporate bonds and preferred stocks receive no safeguards against inflation. Of necessity, moreover, the burden of protecting the investors in utility stocks would fall upon all consumers of service, including those whose incomes do not rise with

[28] See Fred P. Morrissey, "Inflation and Public Utility Regulation," *California Management Review*, Vol. I, No. 4, 1959, pp. 74–88. This article is summarized in 65 *Public Utilities Fortnightly* 625–627 (1960).

rising price levels. There is the further serious question whether the extension of escalator provisions, already widely applied in union wage contracts, to public utility stockholders may not tend to accelerate the inflationary spiral.

Not all writers, however, have shared the doubts on the issue that I have just expressed. Among the academic economists, Professor Morton of the University of Wisconsin has not hesitated to denounce, as grossly unfair and as constitutionally confiscatory, the denial to public utility stockholders of income-earning opportunities that make full allowance for experienced price inflation. In an article published in 1952,[29] his strong plea for inflation adjustments designed to protect utility stockholders against the impairment of their "real" capital was presented with such skill and with such moral conviction that it should be read by everyone concerned with the problem. What makes the article of special interest is its concession, contrary to the position taken by other spokesmen for the public utility investors, that inflation protection is not required on grounds of financial expediency.[30]

Without attempting to follow Professor Morton's reasoning, which insists on the traditional character of a common-stock investment even in a regulated corporation, I may here state my own tentative position, reached in the light of his analysis and that of other, more recent writers. In my opinion, the holders of public utility common stocks have a strong claim for the protection of their "real capital" against price inflation in those situations in which they have already been given reasonable grounds to anticipate such protection through the action or inaction of courts, commissions, and legislatures. If the situation were clearly otherwise—if subscribers to public utility common stocks had been on

[29] Walter A. Morton, "Rate of Return and the Value of Money in Public Utilities," 28 *Land Economics* 91–131 (May, 1952). Professor Morton has subsequently defended his position in appearances as a company witness on rate of return in a number of telephone rate cases.

[30] "We can now quickly dispense with the argument that inflation protection is required to attract capital. The need for such protection, we must agree with Professor Bonbright, has not been established. But we can go further and say that it cannot be established. However desirable such protection may be as a matter of fairness, it certainly is not necessary. . . . So long as the nominalistic historical cost–cost-of-money method of rate-making prevails, nothing stands in the way of continuous expropriation of past investors by means of a fall in the value of money. Furthermore, new inflated dollars will always be forthcoming at a satisfactory current yield, except perhaps during times of galloping inflation." *Ibid.*, pp. 117–118. In my own opinion, this was an incautious concession.

actual or constructive notice that, so far as concerns inflation or deflation, they would be treated exactly like bondholders or preferred stockholders—their later claim for inflation adjustment would have no higher moral standing than would a claim for inflation offsets in favor of these other security holders. In states that have adhered persistently to a fair-value rate base, the argument of reasonable anticipation can be urged with much force. But in states that have long adopted an actual-cost rule of rate making, this argument would need to be modified. Here the claim must rest, not on any asserted right to an inflation offset per se, but merely on the right of public utility stockholders to rely on the continued application of whatever liberality, in rate-of-return allowances, the commissions have actually exercised in the past. As already noted, these allowances have sufficed to create a material excess in the market prices of utility equities over their book values. A later reduction in the allowances, designed to bring market values into equality with book values, might well be held to violate reasonable expectations.

The foregoing remarks on the fairness of claims for inflation protection are concerned only with the asserted injustice of rules of rate making to be applied retroactively to stockholders who were justified in anticipating the application of more favorable rules. But what about the fairness of a denial of inflation protection even in the absence of problems of retroactive equities—even, for example, with respect to a new company, the prospective stockholders of which can be put squarely on notice that, in the future, regulation will adhere strictly to the convention of a stable dollar despite any rise or fall in price levels? This is a more difficult question, since traditional standards of financial ethics tend to support the freedom of individual investors to enjoy or suffer the consequences of their own negotiations, subject only to the proviso, embodied in the Federal Securities Act, that they shall be fully informed about the facts of their proposed investment. If these standards are deemed applicable to the problem now before us, a purchaser of common stock in a utility company who buys the stock on notice that he must take his chances of an impairment of his "real income" or of his investment values in the event of further inflation, has no more claim to restitution than has the buyer of a mining stock to the receipt of indemnity for a failure of the mine to produce gold.

If I understand Professor Morton's position, he would not accept this traditional standard of commercial morality when it comes to exposure of public utility stockholders to the "confiscation" of their property through price inflation. This position is worthy of respect. Yet those persons who take it seem to me to lack the strength of their own convictions when they fail to insist that the same standards of equity be applied, at least in the future, to the purchasers of creditor securities and of preferred stocks.[31] The very public utility companies whose officials and expert witnesses complain of the unfairness of a system of rate control which tends gradually to confiscate the "real capital" of their common stockholders, continue to do most of their financing by public offerings of bonds and preferred stocks which are vulnerable to this form of "confiscation" by the express terms of the indentures or of the corporate charters. Indeed, during times of very low interest rates, utility companies have not hesitated to call bonds bearing higher rates whenever they have seen a financial advantage in so doing.

For the reasons just suggested, I do not find the fairness aspect of the case for or against inflation protection as easy a problem as Professor Morton and other writers appear to find it when they insist, almost as if it were self-evident, that public utility common stockholders should receive protection against the gradual or sudden impairment of their "real" capital through a fall in the value of the dollar. The problem is not easy because price inflation, by its very nature, is unfair and disorderly in its impacts on different classes of people, and because any attempt to save one particular class against its inequities runs the risk of imposing even more severe burdens on unprotected classes. It also runs the risk of adding fuel to the inflationary fire.

[31] The late Professor Sumner Slichter was one of the few responsible economists bold enough to have proposed that the U.S. Treasury be authorized to issue savings bonds payable, not in terms of a fixed number of dollars but rather in terms of a specified amount of purchasing power. Others have condemned the proposal as in danger of accelerating the inflationary spiral. Professor Morton suggests that insurance companies would probably not care to buy bonds payable in variable dollars, since their own liabilities are in fixed dollars. This may be true. But the fixity of liabilities on insurance policies and on annuities may be attributed to the unavailability to insurance companies of investments that provide a sufficiently secure inflation hedge. Note, however, the recent proposals for the issuance by insurance companies of "variable annuities," like those of the College Retirement Equities Fund, to be based on investments in common stocks believed to offer protection against price inflation.

But the fairness issue is a close one—so close that I believe that the merits of proposals for systematic inflation adjustments on behalf of utility stockholders should be judged primarily on the basis of capital-attracting efficiency. While the experience down to date indicates that equity capital can be secured without such adjustments, it also suggests that this capital can be secured only by the concession of rates of return that offer to stockholders a kind of compensation for the very absence of protection. This being the probable situation, the provision of overt adjustments, whether in the rate base or in the rate of return, may prove the more satisfactory arrangement, alike to investors and consumers. The possible advantage of this latter arrangement is that the compensation for inflation would then be paid only if, and to the extent that, inflation actually occurs instead of being paid merely in contemplation of the danger that it *may* occur. But the only way to test this possible advantage is by experiment; and the experiment, in my opinion, is well worth trying out in a number of jurisdictions. Perhaps the most promising opportunities in this direction lie in states in which stockholders are already accorded material but unsatisfactory safeguards against inflation by the crude device of a fair-value rule of rate making.

SHOULD INFLATION ADJUSTMENTS BE MADE VIA THE RATE BASE OR THE RATE OF RETURN?

If adjustment is to be made for inflation and deflation (whether as a matter of limited experiment, as I have suggested, or as a matter of general policy, as Professor Morton insists), the question arises whether it should be made in the rate base or in the rate of return. Contrary to the views of some writers, my own preference would be for the former alternative as a means of avoiding the false appearance of an excessive rate of return during a period of inflation. But for reasons already discussed in Chapter XII, this does not mean the adoption of a fair-value rule of rate making. Instead, it means the acceptance of a rate base measured by depreciated actual cost, restated, however, in terms of dollars equal in purchasing power to the purchasing power of the original capital contributions. Moreover, the restatement should be confined to the common-equity capital if the objective is that of maintaining the integrity of the stockholders' investment. Acceptance of this objective would suggest that the index number by which to measure price changes should be neither an index of construction costs nor an index of wholesale prices but rather the Bureau of Labor Statistics Index of Consumer Prices—the closest available approach to a "cost-of-living" index.

As Professor Morton points out in his above-noted article, stockholder protection against inflation could be secured without any enhancement in the annual allowance for depreciation as an operating deduction. True, the latter allowance would then not suffice to let the utility company recoup its original capital investment in dollars of equivalent purchasing power and might therefore seem to subject the corporate stockholder to an erosion of his "real capital." But, from the stockholder's view, the erosion would be prevented by the nondeduction from the rate base of any allowance for accrued depreciation in excess of the "inadequate" depreciation reserve. The important question, therefore, is whether a larger dollar allowance for accruing depreciation, and hence a heavier build-up of the depreciation reserve, is not called for in order to secure a better balance between the claims of present and future consumers. And here, an affirmative answer would clearly be indicated were the issue not complicated by an income-tax law which refuses to recognize deductions for depreciation other than those based on dollar costs.

With an actual-cost rate base, adjusted in the manner just suggested, rate-of-return determination would follow the same general principles suggested in the first section of this chapter. But the return on the common-equity capital (a capital enhanced by the index-number adjustment) would now be designed to bring the market prices of the stock into balance with the *restated* book values. There would still be a need for an excess of market value over revised book value sufficient to allow for underpricing as well as to allow a margin of safety. But there should no longer be a need to concede a rate of return high enough to send stock market prices to the high premiums over restated book values that they enjoy in the present stock market in comparison to their present book values. If such a concession were continued, and if the inflated book values should become a mere base on which to superimpose correspondingly high excess market values, the experience would discredit the whole attempt to provide stockholders with price-level adjustments.

THE FAIR RATE OF RETURN ON A FAIR-VALUE RATE BASE

It remains for this last section to consider what tests of a fair rate of return are applicable to a "fair-value" rate base which gives important weight to reproduction-cost appraisals.[32] This question has been raised repeatedly in the postwar rate cases and has given rise to disputes between those who insist, and those who deny, that the concession of a fair-value rate base in excess of actual net investment should be wholly or partly offset by a lower allowed rate of return. On this issue the positions of the various commissions have been far from uniform. It is my impression that, in fair-value jurisdictions, the general, though not universal, tendency has been to allow less liberal rates of return than would be allowed to comparable companies subject to an actual-cost rate base.[33] But in most cases, the differentials have not been large enough to result in a complete offset.

As to the question whether, as a matter of "theory," the allowed rate of return on a relatively high, fair-value rate base should be lower than the rate allowed on an actual-cost rate base, no intelli-

[32] For case citations, see Ellsworth Nichols, *Ruling Principles of Utility Regulation—Rate of Return* (Washington, D.C., 1955). The public accounting firm of Arthur Andersen & Co. has digested the return allowances and the corresponding types and amounts of the rate base for a great many cases decided since 1914. *Return Allowed in Public Utility Rate Cases,* and two or more *Supplements.*

[33] In a case involving the intrastate New York rates of a gas company that had recently been through a rate case with respect to its operations in Pennsylvania (a fair-value jurisdiction, as distinct from New York which is an actual-cost jurisdiction with respect to gas and electric companies), the company supported a higher rate of return than it had claimed in Pennsylvania. Said the New York Commission: "The reason for the variance in the claims before the respective commissions was not explained of record, but presumably rests upon the commonly accepted generalization that the fair rate of return on an original-cost rate base may be quite different from the fair rate of return on a fair-value rate base." Pennsylvania Gas Co., Inc., Case 17817, Adopted Dec. 17, 1957.

A study of rates of return actually earned by 103 electric utility companies finds that, in 1957, companies in fair-value jurisdictions averaged a 6.25 per cent return on total capitalization as against a 5.43 per cent average for the original-cost companies—an excess of 15 per cent for the former class of companies. In 1953, the excess came to about 12 per cent. *Electrical World,* May 5, 1958, pp. 97–100.

gent answer is possible in the absence of a more definite rationale of the fair-value rule itself than is forthcoming from the opinions of the courts and commissions. But three, quite different, answers have been suggested by some of the recent rate cases, and these will now be discussed in turn.

THE COUNTERVAILING-FALLACY PRINCIPLE

For reasons suggested in the chapters on the rate base, many participants in public utility rate cases, including the present writer, regret the acceptance of a fair-value rule of rate making and would greatly prefer an actual-cost standard, either in its traditional form or else revised by application of index-number adjustments to the equity capital. This aversion to fair value is true of some, though not of all, of those public service commissions that are nevertheless bound to the standard by legal mandate.

Persons of this persuasion are naturally under strong temptation to go as far as the courts will permit in applying an antidote to the fair-value doctrine in the form of an offsetting rate of return. The perfect antidote, of course, would be a precise offset, with the result that a company which establishes a high, fair-value rate base would be allowed exactly the same annual dollar return to which it would be entitled on an actual-cost basis. But the patent subterfuge involved in this allowance suggests its rejection in favor of a partial offset. This procedure invokes a "countervailing fallacy" principle of rate-of-return determination. That is to say, the assumed fallacy of a fair-value standard of rate making is to be counteracted, at least in part, by what would otherwise be the fallacy of conceding less than a capital-attracting rate of return.[34]

Viewed even from the standpoint of persons who object to the fair-value rule of rate making, the wisdom of any such attempt to circumvent the rule by the introduction of an undisclosed "countervailing fallacy" in the measurement of the percentage rate of return seems to me more than dubious. The serious vice of the procedure, of course, lies not in the mere resort to an offsetting

[34] But if one accepts the contention that, as long as the rate base is restricted to actual dollar costs, public utility stockholders are entitled to an enhancement in the percentage rate of return as a protection against inflation, it follows that this enhancement should be removed in the determination of a fair rate of return on a rate base which itself makes allowance for inflation. The use of a lower rate of return to offset the higher rate base is then entirely logical.

fallacy but rather in the lack of a frank and full disclosure. Indeed, if the procedure were to be upheld by the appellate courts despite full disclosure, it could no longer be criticized as a circumvention of the fair-value law since the law itself would then have ceased to exist except as a meaningless and expensive legal ritual.[35]

THE REPRODUCTION-COST PRINCIPLE

If we accept reproduction cost, not just as one "element" of fair value but as the basic standard of rate control, it follows that the "fair rate of return" should also be measured in terms of reproduction cost and not in terms of those costs of capital which the present company is now incurring because it has done part of its financing in a market different from that now prevailing. This statement is subject to whatever qualification may be applied to the determination of the reproduction cost of the property itself. As to this latter cost, it is customary, though not universal, to base the estimates on prices prevailing on the average over a period of several years rather than on spot prices. A similar procedure of averaging might then apply to an estimate of the current costs of capital. But subject only to this qualification, the fair rate of return would depend on whatever anticipated rate would attract new capital to a new enterprise. The most obvious deviation from the calculation applicable to an actual-cost rate base would be the substitution of current costs of debt and preferred-stock capital for the actual costs now incurred by the company by virtue of its outstanding contractual commitments to its senior security holders. Another deviation would be the substitution of a "typical" or "representative" capital structure for the existing structure unless this latter structure is itself deemed typical.

[35] For an approach to full disclosure, see Re Northwestern Bell Telephone Co., 12 PUR 3d 233 (May 31, 1957). Here, the Nebraska State Railway Commission granted a rate increase designed to produce $600,067 additional earnings per annum, which "would enable the applicant to pay reasonable dividends, meet its debt charges, attract additional capital and provide for additions and betterments to its interstate plant and equipment." The resulting earnings, it declared, would be equivalent to a rate of return of 6.2 per cent on a book-value rate base, 4.64 per cent on the company's computed reproduction-cost rate base, 6.3 per cent on a book-value rate base with a lower allowance for working capital than that for which the company contended, and 6.5 per cent on a book-value rate base with no allowance at all for working capital! But the commission did not state which, if any, of these four percentage rates it deemed important for the purpose of its decision, nor did it refer to any ruling on the measurement of the rate base by the Nebraska courts.

But what becomes of the credit-maintenance test of a fair rate of return if the rate-base is to be derived from estimates of reproduction cost and if the "fair rate of return" is to depend, not on the capital costs incurred by the present company but rather on the costs that *would be* incurred by a hypothetical, new company? The answer is that this test is no longer relevant. For, under a reproduction-cost principle of rate making, regulation attempts to simulate the norm of competitive pricing; and a competitive price is not supposed to be based on the financial needs of any one producer. If the given firm cannot maintain good credit while charging such a price, it must either go out of business or undergo financial reorganization. To be sure, the Supreme Court did not express this position when it accepted the credit-maintenance test of a fair return as one of the two tests applicable even under a fair-value rule.[36] But this great Court has not always taken pains to reconcile what it has said about a fair rate of return with what it has said about a rate base.

At the time of writing this chapter (October, 1958), current interest rates are higher than they have been in earlier postwar years, with the result that public utility companies whose outstanding funded debt is still costing them only a trifle more than 3 per cent are now paying from 4 to 4¾ per cent for new debt capital. Under the actual-cost principle of rate control, these higher current rates of interest would affect the fair rate of return on debt capital only to the extent that they affect, or are anticipated soon to affect, the company's annual interest requirements. But under the reproduction-cost principle of rate control, the effect would be immediate, subject to whatever time-period of averaging may be applied in

[36] But the Ohio Supreme Court seems to have accepted this position in its interpretation of the regulatory law of that state. In rejecting appellants' argument, in a telephone rate case, that the dollar amount of the return should be based upon the company's actual "earnings requirement," the court said: "However, under the Ohio Statutes and the decisions of this court, the percentage return is to be related not to the 'total capitalization' or to the 'net investment' but to the statutory rate base (reconstruction cost new less depreciation) so that *neither the actual capital of or net investment in this public utility nor its actual earnings requirements are really material in a proceeding of this kind.*" (Original italics). The court also indicated that, while consideration may be given to "what would reasonably be required" to provide for taxes, interest, dividends on stock, etc., this summation should be designed to measure the requirements of a *hypothetical* company with a property value equal to the statutory rate base of the company under review. City of Cleveland v. Public Utilities Commission of Ohio, 164 Ohio St. 442, 132 N.E. 2d 216 (1956).

the estimate of reproduction cost. Against this enhancement in the current cost of debt capital there may be an offsetting reduction in the cost of common-equity capital if a study of the behavior of stocks of companies in fair-value jurisdictions proves the willingness of investors to accept lower percentage rates of earnings than they would accept for stocks of companies in actual-cost jurisdictions.[37]

THE INFLATION-PROTECTION PRINCIPLE

The competitive-price rationale of a fair-value rule of rate making still enjoys some support in the literature. But it was never taken very seriously by most commissions and courts. And, especially since the time of the *Hope Natural Gas* case in 1944, emphasis has shifted from a simulated-competition defense of the fair-value rule to the more "practical" argument that a rate base in excess of actual dollar costs is called for as a means of protecting public utility stockholders against price inflation. Indeed, the tendency of commissions to set a so-called fair-value rate base which somehow compromises between actual cost and estimated reproduction cost is more in keeping with the latter objective than with the former. For it is consistent with the view that regulation is powerless to protect senior security holders against inflation and hence that only partial allowance should be made for increases in construction costs between the time of construction and the time of a rate case.

To be sure, the retention of the old fair-value rule is a very crude way of maintaining "the integrity" of common-stock investments against a decline in the purchasing power of the dollar. But it has the support of long-standing legal tradition, which an adjusted-actual-cost rule does not have. Moreover, its very indefiniteness

[37] Comparisons of the earnings-price ratios and dividend-yields on the common stocks of electric companies operating in fair-value jurisdictions, such as Pennsylvania and Ohio, with the ratios on the stocks of comparable companies operating in original-cost jurisdictions, such as New York, have supplied no conclusive evidence that the market rests content with a lower current rate of return on the former class of stocks. One may suspect that the market "discounts" the supposed inflation-hedge advantage of an investment in companies operating under the fair-value rule because of serious doubt as to how the rule will be interpreted in the future, or even as to how long it will continue to live against political opposition from consumer interests. For example, bills to repeal the reproduction-cost mandate of the Ohio statutes have been introduced into the state legislature year after year.

gives it a charm in the eyes of those persons who believe in flexible and noncommittal standards of reasonable utility rates.

In any case, this alternative rationale of a fair-value rate base would call for the acceptance of essentially the same tests of a fair rate of return as those applicable to an actual-cost rate base revised by an index-number adjustment of the common-stock capital. The senior-capital component of the over-all cost of capital would then be based on experienced cost rather than on hypothetical current cost, whereas the common-stock component would receive an allowed rate of earnings designed to keep the market prices of the shares, during periods of prosperity, from falling below their net rate-base values plus allowances for underpricing. But the rate of return should include no allowance for price inflation, realized or anticipated, since any such allowance would be incorporated in the rate base.

HOW ARE FAIR RATES OF RETURN ACTUALLY DETERMINED BY PUBLIC SERVICE COMMISSIONS?

Readers familiar with the actual practice of American rate regulation need no reminder about the uncertain relationship between the supposed "principles" of rate-of-return determination as set forth in this chapter and the considerations that actually lead commissions to allow whatever rates of return they do allow in specific rate cases. In the opinions that accompany their rate orders, commissions seldom attempt to disclose the reasons why they find, say, 5.85 per cent fair in one case and 6.2 per cent fair in another.[38] Especially in fair-value jurisdictions, some of the decisions lead one to suspect that the commissions have first reached a conclusion as to reasonable revenue requirements in terms of dollars per annum and then have proceeded to translate these requirements into whatever combination of a rate base and a percentage rate of return will be likely to pass muster with the appellate courts or with public sentiment.

The most significant generalizations about rates of return under

[38] See, e.g., Re Michigan Bell Telephone Co., 20 PUR 3d 397 (Aug. 6, 1957), in which the Michigan Public Service Commission approved a rate increase designed to yield 6.6 per cent on a net investment rate base. In a dissent contending that a higher rate of return was justified, Commissioner Hunt remarked: "This 6.6 per cent is the return adopted by the majority opinion, but not on the basis of staff testimony. The basis of the majority for arriving at this figure is not clearly stated in their opinion."

American regulation are those based on gross and net income reported as actually earned by the different classes of utility companies. During postwar years, the country's electric utilities, for example, have enjoyed a return averaging about 6 per cent on their invested capital, and a net income after dividends on preferred stock averaging about 10 per cent on their common-stock capital.[39] But the ranges on both sides of the average have been fairly wide, and writers have sometimes classified the electric companies by reference to these ranges. Thus, average rates of return on invested capital have been higher in fair-value states than in actual-cost states, higher for small companies than for very large companies, higher for utilities in fast-growing areas such as Florida than for utilities in slower-growing areas like New England, and higher for companies under the jurisdiction of notably "liberal" commissions like that of Florida than for companies under strict commissions like that of Wisconsin.[40] It is my impression—although I have seen no comprehensive statistical data to this effect—that there is a low correlation between relative rates of return on invested capital and relative prices charged for electric service. What correlation exists may possibly be negative[41]—a situation which would be well worth a careful study (a superficial study would be inconclusive) in an effort to trace any cause-and-effect relationships.

I close this last chapter on rate-level determination by suggesting that the real differences among the different jurisdictions are not based on the textbook distinction between an actual-cost principle

[39] For the period 1948–1956, the rate of return on the composite invested capital on the U.S. private electric utility industry, as derived from Federal Power Commission statistics for Classes A and B companies, has averaged 6 per cent per annum, while the rate of earnings on the common-stock equity has averaged 10.4 per cent. Commonwealth Edison Company Rate Case 44391 before the Illinois Commerce Commission (1957), Edison Exhibits 13.5 and 13.7. The computed return on invested capital includes interest during construction; and the invested capital for each year is taken as the mean of the year-end figures.

For the telephone and gas utilities (pipelines and distribution companies) the average rates of return on invested capital have been somewhat higher, at least during the past several years.

[40] See the study in the *Electrical World* cited in footnote 33, *supra;* also Fred P. Morrissey, "Relation of Growth and Rate of Return for Utilities," 60 *Public Utilities Fortnightly* 361–373 (Sept. 12, 1957).

[41] See "Bills Are Lower with 'Fair Value,'" 149 *Electrical World*, June 16, 1958, pp. 71–74.

of rate making and a reproduction-cost or present-value or competitive-price principle. On the contrary, *all* the commissions administer, with more or less efficiency or laxity, versions of, or minor deviations from, an actual-cost or sunk-cost standard of reasonable utility rates. True, there are important, perhaps critical, differences in administration. But these are not differences in the basic theory of "reasonable" or "optimum" rates. They lie rather in different degrees of liberality in the rates of return that companies are allowed to enjoy *on their actual capital investment;* in the differences between painstaking and lax application of the requirements that rates must be reasonable; in the extent to which the legally imposed rules of rate-base determination place a burden on effective and expeditious rate regulation; and in the degree to which and manner by which allowance is made for price inflation in the determination of a company's current revenue requirements.

PART THREE

The Rate Structure

XVI

CRITERIA OF A SOUND RATE STRUCTURE

Despite its recognized importance as a basis of rate control, the determination of revenue requirements under a fair-return standard, which was the subject of the preceding chapters, by no means exhausts the issues of a rate case. For even if this standard were accepted as the master rule of rate making, overriding all conflicting rules such as that against unjust discrimination—an exalted status which, though sometimes claimed for it, it does not enjoy in fact—there would still remain the question what specific rates will yield a fair return, together with the further question what rates and rate relationships should be chosen when a company's earning power is so high that any one of a variety of tariffs could be made to yield adequate over-all revenues.[1]

[1] As noted in Chap. IX, the Interstate Commerce Commission has given far more attention to rate relationships than to rate levels. But the contrary emphasis has characterized the utility rate cases before the state public service commissions. As to the specific rates, the major concern of these commissions has been to protect the interests of the residential customers. In the words of Russell E. Caywood, "The thought is that the larger customer can protect himself, whereas the small customer requires the help of a third party." *Electric Utility Rate Economics* (New York, 1956), p. 4. But in recent years particularly, very large industrial customers have intervened actively in rate cases, not just with respect to the relative height of their rates, but also with respect to the form: e.g., in the Commonwealth Edison Company Rate Case of 1957–1958 before the Illinois Commerce Commission (Case No. 44391), 24 PUR 3d 209.

Public utility counsel have sometimes argued that once a company's total revenue entitlements have been determined by a commission, the choice of a pattern of rates that will yield the allowed revenues should be left to the discretion of the management, which will then be in an impartial position to make a fair apportionment of burdens among its different classes of customers. This is only a half-truth argument: among other reasons, because a utility company is concerned not just to secure rates that will presently yield the approved "fair rate of return," but to develop a pattern of rates that will promote growth of earnings and that will protect these earnings against business depressions. The better the utility management, the greater is this long-run concern.

A much more plausible reason for caution on the part of a commission in over-

We turn now to principles of rate making designed to throw light on these two other questions, but particularly on the latter. By what basic standards, for example, shall regulation pass judgment on a system of electric-utility rates which allows liberal discounts for incremental blocks of energy; or which levies higher charges, per kilowatt-hour, on residential consumers than on industrial customers; or which concedes lower rates for off-peak consumption than for consumption at peak-time hours or seasons? With the telephone utilities, does public policy justify the practice of the industry in setting higher rates for service in larger communities than for comparable service in small communities even when these differentials are not based on differences in cost of service? And what are the merits of the contentions, advanced by some economists but enjoying no popular support, that rush-hour fares for local-transit service or commuter railroad service should be higher than fares at nonrush periods? These are mere random samples of the many practical issues falling under the subject of rate structure.

But rate-structure problems are far more complex than problems of a fair return even though the latter are by no means elementary; and they are even less amenable to solution by reference to definite principles or rules of rate making. In part, the complexity is due to the mass of technical detail, including the technology of metering, involved in the design and administration of workable rate schedules for different types of utility enterprises. In part it is due

---

riding the rate-pattern policies of a utility company is the one suggested many years ago by Dr. G. P. Watkins, in expressing regret that few American commissions had contributed substantially to the development of principles of electric-rate design. "This situation," he wrote, "is perhaps partly due to doubt as to the possession of adequate powers, but more fundamentally to the diffidence of commissioners when confronted with a subject so complex, both theoretically and practically, as that of electric rates." *Electrical Rates* (New York, 1921), p. 37. The commissions that have given the most attention to rate-structure principles are the stronger commissions, such as those of California, New York, and Wisconsin, which have the aid of relatively large expert staffs.

A strong case can be made for the contention that, as far as feasible, fundamental problems of rate-structure revision should be handled in special proceedings, since they need far more time for satisfactory solution than can properly be given to them in what the Interstate Commerce Commission calls the "revenue cases." Referring to this situation in railroad rate cases, Professor Merrill J. Roberts writes: "The general rate case as presently construed is a veritable farce. Its broad sweep virtually precludes even the most rudimentary consideration of the intimate demand and cost relationships that should govern pricing in specific markets." "Maximum Freight Rate Regulation and Railroad Earnings Control," 35 *Land Economics* 125–138 at 136 (1959). See also footnote 11 of Chap. XVIII.

to the inability of the rate maker to predict the effect of changes in rates on demand for the services and hence on costs of supply—due, in short, to ignorance of demand functions and cost functions. But in part—and this is the most serious theoretical difficulty—it is due to the necessity, faced alike by public utility managements and by regulating agencies, of taking into account numerous conflicting standards of fairness and functional efficiency in the choice of a rate structure. The nature of some of these conflicts will be revealed as the discussion proceeds. But, by way of illustration, we may note the conflict between the desirable attribute of simplicity and the otherwise desirable attribute of close conformity to the principle of service at cost. Here, as with other clashes among various desiderata of rate-making policy, the wise choice must be that of wise compromise; and in reaching this compromise, the practical rate expert would look in vain to any general theory of public utility rates, at least in its present stage of development, for a scientific method of reaching the optimum solution.[2]

In view of this complexity of subject matter, the present study will not undertake descriptions of the typical rate structures of the different types of public utilities; and the reader unfamiliar with these structures is therefore referred to the treatises for background material, in the absence of which the following discussion of general principles may seem hopelessly abstract.[3] Even in its treatment

[2] Certain *approaches* toward a rational solution may be possible. Note, e.g., one economist's attempt to compare the additional costs of time-of-day metering as a device for differential electric rate making based on on-peak versus off-peak use with the possible resulting savings in plant generating capacity. Ralph Kirby Davidson, *Price Discrimination in Selling Gas and Electricity* (Baltimore, 1955), pp. 182–195. But other "intangible" costs of time-of-day metering are not readily assessed. In its investigation of the nationalized electric supply industry, a British inquiry commission concluded that, acting under statutory mandate to simplify rate structures, the Electricity Boards had deviated too far from the principle of rate differentials based on service at cost. *Report of the Committee of Inquiry into the Electricity Supply Industry*, Jan., 1956, Cmd. 9672, pp. 104–105. Needless to say, however, the Committee supplied no formula by which to draw the line between too much and too little simplicity.

[3] In addition to the treatment of rate structures in the general textbooks on public utility and transportation economics, see Caywood's book already cited in footnote 1; J. M. Bryant and R. R. Herrmann, *Elements of Utility Rate Determination* (New York, 1940); L. R. Nash, *Public Utility Rate Structures* (New York, 1933). For a significant critical monograph on modern utility rate structures in the United States, see Ralph Kirby Davidson's study already cited in footnote 2. On many technical issues, no American treatise on electric utility rates can equal that by the distinguished British rate engineer, D. J. Bolton, *Electrical Engineering Economics*, Vol. II: *Costs and Tariffs in Electricity Supply*, 2d ed. (London, 1951).

of principles, these chapters are mere essays on the nature of the more controversial, largely unresolved, problems rather than attempts at systematic development. All of them have one theme in common: the thesis that the most formidable obstacles to further progress in the theory of public utility rates are those raised by conflicting goals of rate-making policy.

CRITERIA OF A DESIRABLE RATE STRUCTURE

Throughout this study we have stressed the point that, while the ultimate purpose of rate theory is that of suggesting feasible *measures* of reasonable rates and rate relationships, an intelligent choice of these measures depends primarily on the accepted *objectives* of rate-making policy and secondarily on the need to minimize undesirable side effects of rates otherwise best designed to attain these objectives. No rational discussion, for example, of the relative merits of "cost of service" and "value of service" as measures of proper rates or rate relationships is possible without reference to the question what desirable results the rate maker hopes to secure, and what undesirable results he hopes to minimize, by a choice between or mixture of the two standards of measurement. Not only this: the very *meaning* to be attached to ambiguous, proposed measures such as those of "cost" or "value"—an ambiguity not completely removed by the addition of familiar adjuncts, such as "out-of-pocket" costs, or "marginal costs," or "average costs"—must be determined in the light of the purposes to be served by the public utility rates as instruments of economic policy. This is a commonplace; but it is a commonplace which, so far from being taken for granted, needs repeated emphasis.

What then, are the good attributes to be sought and the bad attributes to be avoided or minimized in the development of a sound rate structure? Many different answers have been suggested in the technical literature and in the reported opinions by courts and commissions; and a number of writers have summarized their answers in the form of a list of desirable attributes of a rate structure, comparable to the "canons of taxation" found in the treatises on public finance. The list that follows is fairly typical, although I have derived it from a variety of sources instead of relying on any one presentation. The sequence of the eight items is not meant to suggest any order of relative importance.

1. The related, "practical" attributes of simplicity, understandability, public acceptability, and feasibility of application.
2. Freedom from controversies as to proper interpretation.
3. Effectiveness in yielding total revenue requirements under the fair-return standard.
4. Revenue stability from year to year.
5. Stability of the rates themselves, with a minimum of unexpected changes seriously adverse to existing customers. (Compare "The best tax is an old tax.")
6. Fairness of the specific rates in the apportionment of total costs of service among the different consumers.
7. Avoidance of "undue discrimination" in rate relationships.
8. Efficiency of the rate classes and rate blocks in discouraging wasteful use of service while promoting all justified types and amounts of use:
   (a) in the control of the total amounts of service supplied by the company:
   (b) in the control of the relative uses of alternative types of service (on-peak versus off-peak electricity, Pullman travel versus coach travel, single-party telephone service versus service from a multi-party line, etc.).

Lists of this nature are useful in reminding the rate maker of considerations that might otherwise escape his attention, and also useful in suggesting one important reason why problems of practical rate design do not readily yield to "scientific" principles of optimum pricing. But they are unqualified to serve as a base on which to build these principles because of their ambiguities (how, for example, does one define "undue discrimination"?), their overlapping character, and their failure to offer any rules of priority in the event of a conflict. For such a base, we must start with a simpler and more fundamental classification of rate-making objectives.

THREE PRIMARY CRITERIA

General principles of public utility rates and rate differentials are necessarily based on simplified assumptions both as to the objectives of rate-making policy and as to the factual circumstances under which these objectives are sought to be attained. Attempts to make these stated principles subserve all special objectives and cover all specific conditions would be hopeless. Writers on the theory of rates are therefore at liberty to base their analyses on the acceptance of those objectives which are of wide application and the attainment of which may be aided by whatever tests or measures of sound rate structure the analyses suggest.

Among these objectives, three may be called primary, not only because of their widespread acceptance but also because most of the more detailed criteria are ancillary thereto. They are (a) the revenue-requirement or financial-need objective, which takes the form of a fair-return standard with respect to private utility companies; (b) the fair-cost-apportionment objective, which invokes the principle that the burden of meeting total revenue requirements must be distributed *fairly* among the beneficiaries of the service; and (c) the optimum-use or consumer-rationing objective, under which the rates are designed to discourage the wasteful use of public utility services while promoting all use that is economically justified in view of the relationships between costs incurred and benefits received.[4]

In actual rate cases, these three criteria of reasonable rates and rate relationships, and particularly the last two, are by no means always sharply distinguished. But the distinction may be illustrated by the imagined example of a request, submitted to a regulating commission by a group of consumers, that an electric company be ordered forthwith to abandon its present, somewhat elaborate, schedule of class rates, block rates, and two-part or three-part

[4] These three criteria correspond to three of the four "primary functions" of utility rates set forth in Chap. III. The function ignored for present purposes, that of encouraging managerial efficiency, is omitted because of its more direct bearing on the desirable criteria for a fair over-all rate of return. See pp. 262 ff., *supra*. Professor John Maurice Clark had in mind essentially the same three criteria in writing: "The chief purposes of a rate system should be to earn a reasonable total return, to develop the utmost use of facilities so long as every service pays at least its differential cost, and to distribute residual costs fairly according to the responsibility of different users for the amount of these costs." *Studies in the Economics of Overhead Costs* (Chicago, 1923), p. 322. Professor Donald H. Wallace added a fourth possible objective: that of benefiting specific classes of customers, such as customers of substandard income or a submerged industry. Temporary National Economic Committee, Investigation of Concentration of Economic Power, 76th Congress, 3d Session, Monograph No. 32: *Economic Standards of Government Price Control* (Washington, D.C., 1941), pp. 475-478. This fourth objective comes under the heading of "social" principles of rate making as I have used the term in Chap. VII.

tariffs in favor of a uniform kilowatt-hour rate for all customers throughout its franchise territory. Almost certainly this proposal would be held subject to the threefold objection: (a) that no uniform rate, however high, could be made to yield a fair return on the company's invested capital; (b) that, even if it could do so, rate uniformity despite lack of cost uniformity in the supply of different types of service would impose *unfair* burdens on the consumers of the less costly services; and (c) that, quite aside from its unfairness, the uniform rate would result in a serious underutilization of plant capacity because it would cut down the demand for services (especially, for off-peak services) that could be supplied at increment costs materially below average unit costs, while stimulating a wasteful demand for services that can be supplied only at increment costs higher than the average.

Some modern writers who confine their attention to what they call the "economic" principles of public utility rates have ignored the second of these three standards of rate making in their development of these principles, on the ground that fairness questions are beyond the competence of professional economists.[5] Instead, they have centered attention on the third standard, often with special reference to its application under the constraint of a revenue-requirement standard. But a refusal to recognize fairness issues as relevant to the design of a sound rate structure would so far divorce theory from practice that these issues will not be completely ignored in the discussion that follows.

In the remainder of the present chapter as well as in all of the following chapters except Chapter XX ("The Philosophy of Marginal-Cost Pricing"), principles of rate structure will be discussed under the assumption that they are designed primarily to subserve the three above-noted objectives of rate-making policy. But in order to avoid extreme complexities, we shall make three further simplifying assumptions, all of which are implicit in much of the literature on public utility rates.

In the first place, we shall impute an unqualified priority to the "fair-return" standard of reasonable rate levels despite the fact, noted in Chapter IX, that no such priority is accorded either by legal doctrine or by rate-making practice. That is to say, we shall assume that the rates of any given utility enterprise, taken as a

[5] See Chaps. II and VIII.

whole, must be designed as far as possible to cover costs as a whole including (or plus) a fair return on capital investment.

In the second place, we shall assume the availability of a wide range of alternative rate structures, any one of which could be made to yield the allowed fair return on whatever capital investment is required in order to supply the demand for service. This assumption, which implies that the utility enterprise in question enjoys a substantial degree of monopoly power, permits us to center attention on a choice among rate structures, any one of which would be equally fair to investors and equally effective in maintaining corporate credit.

And in the third place, except for incidental references, we shall rule out all of those so-called "social" principles of rate making, discussed in Chapter VII, which may justify the sale of some utility services at less than even marginal or out-of-pocket costs.

IMPORTANCE AND LIMITATIONS OF THE PRINCIPLE OF COST OF SERVICE

Without doubt the most widely accepted measure of reasonable public utility rates and rate relationships is cost of service. In the literature, this measure is generally given a dominant position even by writers who insist upon, or reluctantly concede, the necessity for deviations from cost in the direction of value-of-service principles or of various "social" objectives of rate making. In actual practice there is usually an obvious, marked degree of correlation between the relative charges for different amounts and types of service and the relative costs of rendition. To be sure, local transit rates, with their customary flat fares regardless of distance and (even more important) regardless of time of travel come close to providing an outright exception. But intercity railroad rates, despite their many familiar departures from cost principles [6] and despite their notorious failure to accord well with any other sane principles of rate making, bear important partial correlations with

[6] Referring to railroad rates, the Interstate Commerce Commission said: "Costs alone do not determine the maximum limits of rates. Neither do they control the contours of rate scales or fix the relations between rates or between rate scales. Other factors along with costs must be considered and given due weight in these aspects of rate making." 262 I.C.C. 693, quoted by Justice Douglas in New York v. United States, 331 U.S. 284, 328 (1947).

relative costs. Thus, by and large, Pullman fares are much higher than coach fares; charges for the shipment of ten tons of any given class of freight are much higher than charges for the shipment of one ton; and freight rates from New York City to points in California are far higher than freight rates from New York City to Albany. Electric utility rates deviate from a cost standard much less than railroad rates. But it is a testimony to the prestige of this standard that, whenever actual or proposed electric tariffs are criticized for their asserted unfairness, the criticism usually takes the form of the contention that the rate relationships fail to conform to cost relationships. When this complaint is made before a public service commission, the defenders of the rates are likely to feel in a much stronger position if they can meet it on its own ground, without having to rely on value-of-service arguments in support of preferential rates to favored classes of customers.

The basic reasons in support of a cost-of-service standard of public utility rates and rate relationships have already been discussed at length in the early chapters of this book, particularly in Chapter IV. Here we may recall that the defense rests both on considerations of fairness as among the different customers and on considerations of optimum utilization or "consumer rationing." As to the issue of fairness, a cost-price standard probably enjoys more widespread acceptance than any other standard except for the even more popular tendency to identify whatever is fair with whatever is in one's self-interest. As to the issue of optimum utilization, this same standard (or, at least, a standard of the same name) comports with the "consumer sovereignty" principle, under which public utility consumers should be encouraged to take whatever types of service, in whatever amounts, they wish to take as long as they are made to indemnify the utility enterprise for the costs of rendition.

NECESSARY DEVIATIONS FROM A COST-OF-SERVICE STANDARD

In view of what has just been said, one might suppose that "the theory" of public utility rate structures or rate differentials would call for the acceptance of no basic principle of reasonable or nondiscriminatory rates other than a mere extension of the very principle already accepted in the determination of entire rate levels, namely, the principle of service at cost. Just as, under the fair-return standard, rates as a whole should cover costs as a whole, so the rates for any given class of service (passenger versus freight, residential versus commercial, etc.) should cover the costs of supplying that class, and so the rates charged to any single customer within that class should cover the costs of supplying this one customer. Under this assumption, the theory of rate structures would be reduced to a mere theory of cost determination through the aid of modern techniques of cost accounting and cost analysis.

Unfortunately, however, no such simple identification of "reasonable" rates with rates measured by costs of service is attainable; and this for several reasons, three of which will now be distinguished. The first of these reasons may be called "practical," whereas the other two are theoretical and are based on the non-additive character of the costs attributable to specific classes and units of service.

Excessive complexity of cost relationships. The "practical" reasons lie in the extreme difficulties of cost-of-service measurement together with the fact that, even if all specific costs could be measured, they would be found too complex for incorporation in rate schedules. Most public utility companies supply many different kinds of service even when they confine their activities to nothing but electricity, or gas, or telephone service, etc. In a very real sense, moreover, the supply of any one type of service to thousands of customers at different locations constitutes the supply of a different product to each customer. Equally truly, service rendered at any one time is not the same product as is otherwise comparable service rendered at another time.

But these millions of different service deliveries by a single public utility company are produced in combination and at total costs, most of which are joint or common either to the entire business or else to some major branch of the business. Under these circumstances, the attempt to estimate what part of the total cost of operating a utility business constitutes the cost of serving each individual consumer or class of consumers would involve a hopelessly elaborate and expensive type of cost analysis.[7] For this reason

[7] John Alden Bliss has sent me a quotation from a report by Alex Dow, former chairman of the Detroit Edison Company, to the effect that his company had been obliged to reply on the one hand to the customer who thinks that rates should be uniform per kilowatt-hour and on the other hand to the man "who wants us to determine so exactly the cost of service to each customer that our power plants and distribution systems would become merely unavoidable preliminaries to the operation of a meter department." The TNEC Monograph No.

alone, the most that can be hoped for is the development of techniques of cost allocation that reflect only the major, more stable, and more predictable cost relationships.

But even if, through the miracles of electronic computers and of modern techniques of mathematical analysis, all significant cost differentials could be measured without inordinate expense, they would then be found far too numerous, too complex, and too volatile to be embodied in rate differentials. Stability and predictability of the charges for public utility services are desirable attributes; and up to a certain point—or rather, up to an indeterminate point—they are worth attaining even at the sacrifice of nice attempts to bring rates into accord with current production costs. Indeed, unless rate-making policies are sufficiently stable to permit a consumer to predict with some confidence what his charges will be *if he decides* to equip his home or his factory to take the contemplated service and then to buy the service, a cost-price system of rate making will be self-defeating when viewed as a means of securing a rational control of demand.

These practical considerations leading to the design of rate structures that ignore many cost differentials are illustrated by the general uniformity of rates for gas, electricity, telephone service, and water supply throughout an entire city, despite distances from source of supply, differences in density of population, and other differences that may have a material bearing on relative costs of service. Indeed, in some parts of the country, the rates of large electric power systems are uniform, or almost uniform, throughout the state, no distinction being made between urban and rural areas. Critics of this "blanket rate" policy may well be right in insisting that it carries the principle of uniformity too far.[8] But the criticism is leveled merely against an *excessive* disregard of cost differentials in rate making.

Failure of the sum of differential costs to equate with total costs.

32, cited in footnote 4, *supra*, quotes at page 41 from an opinion by Chairman Maltbie of the New York Public Service Commission reading in part: "In every business, there is always a large percentage of customers, who are served at less than cost, for the reason that it has been found impracticable to devise and apply a system of cost accounting and computation which would carry out the principle literally; and if it were done, it would result in such an elaborate and complicated schedule of rates that the public could not understand it and few could apply it."

[8] See Chap. VII, pp. 112–113, *supra*.

We come now to a further limitation of the cost-of-service principle of rate structures—this one of critical concern when the rates must be made to yield a fair over-all return. It lies in the nonadditive character of the costs allocable, on a cost responsibility basis, to specific classes and quantities of utility service. In view of this failure of "the sum of the parts to equal the whole," the requirement that rates as a whole shall equal costs as a whole cannot be reconciled with a requirement that each consumer shall pay only the costs for which he, and no one else, is causally responsible; nor can it be reconciled with a requirement that each major class of consumers shall pay rates designed to cover the costs of serving that class, no more and no less. In consequence, save under circumstances that could occur only by rare coincidence, one of the two cost principles—the total-cost principle or the specific-cost principle—must give way. And, under the assumption of this chapter, the principle that must yield is that of service at cost as a measure of particular rates and rate relationships.

In stressing this probable conflict between the over-all-cost standard of entire rate levels and the specific-cost standard of the rate structure, the literature on rate theory has attributed it primarily to the distinction between average cost and incremental or marginal cost—a distinction familiar to the economic textbooks on the theory of price determination. This distinction will now be noted, although a second distinction will receive attention later. The point is that, when multiple products, or even multiple units of the same product, are produced jointly or in common, by an organically whole productive process, the only costs allocable solely to any given product or amount of product are *differential* costs. They are measured by a comparison between the total costs of the entire operation with the given output included, and the total costs with that output excluded.[9]

The most familiar and most significant form of a differential cost is incremental cost—the increment in total cost that will result from superimposing the production of the particular amount and type of product under inquiry on the other production. A special

[9] Under limited conditions, however, it is permissible to regard the net cost of one product, among a complex of jointly produced products, as measured by the total cost of producing the whole complex minus the proceeds of the sale of all the other products. These other products are then treated as by-products in the strictest sense of this term.

type of incremental cost, important for the theory of public utility rates, is marginal cost—a concept subject to various definitions but here best defined in a loose way, as the incremental cost, per unit, of producing a relatively small increment of a given product.[10] But these differential or incremental or marginal costs are nonadditive except under special conditions. For the determination of the cost of any particular type and amount of output assumes the continued production of the rest of the output, an assumption which is shifted when the costs of other types and amounts of output are under inquiry.

What has just been said as to the nonadditive nature of differential or incremental costs applies to all public utility companies which produce services of different kinds for many different people and in many different amounts. With an electric utility company, for example, the only cost specifically allocable to the residential service, *and not* to any other service, is the excess in total cost over what would be the cost of supplying all services other than residential. And the same statement would apply to an attempt to measure the cost that a company has actually incurred, or would incur in the future, in supplying a particular amount of service to any single consumer. The usual assumption is that, if the incremental costs of all services, separately measured, were added together, they would fall materially short of covering total costs—an assumption based on the belief that most public utility enterprises operate under conditions of decreasing costs with increasing output. When this assumption is valid, it implies that a public utility company cannot cover its total revenue requirements without charging *more* than incremental costs for at least some of its services.

The nonadditive character of the costs specifically allocable, on a cost-responsibility basis, to the different classes and amounts of public utility services has often been disguised by the acceptance of elaborate full-cost apportionments which begin with total costs and apportion these costs among the various classes of service as one might divide a pie among the members of a dinner party, leaving no residue for the kitchen. These "fully-distributed-cost" apportionments are especially familiar in the railroad field, where

[10] Marginal cost is sometimes defined as the increase in total cost resulting from the production of one additional unit of the product. But a one-unit margin is too narrow for most rate-making purposes.

they have been made under formulas developed by experts in the Interstate Commerce Commission. One such apportionment seems to indicate that the railroads of the United States, taken altogether, have been suffering annual losses of many millions of dollars per year on their passenger business. The usefulness of these apportionments is a debatable subject, which will be discussed in Chapter XVIII. But, in any case, their merits must rest on a claim that they represent, not a finding of the costs definitely occasioned by this class of service rather than that, but rather a *fair* or *equitable* division of total costs or else a statement of relative, not absolute costs. Even the cost analysts who make these full-cost apportionments recognize this fact implicitly when they concede, as they usually do, that a company may find it profitable to sell some classes of service at less than their imputed costs.[11]

The "cost" used as a measure of total revenue requirements is not the same kind of cost as the "cost" most clearly relevant to the design of the rate structure. The source of the previously discussed discrepancy between the total costs of an entire utility business and the sum of the costs causally allocable to the particular amounts and types of service lies in the distinction between *average total* costs and *incremental* or *marginal* costs. Whenever this discrepancy prevails, which it will do if the public utility company is operating under conditions of decreasing unit cost with increasing rates of output, rates set at incremental cost would tend to fall short of total costs. But we must now note another reason why the sum of the costs attributable to the specific services of a public utility company may fail to reflect the total costs of running the entire business.

[11] Public utility companies have sometimes invoked a marginal or incremental cost principle in defense of special rate concessions to very large customers, the defense resting on the contention that the revenues from this favored service will cover, or more than cover, all *additional* costs of its production. The weakness of this defense lies not, as sometimes asserted, in the invalidity of the incremental cost principle, but rather in a company's unsymmetrical proposal to base the preferential rate on incremental cost while basing the other rates on residual cost. Even this latter proposal may be justified in special cases; but the practice constitutes a form of rate discrimination, not a form of cost pricing. Its reasoning has been rejected as a defense against the charge of unlawful discrimination under the provisions of the Robinson-Patman Act. See Herbert F. Taggart, *Cost Justification*, Michigan Business Studies, Vol. 14, No. 3 (Ann Arbor, 1959), pp. 538–539: "The differential cost approach to cost justification is totally unacceptable. This means that a cost cannot be ignored *merely* because a given cost category would not be changed by the acquisition or loss of a certain customer or order or quantum of production." See also Frederick M. Rowe, "Cost Justification of Price Differentials under the Robinson-Patman Act," 59 *Columbia Law Review* 584–617 at 594 (1959).

This reason lies in the important distinction between historical or "sunk" costs and anticipated or "escapable" costs. A company's total revenue requirements, as measured under a fair-return standard, depend on liabilities and quasi liabilities for the payment of operating expenses and capital costs already partly predetermined by earlier transactions, including earlier purchases of plant, land, and other resources. On the other hand, the costs most clearly relevant to the determination of specific rates, at least under an optimum-utilization objective of rate-making policy, are those anticipated costs that can still be escaped or minimized by a control of output. This important distinction between the two types of cost is drawn most sharply when the revenue requirements are determined under an original-cost rule of rate making.[12] But the distinction remains, though in a blurred status, even under a so-called "fair-value" rule as actually applied by courts and commissions.

In short, then, there are two quite different sources of possible conflict between a cost-price system of reasonable rate levels and a cost-price system of specific rates and rate relationships. But, if the revenue requirements of the company are lower than would be the requirements of a new company, as they are likely to be during a period of rising construction costs and rising site values, the two sources of conflict may result in a partial offset. It is with this possibility in mind that some economists, who view with regret the necessity of charging public utility rates in excess of marginal costs, have tended to favor an original-cost type of rate base during a period of price inflation.

THREE WAYS BY WHICH TO RECONCILE THE COST-OF-SERVICE PRINCIPLE OF INDIVIDUAL RATES WITH THE MANDATE OF A FAIR OVER-ALL RETURN

For the reasons just suggested, rates based merely on specific or incremental or marginal costs might well suffice, on occasion, to yield adequate, or even more than adequate, total revenues under a fair-return standard. But the general principles of public utility rates dare not rely on such a convenient harmony. Instead, they

[12] See pp. 75–77, *supra*. In Chap. I of his *Economics of Sellers' Competition* (Baltimore, 1952), Professor Fritz Machlup stresses the impossibility of a rational allocation of the historical costs of standard accounting when the assumed objective is to determine the specific costs of producing any given product among a complex of products.

must apply to situations in which a public utility company, in order to cover its operating expenses plus a fair return, must charge more than incremental costs for much of its service—more, that is, than any costs definitely assignable to specific renditions of service on a cost-responsibility basis. In these situations, which are usually assumed to be typical of American railroads and public utilities, the problem then becomes that of devising rules of rate design under which there can take place a recovery of total revenue requirements without too severe disregard of principles of equitable cost apportionment, and with a minimum sacrifice of standards of optimum utilization of service.

Some of the modern literature of rate theory approaches this problem by treating a rate as composed of two parts: a minimum rate, called a "price," and a surcharge, called a "quasi tax." [13] Subject to important reservations, the "price" component should be set at marginal cost—at the specific cost that can be imputed on a causal basis to a relatively small increment of service. The surcharge, on the other hand, is a charge in excess of specific cost and is designed to yield an appropriate share of the unallocable total costs.

The significance attached to this somewhat artificial distinction between the price component and the surcharge component of a utility rate lies in the different functions performed by the two elements. In setting a minimum rate, the "price" is deliberately designed to create a barrier against wasteful or distorted use of the service in question. That is to say, up to this point any tendency of the rate to restrict consumption is deemed a desirable tendency. On the other hand, the surcharge is imposed merely for the sake of yielding necessary income to the company, and insofar as it also operates further to restrain the use of the service, this result is regarded as an undesirable side effect.[14] Hence, the problem faced by

[13] For example, Marcus Fleming, "Optimal Production with Fixed Profits," *Economica*, Aug., 1953, pp. 215–236. The conception of railroad rates as a form of tax insofar as they exceed out-of-pocket cost, and especially insofar as they are based on "value-of-service" considerations, has had a long tradition. Langdon recalls that the Interstate Commerce Commission expressed this point of view in its first Annual Report, 1 I.C.C. Rep. 31 (1887). Jervis Langdon, Jr., "Regulation of Competitive Business Forces: The Obstacle Race in Transportation," 41 *Cornell Law Quarterly* 57–92 at 86, note 165 (1955).

[14] This statement must be qualified in recognition of the requirement that if a surcharge is imposed on any one service, counterbalancing surcharges must also be imposed on alternative services in order to avoid distortionate substitutions. See Chap. XIX.

rate theory is that of suggesting rules for the determination of surcharges, over and above marginal costs, that will yield the required total revenue with the minimum of incidental harm in the form of curtailment of service otherwise worth rendering. But "minimum harm" cannot be identified with minimum curtailment of consumption in terms of physical intake such as kilowatt-hours, cubic feet of gas, or passenger miles. For, if the curtailment applies to service for which potential consumers would be willing to pay only slightly more than mere incremental costs of production, the loss of benefits is less serious than that resulting from equal curtailment of service which potential consumers would value at far more than marginal cost.

The problem of determining what surcharges will impose the least serious harm in the form of curtailments and distortions of use of service when the rates as a whole must yield total revenue requirements is perhaps the most complex and most difficult problem of modern rate theory. Not only does it require for its solution highly involved mathematical analysis; in addition, the application of this analysis depends on a knowledge of cost functions and of demand functions that practical rate makers simply do not and cannot possess. How far these refinements of theory may lend themselves to practical use in the future is a question on which I venture no opinion. But certain elementary aspects of the problem can be discussed without reference to the tools of mathematics; and they will be discussed in Chapter XIX, on rate discrimination.

Broadly speaking, if one assumes that ideal rates, viewed from the standpoint of the optimum use of service, would be rates set at marginal cost, there are three alternatives, or rather types of alternatives, by which one might seek to approach this optimum while imposing rates designed to meet total revenue requirements.[15] Under the first alternative, rates for different types of service would be made directly proportional to marginal cost, each rate exceeding the correlative marginal cost by the minimum fixed percentage necessary to cover total costs, say 20 per cent. Under the second

[15] A fourth alternative would be the apportionment of the burden of the cost residues, over and above marginal costs, by reference to some "social" principle of rate making in line with the fourth possible criterion of a sound rate structure mentioned by Professor Wallace in the article already cited in footnote 4. But the application of these social principles would not be consistent with the acceptance of marginal or out-of-pocket costs as setting the lower limits of utility rates.

alternative, rate differentials would be designed to cover cost differentials as between substitute classes of service. Under the third alternative, rates would be deliberately "biased" through the imposition of higher surcharges, relative to marginal costs, for those types and amounts of service for which the demand is relatively inelastic, and through the concession of lower surcharges (or even of relief from any surcharge) for types and amounts of service for which demand is relatively elastic. This last-named practice is an application of the familiar "value-of-the-service" principle of rate making, otherwise known as the principle of charging what the traffic (or what the market) will bear.

Before turning to another aspect of the problem of rate making raised by a failure of rates based on incremental or marginal costs to yield adequate total revenues, let me note that the "synthetic" approach just outlined, according to which the rates are conceived of as built up by the superimposition of appropriate surcharges on minimum rates set at marginal costs, is not in widespread use in actual rate making. Instead, the rates are more likely to be derived in part from apportioned total costs, with marginal or out-of-pocket costs serving only to set the lower limits below which no rates will be fixed under a value-of-service principle. The nature and significance of these fully apportioned costs will be discussed at length in Chapter XVIII.

AGGREGATE CLASS COST AS A RATE-MAKING STANDARD INTERMEDIATE BETWEEN TOTAL COST AND MARGINAL COST

In order to draw the distinction between total costs and specific costs in its sharpest form, we have so far contrasted these costs at their extreme ends. That is to say, we have identified total costs with a company's entire revenue requirements, whereas we have identified specific costs with the incremental costs of a very small output of service—marginal costs. But we must now consider the possible relevance for rate-making purposes of costs intermediate between grand-total costs on the one hand and marginal costs on the other. These intermediate costs are illustrated in the railway field by the distinction between the costs of passenger service and the costs of freight service, or by the distinction between the costs of the Eastern Division and the costs of the Western Division. In the electric power field, they are illustrated by the costs of supplying major classes of customers, such as domestic, commercial, industrial, street lighting, electric railway, and rural. Each of these major classes can be divided into numerous subclasses for purposes of cost analysis. But we may ignore this complication in raising the question whether public utility rates should be designed to make domestic consumers as a group pay for domestic-service costs as a whole, and so on through the entire list of service classes.

This is a controversial question, among other reasons because it raises the further question as to the relevant definition of group or class costs. Here economists and utility cost analysts are likely to differ. The analysts have tended to identify class costs with some "reasonably" or "fairly" apportioned share of the total costs; whereas most economists would define the cost of serving any one class of consumers in terms of differential costs. Thus, for a railroad company, the only costs which economists would consider allocable to the passenger business and not the freight business would be whatever costs the railroad would save by dropping this business or avoid by refraining from taking it on. So also, to an electric company, the only cost that can be assigned definitely and solely to the supply of the residential service is the cost of adding this service to the other services, or alternatively the saving in cost that would result from closing down the residential-service department while continuing to supply industrial power, street-lighting service, etc. These two types of differential costs may differ widely in amount; but we may accept incremental costs as the more significant type with respect to an expanding public utility business.

Having defined a class cost as an increment cost, albeit a cost of a very *large* increment, we now face the question whether this is a cost to be covered, as a minimum, by the revenues received from the sale of all services falling within this class. Consistency with the full-cost principle of entire rate levels (unless this principle is adhered to merely for reasons of financial necessity) requires an affirmative answer. That is to say, just as each individual consumer should pay rates at least sufficient to cover the costs allocable expressly to him, so also the domestic consumers as a group should pay rates at least sufficient to cover the incremental costs of supplying that group. Failure to meet this standard of intermediate-cost coverage would involve internal subsidies, contrary to the spirit of financial self-sufficiency.

MAJOR CRITERIA OF A SOUND RATE STRUCTURE ILLUSTRATED BY AN IMAGINARY EXAMPLE OF ELECTRIC-UTILITY RATE MAKING

Despite the early warning that the reader must look elsewhere for concrete examples of public utility rate structures, I have thought it worth while to close this chapter with an imaginary example of an electric-utility enterprise which starts business with a tariff of the simplest form and which faces the need to practice various forms of rate differentiation in pursuit of the objectives of rate-making policy already set forth. Steam generation of power is assumed. The example is grossly oversimplified, and instead of following the actual historical trends of American electric-rate design, it is deliberately tailored for illustrative purposes. In order to follow the simplifying assumption of a rigid, fair-return standard of entire rate levels, we shall suppose that the company, which might be a municipally owned "public corporation," is under a mandate to secure a 6 per cent rate of return on capital investment but seeks to attain this goal by a fair apportionment of total costs among the different consumers and by resort to rate differentials designed to encourage the optimum use of its service.

1. THE INITIAL TARIFF: UNIFORM RATE PER KILOWATT-HOUR

Commencing business at a time when electric supply has but little use other than for illumination, with the anticipated result that use of service will be fairly homogeneous, our company decides in favor of a completely uniform, single rate of charge. This rate, of course, must equal the average total cost of service, inclusive of the 6 per cent rate of return. But there remains a choice as to the unit of service to which the uniform rate shall be applied. Among a variety of alternatives, three receive closest consideration: a uniform charge per customer; a uniform charge per unit of energy (kilowatt-hour); and a uniform charge per unit of the customer's maximum monthly kilowatt demand.

Uniformity of charge per customer (say, $10 per month for any desired quantity of service) has charm in avoiding metering costs. Nevertheless, it is soon rejected because of its utter failure to recognize either cost differences or value-of-service differences between large and small consumers. The choice between a uniform rate per kilowatt-hour and a uniform rate per kilowatt of maximum demand presents a nicer problem, especially so since the engineering consultants disagree as to whether the total costs of the business will vary more nearly with changes in output of energy or with changes in customers' maximum loads. But the decision is finally in favor of the kilowatt-hour charge, not so much for reasons of cost analysis as for reasons of the greater familiarity, and hence the greater public acceptability, of a charge based on amount of service in a popular sense of this term. There is the additional consideration that, if no charge whatever is made for off-peak consumption of energy, consumers will waste the service by carelessness in turning off their lights—a serious source of loss to the company since fuel costs in these early days of the industry are high because of the inefficiencies of available prime movers.

Accordingly, the company commences business with a rate of 20¢ per kilowatt-hour for any amounts of service to any consumers located within its franchise area. The 20¢ rate represents the estimated average total costs of electric service inclusive of the 6 per cent rate of return.

2. THE INTRODUCTION OF QUANTITY DISCOUNTS THROUGH RESORT TO A BLOCK-ENERGY RATE

By an amazingly good stroke of luck, the 20¢ rate per kilowatt-hour actually yields the estimated 6 per cent rate of return. But it obtains this yield in the face of serious deficiencies from the standpoint of the two other major objectives of rate-making policies. In the first place, in treating the total cost of the business as if it varied directly with changes in the kilowatt-hour output of energy —a grossly false assumption—it violates the most widely accepted canon of fair pricing, the principle of service at cost. In the second place, the uniform rate is so high relative to incremental costs that it discourages customers from increasing their consumption even though they would be ready to pay follow-on rates well in excess of incremental costs.

In order to meet these deficiencies, the company decides to introduce a practice familiar in the marketing of commodities at retail, the practice of quantity discounts. However, it departs from the traditional form of this practice by the use of block-energy rates.[16] For the first 25 kilowatt-hours of use per month, a consumer is charged 20¢ per kilowatt-hour as before. For the additional energy up to 50 kilowatt-hours per month he is charged 10¢ per kilowatt-hour. Finally, for all consumption in excess of 50 kilowatt-hours per month he is allowed a 5¢ rate. The lower rates for the second and third blocks are justified partly on cost grounds and partly on value-of-service considerations. As to the cost grounds, it is believed that an increase in consumption per customer will result in a less than proportionate increase in cost of service, chiefly because, within any presently contemplated limits of use, it will not require any expansion of the present capacity of the distribution system.

3. THE INTRODUCTION OF CLASS RATES: DISTINCTION BETWEEN INDUSTRIAL CONSUMERS AND RESIDENTIAL CONSUMERS

The change from a uniform rate per kilowatt-hour to a block rate which allows quantity discounts proves to be a material improvement in actual practice. By reducing its rates for increased consumption the company is able to earn its required 6 per cent rate of return even without any increase in charges to small customers. Despite this fact, the change leaves much to be desired. The chief deficiency of the new tariff lies in its failure materially to improve the company's system load factor. Even at the terminal block rate of 5¢ per kilowatt-hour, electricity is put to only limited use except for lighting. As a result, the company's expensive plant lies largely idle except for a few hours in the evening. What is needed, therefore, is some change in the tariff designed to encourage off-peak use of power. At the present time in the company's development, before the days of many tempting electrical appliances for household use, the most promising source of off-peak use lies in the industrial consumer, who may possibly be induced to stop using his own steam engine and, instead, to buy electric power during the daytime in order to drive his machinery.

[16] "This type of rate specifies certain prices per kilowatt-hour for various kilowatt-hour blocks, the price per kilowatt-hour decreasing for succeeding blocks." Caywood, *Electric Utility Rate Economics*, p. 43. The block-meter rate has almost completely superseded the step-meter rate, which applies to a customer's *entire* consumption of energy the same rate per kilowatt-hour decreasing with increases in consumption.

But the only hope of encouraging widespread industrial purchase lies in the concession of rates low enough to induce factories not only to use electric power but also to purchase this power from the company instead of producing it in their own generating plants.

In order to meet this situation, the company decides to split its service into two classes: residential and small commercial service (which we shall also call residential for convenience) and industrial power service. The residential rates are reduced 20 per cent so that they are now 16¢ for the first block, 8¢ for the second block and 4¢ for the third block. But the industrial power rates are still lower—let us say 4¢ per kilowatt-hour for monthly consumption up to 10,000 kilowatt-hours, 3¢ from 10,000 to 50,000 kilowatt-hours, and 2¢ for all consumption in excess of 50,000 kilowatt-hours. While, at least from a superficial standpoint, the grant of lower rates to power users may be said to constitute discrimination against the residential users, the discrimination—if so it may be called—will actually operate to the advantage of the residential users themselves since they will receive a rate reduction that could not otherwise be given them.

4. THE INTRODUCTION OF A TWO-PART RATE FOR INDUSTRIAL POWER: A DEMAND CHARGE AND AN ENERGY CHARGE

As already noted, the introduction of the low, industrial-power rate was designed to find useful, compensatory work for plant capacity that had been almost idle except during the evening hours. This objective, it was hoped, might be attained by the concession of below-average-cost rates to the one major group of customers that could be expected to use the service largely during the off-peak hours. To a material degree, moreover, these hopeful expectations have proved justified. The new industrial load has increased the system load factor from 15 to 25 per cent, so that an 85 per cent plant idleness has been reduced to a 75 per cent idleness, a striking improvement. Again, as in the shift from step two to step three, the change has been for the better.

But it is still far from good enough, especially so since the class-rate differentiation has one serious flaw: namely, in its identification of industrial use with off-peak use, and of residential use with on-peak use. True, in an early stage of electric-appliance history, the latter identification was not too far from reality. But much of the industrial-power use now overlaps the peak of the residential use, with the result that a material fraction of the industrial power has to be supplied at high, peak-time costs. Indeed, some industrial consumers confine most of their outside purchases of power to seasons and hours when the utility plant is operating close to full capacity, relying on their privately owned generating plants to produce their base-load power.

What is needed, therefore, is an industrial-power rate which makes a nicer distinction among different consumers than the mere distinction between larger and smaller consumers. For this purpose the company, taking its lead from a famous British electrical-utility engineer, Dr. John Hopkinson, introduces a two-part rate composed of an energy charge and a demand charge.[17] The demand charge for each industrial-power consumer is based on the maximum (30-minute period [18]) demand of this customer during the previous twelve months, as measured in terms of kilowatts (rather than kilowatt-hours) by a maximum-demand meter. For any one month, the customer's bill is made up of the sum of the energy charge and of the demand charge. As a result, the consumer with the higher load factor pays a lower charge per kilowatt-hour than the consumer of the same amount of energy at a lower load factor.

The full rationale of this Hopkinson, two-part rate is far from simple. But the rationale usually given (although it will serve only as a first approximation) is that the two-part rate distinguishes between the two most important cost functions of an electric-utility system: between those costs that vary with changes in the system's output of energy, and those costs that vary with plant capacity and hence with the maximum demands on the system (and subsystems) that the company must be prepared to meet in planning its construction program.

But, whatever its rationale from the standpoint of cost analysis, the introduction of the two-part industrial-power rate results in an improved use of existing plant capacity; and it gives such promise

[17] See Caywood, *Electric Utility Rate Economics*, pp. 48–50, 159.

[18] This period may vary from fifteen minutes to one hour or more. The use of a substantial period of time rather than a more nearly instantaneous period is said to have a twofold rationale: first, that electrical generation and distribution systems can stand heavy overloads for short periods; secondly, that the longer the period, the higher is the probability of coincidence between an individual customer's peak and the system or subsystem peak.

of increased power revenues, with only minor increases in the total costs of serving this class of business, that the company finds it feasible again to reduce its residential rates. The favorable effect of the demand charge on the power revenues is expected to result chiefly from three responses on the part of existing or potential power customers: first, from a tendency of existing customers to spread their loads over a longer period in order to minimize their demand charges, instead of bunching them during short periods likely to coincide with the heavy loads of other customers; secondly, from the deterrent effect of the demand charge on those industrial customers who, although operating their own private power plants, have hitherto relied on the company to carry their peak loads; and thirdly, from the fact that the two-part rate, which works to the advantage of a consumer with a high load factor, will offer the more favorable terms precisely to those consumers who are in the best position to resort to private generating plants if unsatisfied with the rates charged by the central station.

To be sure, the Hopkinson two-part rate, in the form here envisaged, is subject to one serious criticism in making the demand charge depend on the maximum demand of the individual customer at any time of day or season, rather than on his demand at the time of the system or distribution peak. But as to this objection, discussion will be reserved for Chapter XVIII.

5. THE INTRODUCTION OF AN OVERT OR CONCEALED THREE-PART RATE: CUSTOMER CHARGE, ENERGY CHARGE, AND DEMAND CHARGE

As already noted, the two-part Hopkinson rate is based on the assumption that one part of the total costs of an electric-utility business is a function of the output of energy of the system whereas another part is a function of plant capacity and hence of all costs related to this capacity. But this twofold distinction overlooks the fact that a material part of the operating and capital costs of utility business is more directly and more closely related to the number of customers than to energy consumption on the one hand or maximum kilowatt demand on the other hand. The most obvious examples of these so-called customer costs are the expenses associated with metering and billing. For residential customers these latter expenses may amount to $1 a month per customer. For industrial power users, the metering and billing expenses may be much higher. In addition, moreover, each industrial power user is likely to have expensive transformer and switching equipment devoted exclusively to his use. The capital costs and maintenance costs of such equipment are properly included as a customer cost (or, perhaps more accurately, as an individual demand cost rather than a system demand cost), in that they reflect a cost imposed by the individual customer even if and when he consumes no power whatever.

Having in mind these customer costs, our company decides to make a more or less direct allowance for them in its rates for charge of service. As to its residential rate, it first contemplates a customer charge of $1 per month, which would be added to the kilowatt-hour charge as a part of the monthly bill. However, this proposal meets with such opposition on the part of customers, who contend (erroneously but bitterly) that it is an attempt to charge them "something for nothing," that the company modifies its plan by introducing a $1.50 minimum monthly bill. Unlike the overt customer charge, this minimum bill permits the customer to consume up to 10 kilowatt-hours of energy without further payment. From the standpoint of cost analysis, it is decidedly inferior to an unqualified customer charge. But this inferiority is held to be outweighed by its greater palatability from the standpoint of the public taste.[19]

As to the industrial power rate, the unpopularity of an outright customer charge is not so serious, since power customers are sophisticated purchasers of electric energy. Nevertheless, even here the charge is incorporated in a minimum demand charge. The introduction of the (concealed) customer charge is significant chiefly for the residential consumer. Here it reduces the number of consumers who, because of their small use of energy, impose an out-of-pocket loss on the company.

The additional revenues derived from the minimum charge,

[19] See Hubert F. Havlik, *Service Charges in Gas and Electric Rates* (New York, 1938). In considering the proposed elimination of a service charge in the nature of a customer charge, the Missouri Commission said: "It is puerile for our commission to enter into the many scientific principles of rate making which support the service charge when the supporting public is unable or refuses to comprehend its scientific structure. Its continuance is and will continue to result in injury to the business of defendant." Stanley v. Harvey Elec. Lt. & Pr. Co., PUR 1921 E, 681. But a commission has some responsibility toward the education of the public in a matter of this kind.

combined with the saving in expenses resulting from the abandonment of service by those unprofitable consumers who decline to pay this charge, makes possible a reduction in the block-energy rates. The company decides to apply this reduction entirely to the tail-end block in order to promote the use of electricity by residential consumers. While the resulting rate for energy in excess of 50 kilowatt-hours per month is well below the average rate for residential consumption, it is justified on economic grounds in that it at least covers incremental costs, so that the residential consumer of large amounts of service is not enjoying an overt subsidy. The low terminal rate is further justified by the probability that the residential consumer who uses unusually large amounts of electricity will be using most of the excess for nonlighting equipment operated at off-peak hours. Unfortunately, this assumption that the larger use means a higher degree of off-peak use will be belied in many individual cases—especially so because the amount of use depends indirectly in part on the size of the consumer's residence. But the company, after considering an allowance for this size-of-house variable, perhaps by resort to a fixed monthly charge based on number of rooms, decides against the proposal—partly in view of its obvious theoretical infirmities and partly in view of its unpopularity.[20]

6. DIFFERENTIAL RATES FOR INDUSTRIAL POWER BASED ON VOLTAGE DIFFERENCES

The last revision in rate design to be introduced in this imaginary illustration takes the form of a rate differential to industrial consumers based upon differences in voltage at which they will accept alternating-current power from the electric company's transmission or distribution system. For reasons of economy, the company must transport its power from its central stations at high voltages, which must be stepped down to safe and workable voltages at or near the customer's premises. But it is feasible to grant to industrial consumers several options as to the voltage at which they shall receive the power from the company's substations or transformers. Hence, if a given industrial consumer is willing to accept the power at high voltage, he is properly given a discount

[20] Aside from the poor correlation between any index of *presumptive* customer load and actual load, such an allowance would fail to motivate a consumer to keep down his actual peak. G. P. Watkins, *Electrical Rates* (New York, 1921), pp. 58–64.

in his demand charges, or in his energy charge, or both, the amount of which should be designed to pass on to him the cost saving which he confers upon the company. With this object in mind and in view of the layout of its transmission and distribution system, the company offers to its industrial consumers an option of three voltages, which we may call "high," "low," and "medium." The company's basic rate (both in the form of demand charges and of energy charges) is the rate for low-voltage consumption. Its rates for medium-voltage and high-voltage consumption are derived from this basic rate with the object of coming as close as feasible to the principle of "rate differentials equal to cost differentials."

COMMENTS ON THE RATE STRUCTURE SUGGESTED ABOVE

The preceding paragraphs have suggested only a few of the many forms of rate differentiation practiced by modern electric utility enterprises, whether privately or publicly owned. They are designed to illustrate five important types of rate differentiation: namely, quantity discounts through block rates; class rates; two-part rates; three-part rates; and rates for alternative types of service related in such a way as to make the differences in charges equal the differences in costs of rendition. Our illustration ignores the fact that the classes of service, instead of being limited to two, are likely to be as many as ten or more; that a company may offer special rates for completely off-peak service including off-peak residential water-heating service; that industrial power is sold subject to many detailed terms including terms which make allowance for differences in the power factor; [21] and that special charges may be imposed by which to reimburse a company for the construction or installation of expensive equipment needed only to serve a specific user of the service.

Nevertheless, even as it stands in this oversimplified form, our illustration suggests the complex nature of the problem of rate design when this problem is envisaged as that of securing revenues which will yield a fair return by means of rate differentials that will not only fairly apportion total costs among the different con-

[21] Caywood, *Electric Utility Rate Economics*, pp. 59–66. Bolton, *Costs and Tariffs in Electricity Supply*, devotes Part IV of his book to this elusive factor in alternating-current electric supply and to forms of tariffs designed to penalize industrial users for power factors that are costly to the supplier, chiefly in compelling it to maintain excess capacity.

sumers, but will also secure the optimum amount of use of each type of service and of all the services taken in the aggregate. Seeking the closest feasible approach to the accomplishment of these three partly conflicting objectives, the rate maker has at his command a wide variety of schemes of differential rate making which he must apply, not singly but in combination. The choice of the best combination is unavoidably in the nature of a compromise, since the particular combination that he chooses is bound to be worse in some respects than many of the combinations which he rejects.

In the United States, for example, most companies, whether of their own volition or through coercion by regulating commissions, impose no maximum-demand charge on residential consumers. Moreover, they usually impose no customer charge except in the concealed and distorted form of a small minimum monthly charge. This rejection of overt multipart rate making as applied to residential use has pretty well restricted residential rate differentiations to the device of the quantity discount in the form of a block-energy rate—a very crude device when judged either from a cost standpoint or a value-of-service standpoint.

With all their deficiencies, the rate design typical of the American electric utilities is probably superior to that of any other regulated business in this country.[22] Along with improved engineering technology and with the development and promotion of electrical

[22] The superiority of so-called public utility principles of rate making, specifically of electric rate making, over traditional railroad-tariff principles was recognized many years ago by Professor John Maurice Clark in an article quoting a statement by Henry L. Doherty, a leading gas-company executive, that "Perhaps no other one factor has contributed so much to the success of the electrical business as the study of the rate problem." Clark, "Rates for Public Utilities," 1 *American Economic Review* 473–487 (1911). In recent years, and under the pressure of competing forms of transportation, this superiority has been recognized by the more progressive railroad executives. Thus, John W. Barriger, Jr., now President of the Pittsburgh and Lake Erie Railroad, insists that the railroads must "price their products to attract mass utilization of their facilities. The electric utilities are the outstanding example of the public and social benefits—as well as the economic and financial soundness—of such pricing." *Super-Railroads for a Dynamic American Economy* (New York, 1956), p. vi.

The primary claim for the superiority of electric-rate principles lies in their far less sweeping resort to "value-of-service" considerations, combined with their more thoroughgoing recognition of multiple cost functions as typified by the distinctions between customer costs, energy costs, and capacity costs. Because railroad tariffs are based so largely on a uniform rate per ton or per hundred pounds for any given commodity and any given movement, with few distinctions for volume except the extremely crude distinction between carload and less-than-carload rates,

appliances, it shares the credit for the amazing success of the industry in reducing rates or keeping them from rising materially during a prolonged period of price inflation. But it is far from ideal, and practical rate makers will do well to consider seriously its alleged infirmities viewed from the standpoint of its critics among the academic economists.

Two of these alleged infirmities, which are really tied together, are, first, the allowance of quantity discounts said to be greatly in excess of any discounts defensible on cost-of-service principles, and secondly, the imposition of demand charges which penalize consumers for high individual demands even though these demands come at hours or seasons that fall well off the peak loads imposed on the system as a whole or even on any major part thereof. More will be said on these points in Chapter XVIII.

shippers are under no inducement to cooperate with railroads in loading cars to their full capacity, not to mention their lack of inducement to supply full-trainload lots. Major credit is due to Mr. John Alden Bliss, a transportation economist, for calling attention to the superiority of utility-rate design over railroad-rate design. See, e.g., "The Picture Is So Different with a Two-Dimensional Price," *The Analysts Journal*, Aug., 1954, pp. 89–95; "Maybe the Utilities Have Got Something in the Way They Make Rates," *Railway Age*, Aug. 18, 1952; "A Model of Make-or-Buy Competition," Contributed Paper (F2) for the Third Annual Meeting of the Operations Research Society of America, June 3 and 4, 1955; "A Short Survey of Non-Linear Pricing Forms," Contributed Paper (D5) for the Eighth National Meeting of the same society, Jan. 9 and 10, 1956; "Some Comments on Tobacco—North Carolina Points to Southern Points (Rail)," *I.C.C. Practitioners' Journal*, Dec. 1951, pp. 281–301.

XVII

MARGINAL COSTS, SHORT-RUN AND LONG-RUN

For reasons already indicated in the preceding chapter, the marginal costs of public utility services, as distinct from average total costs, are of such significance for sound rate determination that some economists have gone so far as to propose their acceptance as measures of rates even when, in consequence, the resulting revenues will fail to cover total costs and must therefore be supplemented by a tax-financed subsidy. The merits of this unorthodox proposal will be discussed briefly in Chapter XX, the last chapter.[1]

But even under the traditional principle that "rates as a whole should cover costs as a whole," marginal cost, or one of its approximate synonyms such as incremental cost or out-of-pocket cost,[2] plays an important role in the design of the rate structure.

[1] In a more logical sequence, Chap. XX, on unqualified marginal-cost pricing, would come first and would then serve as a basis on which to discuss the more complicated and confusing role to be played by marginal cost as a measure of *minimum* rates and as a factor in the determination of rate relationships subject to the requirement that rates as a whole must cover costs as a whole. Somewhat reluctantly, however, I have reversed this procedure in view of the unfamiliarity of marginal-cost pricing philosophy to persons other than academic economists. Incidentally, the relative simplicity of the theory of strict marginal-cost pricing as compared to the theory of "optimum pricing subject to a budgetary constraint" is a spurious simplicity, which the price theorist gains by wishing some of his most embarrassing problems upon the tax theorist!

[2] "Out-of-pocket cost," itself an ambiguous term, is the popular partial equivalent of "marginal cost," especially in railroad parlance. But it is sometimes used to refer merely to the additional *cash* outlay imposed directly by the production of additional output, whereas "marginal cost" also includes any enhancements in noncash costs (such as depreciation due to wear and tear of equipment) attributable to an increase in rate of output. An important study by Professors Wilson and Rose concludes that the Interstate Commerce Commission has in practice (though not often in language) identified "out-of-pocket" cost with average variable cost regardless of the question whether this latter cost remains constant or increases

In fact, it may play a dual role: first, in setting a lower limit below which no rates will be fixed, not even in order to promote the use of service which could not otherwise find a buyer; and secondly, in serving as a basis for relative rates, subject to deviations of a value-of-service nature.

More will be said of these two uses of marginal-cost estimates in the two following chapters. But in the present chapter we must take account of the fact that "marginal cost" is itself a highly ambiguous term, with the result that proposals to base minimum rates or rate relationships on marginal costs mean different things to different people. The most important ambiguity is that suggested by the distinction between "short-run" and "long-run" marginal costs. Indeed, this distinction is of critical import, for most of the really spectacular differences between incremental and average costs of public utility services are those which apply only when the former costs are taken to be of a short-run variety.

The first section of this chapter, which follows immediately, will therefore discuss the distinctions between the two types of marginal cost. Most of these distinctions would apply, with modifications, to short-run versus long-run incremental costs in general and not alone to costs of increments so small that they are called "marginal." The second section will then turn to the relative merits of short-run and long-run marginal costs as a basis of minimum rates. Here it will be noted that, in most public utility cost analyses, major importance has been attached to the longer-term version.

DISTINCTION BETWEEN SHORT-RUN AND LONG-RUN MARGINAL COSTS

GENERAL NATURE OF THE DISTINCTION

In its general sense, the "marginal cost" of a given commodity or service refers to the increase in total cost of production imposed by a relatively small ("marginal") increase in rate of output.

with an increase in the volume of traffic. In consequence, the Commission may well have accepted estimates of out-of-pocket costs, when offered in support of special reduced rates designed to meet competition, that were lower than marginal costs. G. Lloyd Wilson and J. R. Rose, "'Out-of-pocket' Cost in Railroad Freight Rates," 60 *Quarterly Journal of Economics* 546–560 (1946). One may surmise that awareness of this likelihood partly accounts for the Commission's tradi-

Usually this increase is expressed in terms of an incremental cost per unit of increased output. Thus the marginal cost of producing electric energy may be 1¢ per kilowatt-hour, whereas the average cost may be 2¢.

But the amount of the marginal cost, and its relationship to average cost, will depend on many factors, one of which is that of the assumed duration of the enhanced rate of output. To an electric company, for example, the additional unit cost of a 5 per cent increase in its output of energy may be one figure if the enhancement in output is to last five minutes, another figure if it is to continue for five months, and still another figure if it is to be maintained for five years.

Any number of time-duration distinctions may therefore affect an estimate of the marginal costs of a specific kind of utility service. But by a convenient though partly arbitrary convention, economists have applied the words "short-run" or "short-term" to marginal costs estimated under the assumption that the enhanced rate of output will be temporary and will hence be accomplished solely by an increase in the rate of utilization of the existing plant and equipment; whereas they have applied the words "long-run" or "long-term" to marginal costs estimated under the assumption that the enhancement in rate of output will continue indefinitely and hence will be accomplished by an appropriate increase and adaptation of plant capacity.[3] Shortness of time is associated with absence of change in plant and (heavy) equipment partly because a sudden, unanticipated demand for an increased output can be satisfied, if at all, only by a more intensive utilization of the present plant, and partly because a demand for a merely temporary increase in output, even if it could be anticipated long in advance of its occurrence, would not be likely to warrant a plant expansion.

---

tional reluctance to allow railroads to quote competitive rates only slightly above calculated out-of-pocket costs.

[3] This distinction is drawn in nearly all textbooks on economic analysis and is portrayed by geometric curves of the type developed by Jacob Viner in his celebrated article on "Cost Curves and Supply Curves," 3 *Zeitschrift für Nationalökonomie* 23–46 (1931), reprinted in George J. Stigler and Kenneth E. Boulding, eds., *Readings in Price Theory* (Chicago, 1952). For an excellent short exposition, see George J. Stigler, *The Theory of Price*, rev. ed. (New York, 1952), Chaps. 6 to 10; also *The Theory of Marginal Cost and Electric Rates*, an 80-page brochure on European rate making published by the Organization for European Economic Cooperation (Paris, 1958).

CHARACTERISTICS OF SHORT-RUN MARGINAL COSTS

Exclusion of "constant" or "fixed" costs. Reserving a closer analysis of long-run marginal costs for later paragraphs, let us first note the special characteristics of the short-run version. The most outstanding characteristic lies in the vital distinction drawn here between constant costs and variable costs. In short-run analysis the capital costs of plant and equipment are treated as constant and hence are excluded for the purpose of the estimate. Indeed, even a large share of the costs which accountants call operating expenses is treated in the same way, on the ground that many of these costs do not vary, at least not materially, with changes in the rate of plant utilization. This exclusion of a good part even of the operating costs applies notably to a portion, usually the major portion, of the operating-expense deduction for annual depreciation (in reality a capital cost), since only a minor part of this depreciation is deemed to be affected by the degree of use made of the equipment.

It is primarily because of these exclusions of large shares of the total costs of supplying utility service from estimates of short-run marginal costs that the latter costs are often found to constitute mere fractions of average total costs. Thus, if an electric utility company is now operating with a plant of excessive capacity, a substantial increase in rate of output of energy may impose almost no additional cost except for an increase in cost of fuel—an increase of, say, ¼ to ½¢ per kilowatt-hour as compared to average total cost of perhaps 2½¢. Indeed, if the supply of power comes from a hydroelectric plant of redundant capacity and with no opportunity to store the water for future use, the short-run marginal cost of the power may be practically zero. Leaving the plant idle and letting the water run over the spillway would result in almost no saving whatsoever.

Relationship between short-run and long-run marginal costs depends upon the current relationship between rate of output and plant capacity. Because short-run marginal costs exclude many components of total costs on the ground that, for the purposes of the analysis, these components must be deemed constant, one might assume that the former costs would always be relatively low—lower than either long-run marginal costs or average total costs.

Indeed, this assumption is implicit in some of the American treatment of railroad and utility rate making, which refers to "mere" incremental or out-of-pocket or marginal costs as if they were invariably lower than "fully allocated" costs. But no such assumption would be justified. For short-run marginal costs may be lower than, equal to, or higher than long-run marginal costs. And the same statement would apply to a comparison with average total costs.

In the economic theory of price determination, it is a basic principle that, when a plant is of the optimum size for the rate of output which it is called upon to deliver (that is, when the size of the plant is such as to minimize the total cost of continuous production at this rate of output), short-run and long-run marginal costs will coincide.[4] At this optimum size, the additional cost of producing a small enhancement in rate of output by a slightly greater utilization of the present plant will be just equal to the additional cost of producing this same enhancement with the aid of an increase in plant capacity. The increase in capital costs in the latter case would be precisely offset by the greater increase in the operating costs in the former case.

But if an existing plant is put under a strain to produce services beyond the rate of output for which it is well adapted, the resulting disproportionate increase in the variable operating costs may *more* than offset the fixity of the capital costs; and short-run marginal costs, even though still devoid of any capital costs, may nevertheless exceed long-run marginal costs.

With an electric utility plant, three developments in combination are likely to be largely responsible for this tendency of the variable costs sooner or later to increase much more than in proportion to the increase in rate of output. In the first place, the enhanced output must be supplied by those less efficient, obsolescent turbogenerators that would otherwise be kept in stand-by reserve. Fuel cost per kilowatt-hour will therefore increase. In the second place, stand-by reserve will be reduced to and past the danger point —a situation giving rise to an economic cost, although this cost,

[4] This is not to say that, for any desired rate of output, the plant of most economical size is necessarily a plant operating at its minimum unit cost at this rate of output. For, "when a firm is operating subject to, say, economies of scale, it is most economical to build a plant whose minimum cost comes at an output larger than is desired and to operate it at less than this minimum cost output." Stigler, *The Theory of Price*, p. 141. For practical illustrations, see D. J. Bolton, *Electrical Engineering Economics*, Vol. I, 3d ed. (London, 1950), especially Appendix III.

when estimated in advance, must be expressed in probabilistic terms and may never get into the records of the company's accountants. And in the third place, the quality of the service is likely to deteriorate through voltage drops—another economic cost, though one of an intangible nature, which may be partly imposed on the company in the form of impaired good will and impaired standing with a public service commission. But only the first of these three elements of the short-run marginal costs of service produced by an overworked plant can usually be expressed in terms of mills or cents per kilowatt-hour. Hence, a statistical or accounting study of the behavior of operating costs of a given plant in response to changes in rates of output is almost certain to ignore important tendencies toward disproportionate increases in these costs as output approaches the upper physical limits of plant capacity.[5]

From what has been said it follows that the familiar phenomenon of extremely low short-run marginal or incremental costs—low in relationship both to long-run marginal costs and to average total costs—is a phenomenon of redundant plant capacity.

Meaning of short-run marginal costs when a plant is operating at the upper limit of its capacity. Public utility plants are not often called upon to produce services up to the very limits of their physical capacity. As a rule, advance provisions will have been made to increase plant capacity before the growth in consumption of service has encroached too seriously on emergency reserves. But let us suppose that the upper limit has indeed been reached and that any further increase in the rate of output of service would be either literally impossible or else utterly out of the question for reasons of safety. Under this assumption, what is the current, short-run marginal cost of utility service?

[5] "In the continued attempt to minimize load shedding up to and over its maximum safe capacity, risks of widespread failures of supplies had unavoidably to be taken." British Electricity Authority, *Third Report and Accounts* (1950–1951), p. 9. What has just been said may be a partial explanation of the results of a number of empirical studies, notably those by Professor Joel Dean, which tend to show a horizontal (constant) short-run marginal cost curve for a given plant throughout any actually experienced range in the rate of output. Another reason for the observed horizontality may be the paucity of data on the operating costs of plants that were being pushed close to the upper limits of their capacity. See Joel Dean, *Managerial Economics* (New York, 1951), Chap. 5. Compare J. Johnston, *Statistical Cost Analysis* (New York, 1960), section on electricity generation, pp. 44–73.

Here, if "marginal cost" is given its usual meaning, as referring to the marginal cost incurred by a company in *producing* the service, the question just raised is meaningless; for no increase in output is feasible at *any* additional cost. But, for purposes of rate making, an alternative concept and measure of marginal costs may now take the place of marginal production costs: namely, the marginal social cost necessarily involved when the supply of any amount of the scarce service to those consumers to whom the service is rendered means the denial of this same amount of service to those potential consumers who are thereby excluded.[6] Under familiar assumptions of resource allocation, this cost would be measured by the "market-clearing" price for the scarce service —by the price, say 10¢ or even $1 per kilowatt-hour for electricity, that would just suffice to bring demand into equality with potential supply. While such a price would not reflect marginal cost of production to the utility company, it would approximately measure "marginal exclusion cost" and hence would have much the same claim for acceptance, as a measure of rates, that short-run marginal cost of the ordinary variety would have with respect to a plant operated at less than maximum possible load.

Volatility of short-run marginal costs. In view of the above-noted characteristics of short-run marginal costs, it should hardly be necessary to add that these costs are typically of a highly volatile nature. Let the current rate of output be even slightly below the maximum output permitted by plant capacity (after an adequate allowance for emergency reserve), and marginal cost of service may be a mere fraction of average cost. But let the output increase to a rate only slightly in excess of that for which existing capacity is safely adequate, and marginal cost may jump to some multiple of average costs and to another multiple of long-run marginal costs. As will be noted in the second section of the present chapter, it is this volatility which most seriously impairs the usefulness of short-run marginal or incremental costs as measures of minimum rates or as a basis of sound rate differentiation.

[6] See William Vickrey, "Some Objections to Marginal-Cost Pricing," 56 *Journal of Political Economy* 218–238 at 231 (1948).

CHARACTERISTICS OF LONG-RUN MARGINAL COSTS

Indefiniteness of the term except when strictly defined. As already noted, what distinguishes long-run from short-run marginal cost is that the former cost is measured under the assumption of a *sustained* increment in the rate of output—sustained for a period sufficiently long to require, or at least to justify, a change in the capacity and design of the plant and equipment. This means that those capital and operating costs which are treated as constant, and hence are excluded, in short-run cost analysis, are here treated as variable.

But unless the term is restricted to what may be called the "limiting case," it cannot be distinguished *sharply* from short-run marginal cost. For, at the current rate of output, the present plant may have an excess capacity with respect to *some* of its component parts while having no more than an adequate capacity with respect to others. Thus, an electric company might find it economical to supply for the next several years a 10 per cent increase in its output of energy by means of a 10 per cent increase in the capacity of its turbogenerators but with no enlargement whatever of its distribution network. Or a railroad might sustain indefinitely a 10 per cent increase in its freight traffic by a corresponding increase in rolling stock, together with a minor enlargement and improvement of its classification yards, but with no enhancement in its line-haul capacity. To make things even more complicated, one must recognize the likelihood that a given increase in rate of output might feasibly be supplied at first by very minor adaptations in plant and equipment, to be followed by major adaptations in later periods—perhaps after much of the older equipment is in urgent need of replacement.

Aware of this lack of sharpness in the distinction between short-run and long-run cost functions, economists have sometimes sought to give greater precision to the term "long-run marginal cost" by making it refer to those cost increments which would result from a shift from an indefinitely continuous rate of output accomplished by means of a plant of optimum design and capacity for that rate, to an indefinitely continuous higher rate of output to be accomplished by means of a plant of optimum design and capacity for this higher rate. This definition premises a "run" so extremely long that the plant management has the opportunity to carry out all of those adaptations to the changed rate of output which would be economical under a given technology and under a given set of prices for the various "factors of production." [7]

But useful as is this tighter, "limiting-case" definition of long-run marginal cost for purposes of price theory, its acceptance would severely reduce its usefulness as a tool of practical rate making. For, in actual practice, the more significant marginal costs are those costs which can be expected to persist, not forever or even for twenty years, but rather for those shorter periods that are within the horizon of today's rate makers. As a rule, these are the increments in costs that may be anticipated to result, during the next several years, from increases in rates of output to be accomplished by whatever plant additions and improvements will be warranted in view of the actual layout and actual capacity of the present plant. There would be little point, for example, in basing today's railroad passenger rates on the long-run marginal or incremental costs of passenger service if these costs were to be estimated without recognition that an increase in this service may require little or no increase in right of way or in trackage for a long time to come, if ever. For, with most American railroads, partial plant redundancy is not just temporary but chronic, perhaps permanent.

The upshot of these comments on the indefinite meaning of long-run marginal cost is that, when used as a practical standard of rate making, the concept should be defined only in general terms and should be left for whatever nicer definition may be required in the light of the particular rate-making problem. Applied, for example, to a rapidly expanding utility enterprise such as a modern electric or telephone utility, long-run marginal cost may properly be associated more closely with its stricter definition than would be warranted with respect to a declining or slowly growing business, such as that of a railroad.

Qualification of the principle that all costs are deemed variable. We have already noted that a special characteristic of short-run

[7] Thus long-run marginal cost, strictly defined, is a concept of a stationary state. But Professor John Maurice Clark doubts whether, even if the process of change were to be stopped in such an imaginary state, there could be a return of capital to the point that it would reach if starting anew. "Some forms of capital cannot be shifted without loss, no matter how much time is allowed for the piecemeal transfer of the depreciation fund; a railroad embankment, for instance, or a tunnel." *Preface to Social Economics* (New York, 1936), p. 280.

marginal cost determination lies in its distinction between "constant" and "variable costs" and in its complete exclusion of the former costs as components of marginal cost. This statement may seem to imply that, in long-run cost analysis on the other hand, all costs, including all capital costs, should be deemed variable, so that the whole distinction between constant and variable costs should disappear.

Properly interpreted, the above-suggested inference is correct. But the inference must be qualified, since it appears to be belied by the frequent practice of cost analysts in distinguishing between constant and variable costs in their estimates even of long-run incremental or marginal costs.

The first qualification is that the variability of *all* costs is not necessarily assumed except in that *very* long-run marginal-cost determination in which there is time for the *complete* adaptation of plant and equipment to any change in rate of output. If this extreme definition of a long-run marginal cost is not accepted, and if the objective is to estimate the increments in cost, say, that a given railroad would incur during the next five years if it were to enjoy a 5 per cent increase in its volume of traffic during this period, the assumption that some important items of cost will remain almost unaffected by the increment in traffic—say, the costs of the land for right of way—may be quite plausible.

The second qualification, although often not clearly distinguished from the first, is of a different nature. It is called for in recognition of the practice by some cost analysts of treating a given portion of the total costs *as if* it were constant and of treating the remaining portion *as if* its variability with rate of output were of a linear character. This artificial distinction between so-called constant costs and so-called (proportionately) variable costs may permit approximations of nonlinear cost functions in linear terms.[8]

Perhaps the leading examples of this practice are those used in railroad cost analysis. Thus, the Cost Section of the staff of the

[8] See p. 348, *infra*. This technique is illustrated by D. J. Bolton with reference to the relationship between the capital cost and the KVA capacity of a transformer substation. He adds: "The first step, therefore, in simplifying the cost relationship will usually be to express the cost as the sum of two components, one of which is directly proportional to the chosen variable and the other independent of that variable." *Electrical Engineering Economics*, Vol. 2: *Costs and Tariffs in Electricity Supply*, 2d ed. (London, 1951), p. 125.

Interstate Commission, in its estimates of the "long-run out-of-pocket costs" of railroad full-carload-lot freight traffic, "takes the out-of-pocket costs at 80 per cent of the total operating expenses, rents and taxes, plus an allowance for the cost of the long-run variable capital investment taken at 4 per cent (after income taxes) on all the equipment and about half the road property." [9] That is to say, 20 per cent of the operating expenses at any given rate of traffic density, and about half of the assumed 4 per cent capital costs of the road property, are treated as if they will not increase with small increments in traffic. In consequence, a 1 per cent increase in traffic would be assumed to increase operating expenses, etc. by $\frac{8}{10}$ of 1 per cent, to increase the annual capital cost of the equipment by 1 per cent, and to increase the annual capital cost of the road property by $\frac{1}{2}$ of 1 per cent. Referring to this procedure of long-run out-of-pocket cost analysis in 1954, Mr. Ford Edwards, who directed the development of the procedure in the late 1930s, writes: "The long-run rail freight out-of-pocket costs in the aggregate, as computed by the Cost Section, run to some two-thirds of the rail carriers' total revenue requirements, including the going rate of return and the passenger and LCL deficits when they occur. This leaves about one-third of the aggregate revenue requirements from freight to be apportioned as 'burden' on a value-of-service basis." [10]

I stress the above-noted distinction between so-called "constant" and "variable" costs even in long-run marginal-cost determination because of the pronounced tendency of recent writers on cost analysis to insist that no costs of producing utility or railroad services will fail very long to respond to changes in rates of output.[11] This insistence might be taken to discredit any long-run marginal cost analysis which makes use of the concept of constant costs. But no such wholesale condemnation of the practice would be warranted, since it would fail to recognize the convenience of the mathematical device of a "pseudo constant cost," for purposes of rough approximation.

Long-run marginal costs may vary with changes in rates of out-

[9] Quoted from Ford K. Edwards, "Transportation Costs, Value-of-Service and Freight Rates," 21 *I.C.C. Practitioners' Journal* 494–510 at 494 (1954).

[10] *Ibid* at 494–495.

[11] See, e.g., H. A. A. De Melverda, "The Illusion of Fixed Costs," *International Economic Papers*, No. 2 (London and New York, 1952), pp. 155–177.

put measured along more than one dimension. Although, with the two qualifications noted in the preceding paragraphs, all production costs of utility services must be deemed variable in the long run, it by no means necessarily follows that all of these costs vary with changes in rates of output or volumes of service measured in the same units, along one single dimension. In electric and gas utility cost analysis, for example, all or nearly all of the total operating and capital costs of production are often divided into three categories: those costs held to vary with the *number* of customers (customer costs); those costs held to vary with the supply of energy ("energy" or "volumetric" or "commodity" costs as measured in kilowatt-hours or in cubic feet of gas); and those costs held to vary with the maximum load imposed on the utility system (demand-related costs or capacity costs). But some cost analyses make use of a much finer and more detailed breakdown of cost functions.

A brief review of multidimensional, "functional-cost analysis" will be presented in the following chapter. But I mention the subject here merely in order to note the fact that marginal costs, especially though not exclusively those of the long-run variety, are by no means necessarily limited to marginal energy costs or to any other singly measured unit costs.[12] Thus, the long-run marginal costs of electric power supply at time of system peak may include allowances for marginal capacity cost, marginal energy cost, and marginal customer cost. And even estimates of short-run marginal costs may include allowances for costs other than energy costs, such as costs of meter reading and billing, which vary with number of customers.

MARGINAL COSTS OF PEAK VERSUS OFF-PEAK SERVICE UNDER SHORT-RUN AND LONG-RUN COST ANALYSIS

Unfortunately for the purpose of simple exposition, the distinction between short-run and long-run marginal cost, seldom very sharp at best, is further complicated by a situation that will receive much attention in the next chapter: that of a plant which operates chronically at, or close to, full capacity at certain times of day, week, or season but which has partly idle capacity at other

[12] See p. 219 of Vickrey's article already cited in footnote 6.

times. Here, the incremental or marginal cost of the off-peak service is of the character which we have called "short run" in that it embodies no incremental capacity cost. But here the plant redundancy is periodic and may even be permanent, since any increase in off-peak consumption may never catch up with whatever increase in plant capacity is provided in order to keep pace with increases in peak loads.

Even so, however, one may draw a distinction here between short-run and long-run marginal cost determination, although the distinction is now of a "mixed" nature. As to any clearly and permanently off-peak service (off-peak with respect to every part of a utility plant), so-called short-run analysis is the only applicable analysis. No component of plant capacity cost or capital costs should here be included in the estimate of marginal cost of service. But as to the peak-time service, there is a significant distinction between short-run and long-run marginal cost. Under short-run analysis, the marginal cost even of peak service includes only the incremental operating costs and excludes all capital costs or "capacity costs." But, for reasons already noted in the discussion of short-run marginal costs, these incremental operating costs may possibly be very high since they will include the incremental cost of production by the use of relatively inefficient equipment and since they may even include the "hazard costs" of operation with inadequate reserves or with temporary overloads. On the other hand, the *long-term* marginal costs of service supplied at times of system peak will include full allowances for whatever increments in capacity costs may be warranted in order to add to the supply of this type of service.

In the following chapter we shall note that the allocation of capacity costs respectively to on-peak and off-peak service illustrates a "limiting case" of joint-product pricing under conditions of pure competition—the case in which the one product (peak-time service) is the main product and the other product (off-peak service) is a by-product in its strictest, most extreme sense. Here, under the principles of competitive pricing, the price of the by-product would cover only its separable costs and would not share even to a slight extent in the coverage of the common costs.

COMPLICATION OF PLANT INDIVISIBILITIES

One further complication of long-run marginal-cost determination is that which arises because of the literal or practical indivisibility of some of the "factors of production"—particularly, of some of the larger units of plant and of heavy equipment.

In the *simpler* expositions of the behavior of the costs of a business enterprise in response to increases in rates of output, *continuously* rising output curves and total-cost curves are assumed—curves in which total cost rises smoothly with, though not necessarily proportionately to, the increasing rates of output. With rapidly expanding public utility systems, this assumption of continuity may give a sufficiently close approximation to the real-life uptrends in total costs to serve the purposes of rate determination. But at times the discrepancies between this assumption and actual cost behavior become serious enough to be of concern to the rate maker. These are times when plant expansion, either as a matter of utter necessity or as a matter of good economy, takes place in fairly large jumps.

The classic example of such a jump is that of a railroad which expands its line-haul capacity by shifting from a single to a double track. But many similar examples apply to the expansion of the urban utilities. Thus, in order to take advantage of the recognized economies of large-scale turbogenerators, an electric company may plan its program of expansion so that its output will first encroach on its reserves pending the completion of a gigantic generating unit, after which time the reserve may be excessive for the next year or more. And thus a telephone, electric, or gas company, in order to minimize the long-run costs plus the inconveniences of network installations involving the tearing up of city streets, may install all at once distribution capacity adequate for an estimated growth in load over the next several years.

Under these circumstances of a "jumpy" program of plant expansion, how does one measure the marginal costs of supplying the service? If the question concerns the determination of short-run marginal costs, the answer is relatively simple. Marginal costs of this type ignore capital costs in any event, regardless of the rate of expansion. Hence, if an electric utility company has just put into operation a gigantic turbogenerator which makes its power capacity temporarily excessive, marginal cost will drop correspondingly, only to rise again as the increase in plant utilization shifts from one of redundancy to one of shortage of reserve capacity. If utility rates were to follow these changes in short-run marginal costs, they too would rise and fall like waves—clearly an impractical situation.

But if our concern is to determine *long-run* marginal costs—incremental unit costs that may be expected to remain relatively stable despite temporary changes from situations of plant overcapacity to plant undercapacity—the problem is not so easy. Here, one must attempt some estimate of *average* incremental costs per unit of output over the life of the indivisible asset. This possible solution is fraught with difficulties. But so, for that matter, is the solution of *any* problem of long-run utility cost imputation.

SHORT-RUN VERSUS LONG-RUN MARGINAL COSTS AS MEASURES OF MINIMUM RATES

Having noted in the preceding section the very striking differences that are likely to prevail at any given time between short-run and long-run marginal costs, we may now consider the relative importance that should be attached to these two types of cost in the design of the rate structure. The question takes on a sharper form when raised under a proposal that all rates be fixed at mere marginal cost—a proposal to be discussed in Chapter XX—than it does when raised subject to the constraint that rates as a whole must be made to cover total costs. But we may discuss the question here under the assumption of this constraint and with reference to the use of marginal cost or incremental cost as a measure of *minimum* rates. Unfortunately, no simple answer to this question of choice is acceptable, since it presents one of the many dilemmas of rate-making policy.

THE CASE FOR SHORT-RUN MARGINAL COSTS

The argument in favor of short-run marginal costs as a basis of minimum rates can be stated briefly by the proposition that the costs which should govern the rates to be charged at any given time are the costs that actually prevail at this time and not the costs that will or would prevail on the average during an indefinite period in the future. These current costs are governed by the relationship between the present output of service and the present capacity of the plant. If this capacity is temporarily excessive, the rates for services in elastic demand should be brought down toward the temporarily low marginal costs in order to encourage consumers to make full use of the excess as long as it lasts. On the other hand, if plant capacity is inadequate, rates should be raised temporarily to the high short-run marginal costs of services in order to put the limited capacity to its most urgently demanded use and in order to avoid any overt rationing of service or any waiting list of unsatisfied potential customers.

If the present capacity of the utility plant and equipment happens to be optimum for the existing output of service, the problem of a proper choice between the two types of marginal cost will not arise, since under this condition the two costs will coincide. But when capacity and output are not in ideal balance, which is likely to be either a chronic or a frequently recurring situation, rates should then be related to the temporarily and "abnormally" high or low marginal costs, partly for the purpose of adapting the current demand for service to available supply and partly for the purpose of encouraging a speedier upward or downward adjustment of plant capacity to the anticipated future demand.

This view that utility rates should approximate short-run marginal costs, at least to the maximum extent permitted by the requirement that rates in the aggregate must cover total costs, is in accord with the view that public utility rate making should accept competitive-price standards of reasonable rates and rate differentials. For, under the theories of pure or perfect competition, prices are supposed to tend to come much more quickly into accord with short-run marginal costs than into accord with long-run marginal costs.

THE CASE FOR LONG-RUN MARGINAL COSTS

Even defenders of short-run marginal costs as the proper measures of minimum utility rates and rate relationships would concede the practical objections to cost determinations of such a volatile, extremely short-run character that changes in rates could not feasibly be made to keep pace with changes in costs. Public utility and railroad rate making, especially when subject to regulation by administrative commissions, is a notoriously expensive and cumbersome procedure, with the result that new rates cannot be made to supersede old rates with anything like the rapidity with which an unregulated enterprise may find it feasible to change its quoted prices.

But the defenders of long-run marginal-cost or incremental-cost rate making would not be satisfied with this minor concession to the need for a leisurely procedure of rate regulation. For they would insist on the need for even greater stability in rate levels and in rate structure than that which would be imposed by the duration of a rate case. This asserted need rests on the view that the rates which play the major role in controlling the types and amounts of use of public utility services are those rates which are expected to prevail over a considerable period of time—probably over the next several years. It is these longer-run, anticipated rates, when compared with anticipated prices for substitute products or services, on which individuals must rely in making rational decisions whether to install oil-heating or gas-heating furnaces; whether to buy gas ranges or electric ranges for the kitchen; whether to locate an aluminum-reduction plant near the source of hydroelectric power on the St. Lawrence River or to locate it instead near the source of low-cost steam-electric power in the Ohio Valley, etc., etc. Once these commitments have been made, the demand for utility services consequent thereon will be largely predetermined by the consumers' investment in equipment and will depend only to a minor extent on any temporary changes in rates of change. In other words, the demand for public utility services is likely to be much less elastic in the short-run than in the longer run.[13]

To be sure, the rate schedules now on file are likely to have a

[13] On the relatively low elasticity of short-run demand curves, see Stigler, *The Theory of Price*, p. 47. In support of long-run marginal cost as a factor in rate making, see Donald H. Wallace, "Joint and Overhead Cost and Railway Rate Policy," 48 *Quarterly Journal of Economics* 583–619 (1934). Commissioner Joseph B. Eastman of the Interstate Commerce Commission said: "While some shippers gain a temporary advantage from unstable and unpublished rates, the greater present good and the long-run good of all require that the transportation factor in the cost of doing business be known and predictable." *Regulation of Transportation Agencies*, Sen. Doc. No. 152, 73d Congress, 2d Session (1934), p. 22. But on the allegedly exaggerated value of stable prices, see Frederick V. Waugh, "Does the Consumer Benefit from Price Stability?" 58 *Quarterly Journal of Economics* 602–614 (1944).

decided effect on those decisions of potential customers which will govern their future uses of utility services. But this effect will depend primarily on consumers' assumptions that the currently published rates, even if subject to fractional increases or decreases as a result of a new rate case, will not undergo a change in their general orders of magnitude or in their general relationship to the prices of substitute services or commodities.

In view of the above-noted importance attached to existing utility rates as indicators of rates to be charged over a somewhat extended period in the future, one may argue with much force that the cost relationships to which rates should be adjusted are not those highly volatile relationships reflected by short-run marginal costs but rather those relatively stable relationships represented by long-run marginal costs. The advantages of the relatively stable and predictable rates in permitting consumers to make more rational long-run provisions for the use of utility services may well more than offset the admitted advantages of the more flexible rates that would be required in order to promote the best available use of the existing capacity of a utility plant.

The history of railroad and utility rate regulation in this country would supply numerous examples of the dangers of especially low "incremental cost" rates which, at the time of their establishment, seemed well justified by their compensatory character but which, at a later time, failed to cover even their out-of-pocket costs.[14] What happened here is that the low rates, originally designed to promote greater use of a temporarily redundant plant, helped to stimulate demand to such an extent that the plant became inadequate, with the result that the incremental or marginal costs of the service rose to levels equal to or even above average total costs. Such a situation arose in New York State and elsewhere

[14] On the danger of promotional rates based on temporarily low incremental costs, see Hubert F. Havlik, *Service Charges in Gas and Electric Rates* (New York, 1938), pp. 59–60. This danger was emphasized by Milo R. Maltbie during his chairmanship of the New York Public Service Commission, e.g., in N.Y. Central R.R. Co. Rates (Commutation Fares), Case 6533, P.U.R. 1932C 75. Mr. John Alden Bliss, a transportation economist, calls my attention to a classic example of this danger in railroad rate making: namely, to James J. Hill's low eastbound lumber rates, which finally created a new eastbound peak and a consequent need to haul empties westbound. "Such low rates," writes Mr. Bliss, "are difficult to correct, even without presuming regulation." The influence of vested interests in established railroad-rate relationships is discussed by Professor I. L. Sharfman, *The Interstate Commerce Commission*, Vol. III B (New York, 1936), pp. 667 *et seq.*

with respect to the manufactured-gas companies, which at one time offered sharply reduced rates to users of the gas for house heating in order to put to fuller use plants that had been made redundant by the competition of electricity. Even before the coming of natural gas, house-heating use in some areas had enhanced the demand for manufactured gas beyond the capacity of the existing plants, with the result that the heating rates ceased to be compensatory. The companies sought, and finally secured, permission from the public service commission to raise these rates; but only after a considerable delay, during which time consumer spokesmen complained that they had installed gas furnaces "on the faith" of the persistence of the favored rates.

But one should not conclude from the foregoing remarks that the danger from rates based largely on temporarily low, short-run marginal costs is a danger never worth running. Indeed, something can be done to minimize the danger through the issuance by a public utility company and by a regulatory agency of clear-cut warnings that especially low rates, designed to make the best feasible use of temporarily excessive plant capacity, are subject to cancellation on very short notice. Thus, during the 1930s, the Ontario Hydro-Electric Power Commission, which supplies the Province of Ontario with most of its electric power and which was then faced with a gross excess in water-power capacity, granted phenomenally low temporary rates to industries which used the power to heat boiler water and to serve other "low-grade" purposes. Indeed, the familiar American use of low rates for "interruptible power" (interruptible within limitations) and of still lower rates for "dump power" (interruptible with few if any limitations) serves the same general purpose.

As to the possible use of a high short-run marginal cost as a measure of rates when the existing capacity of a utility system will not suffice to supply all service that would be demanded at "normal" rates of charge, the practical and political objections to this practice have been deemed so serious that resort to overt rationing or to the policy of first come, first served has been the accepted alternative. For reasons suggested in an earlier chapter,[15] these objections seem to me not merely serious but almost fatal if the public utility in question is operating under private ownership.

[15] Pp. 98–99, *supra*.

I conclude this chapter with the opinion, which would probably represent the majority position among economists, that, as setting a general basis of minimum public utility rates and of rate relationships, the more significant marginal or incremental costs are those of a relatively long-run variety—of a variety which treats even capital costs or "capacity costs" as variable costs. Short-run marginal costs should not be ignored. But they should be used with caution, and with special warnings of the liability of rates based thereon to cancellation or revision on short notice.

XVIII

FULLY DISTRIBUTED COSTS

As already noted in Chapter XVI, writers on the economic principles of public utility rates have suggested that, when the rates of any given utility enterprise must be made to cover total costs of production even though the enterprise is operating under conditions of declining unit costs (of unexhausted economies of scale), each individual rate should be made up of two components: a minimum price set at the marginal cost of the service, and a surcharge or quasi tax designed to contribute some appropriate share of those additional revenue requirements which would fail to be covered if all rates were to be held down to their minima. While even the surcharge would not be independent of marginal cost, its relationship thereto would not necessarily be a simple one and might well be deliberately "biased" by value-of-service considerations. The same idea is implicit in the more popular but cruder assertion that public utility and railroad rates should be set somewhere between "cost of service" (that is, out-of-pocket or marginal costs) as a lower limit and "value of service" (that is, what the traffic or market will bear) as an upper limit.

In actual practice, however, rate structures are seldom built up in this two-step manner. Instead, if based on any comprehensive cost analysis at all (which appears to be true in only a tiny minority of cases) they are derived analytically, not synthetically, from apportioned total costs of service. Thus, with an electric utility company, the analyst may first distribute total annual costs among nine classes of service, more or less: residential, commercial, industrial power, street lighting, etc. He may then redistribute the costs of each class among the units of service within this class, distinguishing among customer units, energy units (kilowatt-hours), and maximum-demand units (kilowatts). The first apportionment is supposed to indicate the aggregate revenues that would be due from each class of service if rates were to be based solely on costs of production. The second apportionment is supposed to serve as a guide to the determination of the pattern of each class rate—a pattern that may be composed of a minimum monthly charge per customer, a set of declining block-energy charges, and (for larger consumers) a set of declining block-demand charges.

Even those experts who make and defend these apportioned total costs in rate cases before public service commissions or courts seldom, if ever, offer them as final measures of reasonable rates and rate relationships. Instead, they concede that rates which deviate substantially from the cost apportionments may be justified by a variety of noncost considerations. This concession goes to the point of recognizing the validity and compensatory character of "competitive" or "promotional" rates, such as one for large industrial power, which fail to cover the very costs which the analysts have imputed to the class of service in question.

But there remains the question what, if any, significance should be attached to these fully distributed costs even as guides, or even as points of departure for rate determination, in view of the admitted fact that they fail to mark the dividing line between compensatory and noncompensatory charges for particular classes or quantities of service. And to this question the customary answers are woefully inadequate. The reply most frequently offered is that cost of service is only one of several factors to be considered in rate-structure determination. But this assertion, while quite valid, is also quite beside the point. For the question at issue concerns the doubtful meaning and significance of apportioned total costs and not the weight to be given to a clearly defined specific cost as a basis of rate making.

Mindful of the widespread failure of the cost analysts themselves to supply a really satisfactory answer to this critical question, and mindful also of the notorious disagreements among the experts as to the choice of the most rational method of overhead-cost allocation—a disagreement which seems to defy resolution because of the absence of any objective standard of rationality—public utility managements and public service commissions have often denied or doubted the value of comprehensive total-cost apportionments even as useful guides to rate-structure design. Their doubt is fortified by recognition of the heavy expense and time-consuming character of any thoroughgoing cost analysis as well as by an awareness of the danger that unsophisticated participants in a rate case may be under the illusion that the apportioned costs mean what they seem to mean.

This adverse or skeptical attitude of many public utility companies and public service commissions toward fully distributed cost apportionments may well be justified.[1] But one should not condemn the procedure too hastily, for it is not devoid of at least a *plausible* rationale. What, then, is this rationale? This is the primary question for discussion in the present chapter.

FULLY DISTRIBUTED COSTS AS FIRST APPROXIMATIONS OF MEASURES OF REASONABLE RATES

Fully distributed cost analysis is of many different types, and only within limits do these types have the same rationale. But common to most of them when used as a basis of rate making is the view that the

[1] The extent to which electric companies make such apportionments for their private use, or in order to have them available if they should be demanded by a commission, is not a matter of record. But comprehensive apportionments have been presented officially in only a tiny fraction of the rate cases. In Re Pacific Gas and Electric Co., Oct. 15, 1952, 96 PUR NS 493, 529–530, where the company presented an analysis of class costs of electric service, the California Commission remarked: "Here for the first time in recent years in a formal rate case the Commission has before it functional cost computations to aid in the determination of proper rate levels." Yet this commission is a recognized leader in its review of rate structures. One of the most elaborate cost-apportionment studies was made in Re Consolidated Edison Co. of New York, Inc., 96 PUR NS 194, a case instituted in May 29, 1946, and not decided until July 14, 1952! Even here, however, the "allocations were concerned primarily with the allocation of cost between different classes of service rather than the rate structure within each class." (Separately printed opinion, p. 269). In a now-pending electric rate case of the same company, the New York Commission has ordered the company, over its objections, to make a comprehensive cost breakdown. But in 1953 and 1958 (24 PUR 3d 209, 214), the Commonwealth Edison Company successfully resisted intervenors' appeals to the Illinois Commerce Commission for a ruling requiring a similar breakdown. See n. 11, *infra*. In Matter of Suspension of Rates of the Union Electric Co., Case 14,039, July 16, 1959, the Missouri Public Service Commission said: "Cost analyses . . . which depend on arbitrary bases of apportionment of investment and expenses . . . are subject to erroneous assumptions and conclusions and merit only limited consideration. . . ." Two recent decisions declining to require cost analyses in support of rate-increase applications are Produce Terminal Corp. v. Illinois Commerce Commission, 414 Ill. 582, 112 N.E. 2d 141 (1953); and City of Pittsburgh v. Pennsylvania Public Utility Commission, 171 Pa Super 187, 90 A 2d 607 (1952). But the Federal Power Commission uses full-cost apportionments in setting or approving wholesale rates of natural-gas pipeline companies. See p. 354, n. 16, *infra*.

apportioned costs are useful as first approximations of reasonable rates—approximations based on the provisional assumption that rate relationships should depend entirely on cost relationships. Carefully to be excluded, therefore, are all noncost factors—factors such as those based on statutory mandates (for example, a mandate that preferential rates must be given to residential consumers); or on political or social considerations (for example, considerations favoring low rates for churches or for low-rental housing); or on the vested interests of existing customers who may have acted in reliance on the continuance of the old rates; or, above all, on value-of-service differentials. While all of these factors are arguably entitled to some weight in the design of the rate structure, their recognition is beyond the purview of the cost analyst, who should seek to confine his estimates to the determination of the behavior of costs in response to the changes in rates of output of various services, without incorporating "value judgments" about the bearing of this observed behavior on the fairness or reasonableness of the charges for the service.

But what, then, is the meaning of total-cost apportionments which admittedly do not reflect differential or incremental costs and which therefore fail to mark the dividing line between compensatory and noncompensatory charges for different types of service? The only plausible answer, in my view at least, is that these apportionments should be designed to reflect *relative* differential or incremental or marginal costs, not absolute costs.[2] If, for example, the apportionment of the total annual costs of supplying electric utility services were to impute $5,000,000 of these costs to residential service and $2,500,000 to small-commercial service, this imputation should imply that the incremental costs of the former class of service, whatever they may be, are much greater than the incremental costs of the latter. And it may even be de-

[2] Supporters of fully distributed cost apportionments have sometimes defended them as measures of relative rather than absolute costs. In Re Consolidated Edison Co., July 14, 1952, the New York Public Service Commission said: "Cost of service studies, even when they include all classes of service, are not acceptable as representing absolute, precise values; they are, rather, relative indications of costs." 96 PUR NS 194 at 395. For this reason, the Commission belittled any studies limited to a single class of service. Compare Donald M. Henry's statement that, while cost apportionments are "more or less arbitrary," they have some value as showing relative changes in costs of the same classes of service over periods of time: "What is the Cost of Service: What is Equitable Allocation?" 93 *Gas Age*, May 4, 1944, pp. 58–60 and 106.

signed to indicate (although by no means all cost analyses are so designed) that the former incremental costs are twice as high as the latter. But what it should not be assumed to assert is that the respective sums of $5,000,000 and $2,500,000 per annum represent the respective annual costs of the two classes of service. These costs, were they to be measured, would be differential costs and hence, save under exceptional conditions, would be nonadditive.

Fully apportioned costs, then, should reflect cost relationships, not absolute costs. But beyond saying that the relationships should be among incremental or marginal costs, one cannot generalize as to their precise nature, since in this respect the analyses are not uniform. A relationship of direct proportionality suggests itself and is perhaps the most generally useful one for rate-making purposes.[3] But it is rejected sharply, for example, by the Cost Section of the Interstate Commerce Commission in its apportionment of so-called "constant costs" or "burden." [4]

The particular cost relationship apparently sought for by most cost analysts is one that would measure those rate relationships which could be called "completely nondiscriminatory." These hypothetical, cost-related rates could then be used as points of departure from which to derive actual rates which would incorporate desirable types and degrees of discrimination while avoiding discrimination that could be deemed "unjust" or "undue."

[3] See p. 375, *infra*.

[4] Interstate Commerce Commission, Bureau of Accounts and Cost Finding, "Explanation of Rail Cost Finding Procedures and Principles Relating to the Use of Costs," Statement No. 4, p. 54; Samuel A. Towne, Chief, Cost Finding Section, I.C.C., "Cost Level Guides to Rate Making," 21 *I.C.C. Practitioners' Journal* 697–707 (1954); Ford K. Edwards, "Transportation Costs, Value-of-Service and Freight Rates," 21 *I.C.C. Practitioners' Journal* 494–510 (1954); Edwards, "Cost Analysis in Transportation," 37 *American Economic Review, Proceedings* 441–461 (May, 1947). Dudley F. Pegrum discusses the uses made of these fully distributed costs in an article on "The Economic Basis of Public Policy for Motor Transport," 28 *Land Economics* 244–263 (1952). And see the reference to them by Justice Douglas, speaking for the Supreme Court in New York v. United States, 331 U.S. 284, 316 (1947), footnote 21: "The sum of the out-of-pocket costs plus a pro rata distribution of the constant or fixed costs is referred to as fully distributed costs." In a leading rate case, Northern Pacific R. Co. v. North Dakota 236 U.S. 585, 597 (1915), the Supreme Court said: "The outlays that exclusively pertain to a given class of traffic must be assigned to that class, and the other expenses must be fairly apportioned. It may be difficult to make such an apportionment, but when conclusions are based on cost the entire cost must be taken into account." This oft-quoted statement is very confusing from the standpoint of cost analysis. For the apportionment of "the other expenses" in such a way as to portray true costs, instead of being "difficult," is literally impossible.

Unfortunately, however, and for reasons to be discussed in the following chapter, no rate relationships can be made completely nondiscriminatory as long as all or some of the rates must be set above marginal costs in order to yield adequate revenues. And this fact may explain some of the disagreements among the experts as to the more rational formulas for the apportionment of total costs among different units of service. One such disagreement, which will receive attention in this next chapter, concerns the question whether rates for different kinds of service, in order to avoid the attribute of discrimination, must be made directly *proportional* to marginal costs, or whether they should be based instead on *differences* in marginal costs. Here, the choice is that between the horns of a dilemma.

TWO MAJOR TYPES OF FULLY DISTRIBUTED COST ANALYSIS

1. THE DOUBLE-STEP TYPE

Despite an ambiguity due to its failure clearly to define "relative costs," the above exposition of fully distributed costing goes about as far as one can go toward expressing the basic philosophy of the practice. For more explicit expositions, one must distinguish different types of analyses. By all means the most important distinction is that between those total-cost apportionments which superimpose a distribution of admittedly unallocable cost residues on estimates of incremental or marginal costs, and those other apportionments which recognize no difference between true cost allocation and mere total-cost distribution.

The first, or double-step, type might also be called the "railroad type" because of its application to railroads (and other transportation agencies) by the Cost Section of the Interstate Commerce Commission. The Cost Section distinguishes between (directly) variable costs and constant costs in a manner noted in the preceding chapter. The variable costs alone are assigned to the different units of freight traffic as representing "long-run out-of-pocket costs"—a term with a meaning here not distinctly different from that of the economist's "long-run marginal costs." There remains a residue of total costs, or total "revenue requirements" which, since it is found to behave as if it were constant over substantial variations in traffic density, is strictly unallocable on a cost-finding basis. Nevertheless, because the Cost Section has felt impelled to make some kind of a distribution of total costs, it has apportioned this residue, which it sometimes calls "burden," among the units of carload traffic on a basis (partly ton, partly ton-mile) which is concededly quite arbitrary from the standpoint of cost determination. In recent years, this burden (which includes allowances for revenue deficiencies in the passenger business and in less-than-carload freight traffic!) has amounted to about one third of those total revenue requirements which the carload freight business is supposed to be called upon to meet.

Since this book is concerned only incidentally with railroad rates, it will not attempt to analyze the methods by which the staff of the Interstate Commerce Commission has estimated out-of-pocket costs and apportioned residue costs. Suffice it to say that the usefulness of the latter apportionment is questionable. But in any event, full credit should be given to the Cost Section for its express and overt recognition of a vital distinction too often ignored in utility-cost analyses: namely, that between a cost allocation designed to reflect the actual behavior of costs in response to changes in rates of output of different classes of utility service; and a mere cost apportionment which somehow spreads among the classes and units of service even those costs that are strictly unallocable from the standpoint of specific cost determination.[5]

2. THE SINGLE-STEP TYPE

We turn now to a type of fully distributed cost analysis which, unlike the "railroad type," draws no distinction between cost allocation and cost apportionment: the single-step type.[6] It might be called the "public utility" type because of the considerable use to which it has been put in gas and electric utility rate cases. Here

[5] But, in the actual design of rate structures, the local public utilities have made far better use of their cost analyses, despite deficiencies, than the railroads and the Interstate Commerce Commission have made of whatever analyses have been at their command.

[6] See Russell E. Caywood, *Electric Utility Rate Economics* (New York, 1956), Chap. 11; L. R. Nash, *Public Utility Rate Structures* (New York, 1933), Chap. 11; and the much more elaborate treatment by D. J. Bolton, *Electrical Engineering Economics*, Vol. 2: *Costs and Tariffs in Electricity Supply*, 2d rev. ed. (London, 1951), Chaps. 4 to 8.

no attempt is made, first to determine out-of-pocket or marginal costs and then to superimpose on these costs "reasonably distributed" residues of total costs. Instead, all of the total costs are treated as variable costs, although these costs are divided into costs that are deemed to be functions of different variables. Moreover, whereas in Interstate Commerce Commission parlance "variable cost" means a cost deemed to vary in *direct* proportion to changes in rate of output, in the type of analysis now under review "variable cost" has been used more broadly, so as to cover costs which, while a function of some one variable (such as output of energy, or number of customers), are not necessarily a *linear* function.

As already noted in an earlier paragraph, the more familiar cost analyses of utility enterprises or utility systems divide the total costs among a number of major classes of service, such as residential, commercial, industrial power, street lighting, etc. This "grand division" permits many costs to be assigned in their entirety to some one class, such as street lighting, or at least to be excluded completely from some important class or classes. High-tension industrial power service, for example, would not be charged with any share of the maintenance costs or capital costs of the low-tension distribution lines. But the major portions of the total costs of a utility business are common or joint to all, or nearly all, classes of customers; and these costs must somehow be apportioned among the various classes and then must somehow be reapportioned among the units of service in order to report unit costs that can serve as tentative measures of reasonable rates.

The general basis on which these common costs are assigned to differently measured units of service will be illustrated by the following highly simplified problem of an electric-utility cost analysis. But before turning to this example, we must distinguish two subtypes of analysis, both of which belong to the single-step type rather than to the double-step type.

In the first subtype, the analyst (following the practice of railroad analysis in this particular respect) distributes both total operating costs and total annual capital costs (including an allowance for "cost of capital" or "fair rate of return") among the different classes and units of service. Here, an apportionment, say, of $5,000,000 of the total costs to residential service as a class would include an allowance of perhaps 6 per cent as the cost of whatever capital is deemed to have been devoted to the service of the residential consumers.

But in the second subtype, which I take to be the one more frequently applied, only the operating expenses and not the "cost of capital" or "fair return" are apportioned directly among the various classes of service. To be sure, the capital investments in (or, alternatively, the estimated "fair values" of) the plant and equipment are apportioned among the different classes, as are also the gross revenues received from the sales of the different services. But any resulting excess of revenues received from a given class of service over the operating costs imputed to this class is reported as a "return" realized on the capital investment attributed to the same service. Thus, during any given year (a) if the revenues from the residential service are $7,000,000, (b) if the operating expenses imputed to this class of service come to $5,000,000, and (c) if the net investment in (or value of) the plant and equipment deemed devoted to this service amounts to $30,000,000, the cost analyst will report that residential service, in the aggregate, has yielded a return of $2,000,000 or 6⅔ per cent. Other services will show different rates of return, some probably much lower and some higher.

There are obvious reasons of convenience for this practice of excluding "cost of capital" from the direct apportionment of annual costs among the different classes of service—notably, the avoidance of the controversial question what rate of return should be held to constitute "cost of capital" or "fair rate of return." But the practice is likely to be misleading, since it may seem to support a conclusion that, as long as the revenues from any class of service cover the imputed operating expenses plus *some* return on capital investment, however low, the rates of charge for this service are compensatory. Needless to say, any such inference would be quite unwarranted.

For the reason just suggested, I shall assume the use of the first subtype of fully distributed cost apportionment in the following simplified example. That is to say, an allowance for "cost of capital" will be assumed to be included directly in the cost apportionment.

THREE-PART ANALYSIS OF THE COSTS OF AN ELECTRIC UTILITY BUSINESS

In order to simplify the exposition of a typical fully apportioned cost analysis, let us assume the application of the analysis to an electric utility company supplying a single city with power generated by its own steam-generation plant. Let us also assume the existence of only one class or type of service, all of which is supplied at the same voltage, phase, etc. to residential, commercial, and industrial customers. This latter assumption will permit us to center attention on the most controversial aspect of modern public utility cost analysis—the distinction among costs that are functions of outputs of the same service measured along different dimensions.

Since the company under review is supplying what we are here regarding as only one kind of service, we might suppose that the problem of total cost apportionment would be very simple; indeed, that it would be limited to a finding of the total annual operating and capital costs of the business, followed by a calculation of this total in terms of annual cost per kilowatt-hour of consumption. In fact, however, the problem is not so simple. For a statement of costs per kilowatt-hour would ignore the fact that many of these costs are not a function of kilowatt-hour output (or consumption) of energy. A recognition of multiple cost functions is therefore required.

The simplest division, and the one most frequently used (with subdivisions) in gas and electric rate cases, is a threefold division of the total operating and capital costs into "customer costs," "energy" or "volumetric costs," and "demand" or "capacity" costs.[7] If this threefold division of costs were to have its counterpart in the

[7] Other cost breakdowns, such as those allowing for the power factor, for voltage differences, for distances between points of generation and points of consumption, and for the customer-density factor, have been used to a limited extent. Compare Vickrey's selection of six parameters in order to approximate the response of the operating costs of the New York City Rapid Transit System to various changes in service and traffic: Train miles; car miles; maximum number of cars in service; number of passengers carried; number of passengers carried during the peak hour; and the layout of the system, consisting of the number of route miles, number of stations, etc. William S. Vickrey, *The Revision of the Rapid Transit Fare Structure of the City of New York.* Technical Monograph No. Three, Finance Project, Mayor's Committee on Management Survey of the City of New York, Feb., 1952, p. 8.

actual rates of charge for service, as it actually does have in some rates, there would result a three-part rate for any one class of service. For example, the monthly bill of a residential consumer might be the sum of a $1 customer charge, a $5 charge for 250 kilowatt-hours of energy at 2¢ per kilowatt-hour, and a $2 charge for a maximum demand of 2 kilowatts during the month at the rate of $1 per kilowatt—a total bill of $8 for that month. But our present interest lies in the measurement of costs of service, and only indirectly in rates that may or may not be designed to cover these costs. Let us therefore consider each of the three types of cost in turn, recognizing that this simplified classification is used only for illustrative purposes; costs actually vary in much more complex ways.

1. THE CUSTOMER COSTS

These are those operating and capital costs found to vary with number of customers regardless, or almost regardless, of power consumption. Included as a minimum are the costs of metering and billing along with whatever other expenses the company must incur in taking on another consumer. These minimum costs may come to $1 per month, more or less, for residential and small commercial customers, although they are substantially higher for large industrial users, who require more costly connections and metering devices. While costs on this order are sometimes separately charged for in residential and commercial rates, in the form of a mere "service charge," they are more frequently wholly or partly covered by a minimum charge which entitles the consumer to a very small amount of gas or electricity with no further payment.

But the really controversial aspect of customer-cost imputation arises because of the cost analyst's frequent practice of including, not just those costs that can be definitely earmarked as incurred for the benefit of specific customers but also a substantial fraction of the annual maintenance and capital costs of the secondary (low-voltage) distribution system—a fraction equal to the estimated annual costs of a hypothetical system of minimum capacity. This minimum capacity is sometimes determined by the smallest sizes of conductors deemed adequate to maintain voltage and to keep from falling of their own weight. In any case, the annual costs of this phantom, minimum-sized distribution system are treated as customer costs and are deducted from the annual costs of the existing system, only the balance being included among those demand-related costs to be mentioned in the following section. Their inclusion among the customer costs is defended on the ground that, since they vary directly with the area of the distribution system (or else with the lengths of the distribution lines, depending on the type of distribution system), they therefore vary indirectly with the number of customers.

What this last-named cost imputation overlooks, of course, is the very weak correlation between the area (or the mileage) of a distribution system and the number of customers served by this system. For it makes no allowance for the density factor (customers per linear mile or per square mile). Indeed, if the company's entire service area stays fixed, an increase in number of customers does not necessarily betoken any increase whatever in the costs of a minimum-sized distribution system.

While, for the reason just suggested, the inclusion of the costs of a minimum-sized distribution system among the customer-related costs seems to me clearly indefensible,[8] its exclusion from the demand-related costs stands on much firmer ground. For this exclusion makes more plausible the assumption that the *remaining* cost of the secondary distribution system is a cost which varies continuously (and, perhaps, even more or less directly) with the maximum demand imposed on this system as measured by peak load.

But if the hypothetical cost of a minimum-sized distribution system is properly excluded from the demand-related costs for the reason just given, while it is also denied a place among the customer costs for the reason stated previously, to which cost function does it then belong? The only defensible answer, in my opinion, is that it belongs to none of them. Instead, it should be recognized as a strictly unallocable portion of total costs. And this is the disposition that it would probably receive in an estimate of long-run marginal costs. But the fully-distributed cost analyst dare not avail

[8] This is in accord with the views of Hubert F. Havlik: *Service Charges in Gas and Electric Rates* (New York, 1938), Chap. 8 and Appendix A. Allocation, in whole or in part, would be at least theoretically possible if a customer-density parameter were added to the three traditional cost components. See G. P. Watkins, *Electrical Rates* (New York, 1921), p. 212. But if this factor were embodied, not only in cost analysis but in the resulting rate differentials, rates would not be uniform throughout a given community and hence would violate a generally accepted tradition.

himself of this solution, since he is the prisoner of his own assumption that "the sum of the parts equals the whole." He is therefore under impelling pressure to "fudge" his cost apportionments by using the category of customer costs as a dumping ground for costs that he cannot plausibly impute to any of his other cost categories.

2. THE ENERGY COSTS

The energy-cost component of this threefold division of total annual costs is supposed to consist of those costs which would vary with changes in consumption of energy, measured in kilowatt-hours, even if the number of customers should remain constant and even if there were no change in maximum load upon the system or subsystem as measured by kilowatts or kilovolt amperes.[9] The most obvious costs of this character are fuel costs, although a small portion even of these costs may be regarded as demand-related on the ground that some fuel is required in order to maintain a "spinning reserve." But other operating costs may also be deemed to vary with output of energy and hence with consumption of energy, including whatever depreciation of the equipment may be regarded as a function of use rather than of obsolescence and aging.

Reduced to costs per kilowatt-hour, the imputed energy costs may be only a fraction of total average costs. It is this relative smallness which is often held to justify a company in conceding very low rates for off-peak or interruptible services, on the ground that these services impose upon the company little or no additional capacity costs.

The treatment of energy costs as a separate cost function is subject to one serious deficiency: namely, in its assumption that the

[9] Estimates of the ratio of energy-related costs to total costs of electric supply (including capital costs) have ranged from ⅓ down to only ¼. Referring to British conditions, Bolton writes: "More accurate costing has shown that, on the average, only one-quarter of the total costs of electricity supply are represented by coal or items proportional to energy, whilst three-quarters are represented by fixed costs or items proportional to power, etc." D. J. Bolton, *Costs and Tariffs in Electricity Supply* (London, 1951), p. 59. But he notes two practical reasons, among others, why this situation does not justify a corresponding dominance of demand charges rather than energy charges in electric rate structures: (a) that the effective power demand imposed upon the system by any given individual is very difficult to determine, and (b) that a pure demand-charge rate would probably lead to a more serious waste of energy than a pure energy rate would lead to a waste of power capacity. The latter reason invokes a "value-of-service" or "demand-elasticity" principle of rate making rather than a cost principle.

cost to the company of producing any given amount of energy, measured in kilowatt-hours, is independent of the system load factor. For such an assumption may be belied by the fact that the turbogenerators which carry the company's base load will be much more efficient than those generators which are relied upon to carry the peak loads. Hence, the cost analysis may be in danger of overstating the relative energy costs of off-peak service and of understating the relative energy costs of on-peak service. Recognizing this danger, the analyst may undertake to offset it by imputing to the peak-time consumers a lower capacity cost than would otherwise be deemed justified—a capacity cost based on a steeply written-down net investment in, or appraised value of, those older, less efficient turbogenerators that will be operated only for a few hours per day.

3. THE CAPACITY COSTS OR DEMAND COSTS

We come now to that category of costs, the treatment of which has made a nightmare of utility cost analysis.[10] For the problem which it presents is that of imputing joint costs to joint products or by-products, and not merely that of distributing those common but nonjoint costs which vary more or less continuously with number of consumers or with rates of output.

Here, as with the other two categories of cost, there is no general agreement as to what items or portions of total costs should be included among the demand-related costs, perhaps because cost functions are far too complex to be reflected by the arbitrary, three-way classification of customer, energy, and demand. But we

[10] In addition to the references already cited in footnote 6, see the two technical reports (K/T 106 and 109) on demand-related cost allocations by special committees of the British and Allied Industries Research Association, 15, Savoy Street, London, W.C. 2, dated 1943 and 1945 respectively; W. Arthur Lewis, *Overhead Costs* (London, 1949), Chaps. 1 and 2; Ralph Kirby Davidson, *Price Discrimination in Selling Gas and Electricity* (Baltimore, 1955), Chap. 8; and a series of discussions on "Peak Loads and Efficient Pricing" in the *Quarterly Journal of Economics* by Professors Jack Hirshleifer, H. S. Houthakker, and Peter Steiner: 71 *Q.J.E.* 585–610 (1957); 72 *ibid.* 451–468 (1958). In one of these articles, Professor Hirshleifer refers to a masterly theoretical treatment of the subject by Marcel Boiteux, a distinguished engineer of Électricité de France: "La Tarification des demandes en pointe: Application de la théorie de la vente au coût marginal," 58 *Revue générale de l'électricité* 321–339 (1949). This article, with modifications, has been translated: "Peak-Load Pricing," 30 *The Journal of Business of the University of Chicago* 157–179 (1960).

may ignore this lack of agreement in order to turn attention to the far more critical problem of apportioning whatever costs have been found to be demand related. Suffice it to say that this category of costs includes the major part of the total allowance for depreciation, for property taxes, and for return on investment, together with a substantial part of the operating and maintenance expenses. Whether or not corporate income taxes should also be included, on the ground that they tend to vary with the earned return on investment in plant and equipment and hence indirectly with the capacity of the plant, is a question sometimes debated in the rate cases. But we may here pass this question with the note that the asserted correlation must be far from close.

Assuming, then, that an estimate has been made of total capacity costs, probably expressed in terms of costs per annum, the question now arises as to the proper apportionment among, or allocation to, services supplied at different load factors and at different times of day or season. These services share responsibility in different degrees, if at all, for the creation of the system and subsystem peak loads and hence indirectly for the capacity costs that must be incurred in advance in order to meet these loads.

In attempting to assess these relative responsibilities, the analyst is offered a wide variety of alternative formulas of apportionment, each of which has received support from some rate experts. Testifying before the Illinois Commerce Commission in a recent rate case, Vice President Gordon Corey of the Commonwealth Edison Company noted the existence of twenty-nine such formulas.[11] Most of them have no claim whatever to validity from the standpoint of cost determination, and only a dubious claim to acceptance as compromise measures of reasonable rates. Hence they will not be reviewed in this book. But, for illustrative purposes, we may mention

[11] Case No. 41130 before the Illinois Commerce Commission, *Objections of Commonwealth Edison Company to Motion of City of Chicago, Dated September 15, 1953, Respecting Cost of Service Studies.* In this brief the company stated that the making of a complete class-of-service cost study would require detailed analyses to determine the proper apportionment of over 6,000 different kinds of costs which the company had segregated according to functions or activities that cause such costs to be incurred. As a witness for the company in this case, I supported its objections to a class-cost analysis on the ground that, in order to be of any value, its preparation and critique would be formidably time consuming and would introduce controversial issues not essential to the timely solution of what was essentially a "revenue" or "rate-level" case.

three formulas that receive special attention in the latest American treatise on electric rates, by Mr. Russell E. Caywood of the West Penn Power Company.[12]

The first is the peak-responsibility formula. Here, the entire capital costs are imputed to those services that are rendered at the time of system (or subsystem) peak (probably at the time of the *annual* peak) and in proportion to the kilowatt demand imposed at this time—an integrated demand rather than an instantaneous demand, measured over some short period such as thirty minutes or longer. Service rendered completely off-peak would be assigned no responsibility whatever for the capacity costs.

The second formula is called the "noncoincidental-demand" formula.[13] Like the first, it apportions capacity costs entirely on the basis of kilowatts of load rather than on the basis of kilowatt-hours of energy. But it apportions these costs in proportion to the maximum demands of the different classes and of different individual consumers of service even though some of these demands, so far from coinciding with the peak load of the company, may be completely off the system peak. This type of cost apportionment, whatever merit may be claimed for it as a basis of rate making on grounds of "fairness," or on grounds of presumptive value-of-service considerations, is not really a cost analysis at all and should not be allowed to masquerade as such. Yet precisely this method of apportionment is said to have a wider currency than does any alternative method in American rate making.

The third formula is called the "average-and-excess-demand" method by Caywood, who points out that it reaches the same result, though by a different mathematical technique, as that reached by

[12] *Electric Utility Rate Economics* (New York, 1956), pp. 156–167.

[13] Economists have been particularly critical of this method. In the words of Professor W. Arthur Lewis, referring to attempts by rate engineers to offset its obvious infirmities by allowances for the different diversity factors of different groups of consumers, "no amount of correction can alter the fact that the standing costs of the undertaking are related not to the maximum rate at which the individual consumer takes, but to the amount he takes at the time of the station peak." *Overhead Costs* (London, 1949), p. 52. (But this sentence should have been amended to include not only station peak but distribution-system peak, which may be an even more critical factor.) In industrial-power rate making, the use of noncoincidental demand as the basis of the demand charge has been defended on the ground that the resulting two-part, Hopkinson rate follows the cost behavior of an isolated, private steam plant. Caywood, *Electric Utility Rate Economics,* p. 32. But this is a value-of-service argument and is not germane to an analysis of the costs actually incurred by the public utility company itself.

another method called the "Greene method." Here, the assumed cost of that portion of the company's plant capacity which would be needed even if all consumers were taking their power at 100 per cent load factor is apportioned among consumers in proportion to their average loads—that is, in proportion to their kilowatt-hour consumption of energy during the time period in question. But the assumed cost of the excess in actual plant capacity over this lower, hypothetical capacity is apportioned "by applying the non-coincidental peak method to the difference between maximum loads and average loads."

Like other formulas which, overtly or in effect, apportion a part of the capacity costs among kilowatt-hours of energy rather than entirely among kilowatts or kilovolt amperes, the "average-and-excess-demand method" has a certain degree of justifiable support from the standpoint of cost analysis. This support lies in the fact that, when the extent of coincidence between the maximum load of any given consumer and the peak system load cannot be measured or prophesied directly, one may be justified in assuming a greater *probability* of coincidence if the customer is operating on a high load factor than if he is operating on a low load factor. Thus, the maximum load of a 100 per cent load-factor consumer is *certain* to coincide at some point of time with system peak, whereas the maximum load (or, for that matter, even the entire load) of a 10 per cent load-factor customer may be entirely off the system peak. But when resort must be had to these "stochastic" or "probabilistic" methods of capacity-cost imputation, no general formula is worthy of much respect. Instead, empirical studies of the relationship between load factors and coincidence factors for different uses of service (air conditioning, water heating, elevator operation, etc.) are required. Such studies have been made by a number of companies, notably, by the Philadelphia Electric Company under the direction of its distinguished rate engineer, Mr. Constantine Bary.[14]

[14] See, e.g., "Economic Significance of Load Characteristics as Applied to Modern Electric Service," presented at 53d Annual Meeting, Association of Edison Illuminating Companies, Jan., 1938; "Coincidence-Factor Relationships of Electric-Service Load Characteristics," *Transactions*, American Institute of Electrical Engineers, Vol. 64, 1945; "Load-Structure of a Modern Electric Utility System," *Electric Light and Power*, May, 1954; "Load Characteristics of 'Small Light and Power' Customers—Their Economic Significance," *Electric Light and Power*, Oct. 15, 1955; "Dollars and Sense of Electric House Heating," presented at 75th Annual Meeting, Association of Edison Illuminating Companies, Nov., 1959.

Limitations of the various capacity-cost apportionment formulas. Of the three formulas just described, the one that would probably come closest to receiving support from the economists, at least viewed from the standpoint of cost analysis, is the system-peak responsibility method. At least there would be general agreement that, with qualifications noted in the footnote below,[15] the cost to a company of rendering any type of service which can be counted upon positively to stay off the system (or subsystem) peak does not include any capacity cost. Whether or not the rates actually charged for such a service should nevertheless attempt to recover parts of the company's capacity costs because of "value-of-the-service" factors, or else because of a widely held view that even off-peak users and interruptible power users should make some *fair* contribution to the costs of a plant which confers upon them a benefit, is another question—a question which, while related to that of cost imputation, is by no means necessarily tied thereto.[16]

But what, then, makes capacity cost allocation or apportionment such a highly controversial problem? The answer lies in the fact that capacity costs, instead of being ordinary overhead costs, common to different kinds of amounts of service, are *joint* costs—the costs of producing services which are joint products when they are rendered at different periods of time. So important for the purpose

[15] If an electric power station were constructed for the *sole* purpose of supplying a peak demand occurring, say, only one hour per day, less efficient and hence less expensive turbogenerators (possibly with gas turbines) would be installed for the sake of maximum economy. Hence, when stations are designed to supply a 24-hour variable load, the additional costs of the more efficient generating units are theoretically chargeable to off-peak use. Moreover, with run-of-the-river hydroelectric plants the system peak load is not necessarily the load encroaching most seriously on capacity.

[16] The cost apportionments used by the Federal Power Commission in its determination of reasonable rates for natural-gas pipeline companies, such as those which treat 50 per cent of the capacity costs as a demand cost and 50 per cent as a volumetric cost, would be utterly ridiculous if viewed as attempts at actual cost determinations. See, e.g., Matters of Atlantic Seaboard Corporation and Virginia Gas Transmission Corporation, 11 F.P.C. 43 at 56 (1952). These apportionments can be justified, if at all, only on "fairness" or "value-of-service" considerations. Unfortunately, the Commission is bound by the provisions of the Natural Gas Act, which compel it to make arbitrary cost apportionments by restricting its rate-making jurisdiction to the sale of gas for resale, as distinct from sales made directly to industrial customers. On the general subject of gas-cost allocation, see Hans Nissel, "The Impact of Cost Allocations upon Future of the Natural Gas Industry," 66 *Public Utilities Fortnightly* 512–524 (1960); Larry Shomacher, "Is FPC Gas Cost Allocation Equitable?" 50 *Public Utilities Fortnightly* 680–688 (1952); Ott Eckstein, "Natural Gas and Patterns of Regulation," 36 *Harvard Business Review* 126–136 (1958); and Ralph Kirby Davidson's book cited in n. 10.

of the present inquiry is this distinction between joint production and merely common production of multiple products that a brief digression on the subject is in order.

Joint costs as distinguished from nonjoint common costs. Many years ago, the importance of the distinction between joint costs and nonjoint common costs was brought dramatically to the attention of the academic economists by a now famous debate on railroad rates between Professor Frank W. Taussig of Harvard University and Professor A. C. Pigou of Cambridge University.[17] Writing for the *Quarterly Journal of Economics,* Professor Taussig had taken issue with those earlier writers who had attributed the familiar railroad practice of "charging what the traffic will bear" ("value-of-service" rate making) to the possession of monopoly power. This practice, he contended, was quite in keeping with the recognized principles of joint-product price determination under conditions of strict competition. For even under competition, the relative prices of joint products are not based on relative costs of production. Instead, the price differentials are based on differences in relative demands for the respective products.

Classical examples of joint products are that of cotton fiber and cotton seed, and that of beef and hides. These sets of products are produced in more or less fixed proportions, with the result that much of the costly action needed in order to produce any given amount of the one (beef) will also go a long way toward the production of a corresponding amount of the other (hides). In competitive equilibrium, the prices of the joint products in the aggregate will equal their combined costs of production. But aside from the fact that each product will be priced at least at its separable costs of production (since otherwise its production would not be brought to completion), the respective prices will not be based on respective costs.[18] Instead, the sharing of the joint costs will depend entirely

[17] 27 *Quarterly Journal of Economics* 378–384, 535–538, 687–694 (1913). Professor Pigou amplified his argument in his *The Economics of Welfare,* 4th ed. (London, 1932), Chap. 18, "The Special Problem of Railway Rates." For a much cited later article on the subject, see Donald H. Wallace, "Joint and Overhead Cost and Railway Rate Policy," 48 *Quarterly Journal of Economics* 583–619 (1934).

[18] But the marginal cost of a by-product (in its extreme form) is determinate for reasons to be noted presently. In theory, it is possible to determine the marginal costs even of joint products if these products, so far from being producable only in fixed ratios, can feasibly be produced in varying proportions. The marginal cost of Product A is then the increase in total cost resulting from the output of an additional unit of this product, output of the other products being kept

on the relative forces of demand. Hence, from the standpoint of cost analysis (though not from the standpoint of competitive price determination), a joint cost is an unallocable cost.

These classical principles of joint-product pricing were believed by Professor Taussig to be widely applicable to railroad rate making. Thus, he argued, passenger service and freight service might be considered joint products, as might also be lumber-transport service and wheat-transport service, etc. And the familiar practice under which the charges for the transport of high-valued commodities would be several times as high as charges for the transport of low-valued commodities was, therefore, in his opinion, quite consistent with competitive-price theory.

To this position Professor Pigou took sharp exception. While he conceded the existence of some examples of jointness in the production of different kinds of railroad service, he believed that these examples were very limited and that they did not include those cited as such by Professor Taussig. The most outstanding example, he said, was that of the forward and return haul of a railroad train. But passenger and freight service are not joint products, since the addition by a railroad of another passenger train or passenger car does not reduce the cost of adding freight trains or freight cars—quite the contrary, in fact, if the road is being operated close to the limit of its line-haul capacity. And the same statement applies to the transportation of different kinds of freight.

Pigou therefore concluded that the familiar "value-of-service" rate-making practices of the railroads belong, for the most part, in the category of monopolistic price discrimination and not in the category of competitive price determination under conditions of joint production. And on this main issue, Pigou's position rather than Taussig's has won general acceptance among the economists, although later rate experts have attributed to the multiple freight and passenger services of a typical railroad system more widespread elements of jointness (particularly, "time jointness") than Pigou was ready to concede.

---

constant. See the article "Cost" by Jacob Viner, in 4 *Encyclopaedia of the Social Sciences* 466–475 at 473. But such "fine" measurements of marginal cost are of limited usefulness in utility rate making, at least in its present stage of development. Moreover, "while cases of *absolutely* fixed proportions may be nonexistent, the proportion in which some products are produced may be variable only within very narrow limits and at great expense." National Bureau of Economic Research, Committee on Price Determination, *Cost Behavior and Price Policy* (New York, 1943), p. 177.

Capacity costs of utility services as a kind of joint costs. Electric and gas companies do not supply the particular examples of joint products and joint costs emphasized by Pigou with respect to transportation companies, namely, those of forward and backward haul. But they supply many examples of another, and even more important, kind of joint production, that of "time jointness." This factor of jointness is especially prominent for the electric utilities, since electricity cannot feasibly be stored for future use either by the producing company or by the consumer. Hence, electricity supplied at any given time is, in a significant sense, a different product from electricity supplied at any other time. And since, with exceptions that we may here overlook, an increase in plant capacity, if made for the purpose of enhancing the rate of output in any one period of time (say, between 5 P.M. and 10 P.M. during the winter, or between 10 A.M. and 4 P.M. during the summer), is also available for an enhancement in the rate of output at other times, services rendered at different times of day, week, or year are true joint products. This being the case, any apportionment of capacity costs, say, as between morning service and evening service, or as between winter service and summer service, or even as between all services rendered in the year 1959 and all services rendered in the year 1960, is a partly arbitrary apportionment from the standpoint of cost determination, however justified or convenient or rational it may be from the standpoint of reasonable rate determination.

In the determination of rate differentials, an "ideal" apportionment of capacity costs would be whatever apportionment would result in an approximation to a 100 per cent system load factor. Were it not for circumstances to be noted below, an "ideal" [19] structure of utility rates would solve the problem of cost imputation as between on-peak and off-peak service by preventing the

[19] Not *quite* ideal, however, since an electric utility company needs time to shut down its generating units for major repairs. In its 1957–1958 rate case before the Illinois Commerce Commission, the Commonwealth Edison Company stated that its seasonal load curve had now become so relatively level (due largely to its summer air-conditioning load) that it was planning its equipment overhauling "so that the sum of the following items remains constant for each month of the year: (1) the monthly peak load, (2) the capability of units scheduled to be out for overhauling in that month, and (3) limitations on capacity due to temperature of circulating water." Testimony of Murray Joslin, Vice President, Edison Exhibit 2, I.C.C. Case No. 44391. In consequence, "Edison's load pattern has materially changed, so that, currently, each month's peak load is as important as the annual system peak load in determining capacity requirements." Company's Reply Brief, p. 38.

problem from arising. A skillfully designed system of rate differentials would so distribute the burden of paying for capacity costs among consumers of services rendered at different periods of time, that the company's load valleys would be raised and its peaks would be lowered to the level of a plain. These differentials would be based on relative demands, on market-clearing forces rather than on cost analysis.

In actual practice, modern public utility rate making has gone a certain distance toward the realization of this ideal through the use of rates (such as those for interruptible power or for controlled water heating) which give to consumers a financial incentive to take their service at what is to them less convenient times of day, week, or season.[20] But it has not gone full way in this direction; nor can it be expected to do so, although further progress may be hoped for as a result, among other things, of further refinements in the art and science of rate making and in the promotion of new uses of service.

The most obvious restriction against valley-filling policies of differential rate making is that imposed by the limited potential demands for off-peak services—demands so limited that plant capacity would not be fully utilized even if off-peak rates were cut to bare incremental energy costs. This limitation may well apply to most electric systems during the dead of night, during holidays, and possibly even during seasons of the year intermediate between periods of maximum winter loads and maximum summer loads.

But even if an electric company were able to keep its plant running at full capacity through the entire year by cutting its charges for what would otherwise be off-peak service to bare energy costs, it might still not be warranted in doing so. For, as long as the com-

[20] A book on electric rate making in Switzerland states that, in that country, tariff policy has stressed its role as the most effective means of regularizing consumption. Newer uses of power have been so well developed that the December evening peak has become "only a minor peak." Pierre Devantéry, *Les Prix de l'électricité* (Lausanne, 1950), p. 105. But, as the more perspicacious writers on utility rates have long pointed out, "The practical improvements resulting from scientific rate systems have come not so much from limiting peak demands by penalizing them with heavy charges as from developing off-peak business by freeing it from the burden of overhead." John Maurice Clark, *The Economics of Overhead Costs* (Chicago, 1923), p. 328, citing examples and referring to Watkins, *Electrical Rates* (New York, 1921), p. 22. Bolton, taking a similar position, writes: "Of the two ways of improving system load factor, namely, by raising the individual load factor and by increasing the diversity, there can be little doubt that the second holds out far greater possibilities." *Costs and Tariffs in Electricity Supply*, p. 87.

pany operates under conditions of declining unit costs, it *must* charge more than incremental or marginal costs for some of its services in order to cover its total costs. And off-peak service is likely to be among those services that can best stand a charge of this nature.

At first thought one might assume that the price elasticity of demand for services at unpopular periods of time would always be high, with the result that these services should be offered for sale at "bargain" prices only slightly above their marginal costs. But this assumption would overlook the fact that the demand for certain amounts of off-peak service is a "convenience" demand—a demand that is highly inelastic over substantial but limited ranges of output. Hence, a company may well be justified in making no general concessions for nighttime service even though it may offer special, incentive rates for electricity to be used at night solely for the use of water heating.

The continued presence of peaks and valleys in public utility plant utilization gives qualified support to the system-peak responsibility principle of capacity-cost allocation. Regardless of the reason for the persistence of peaks and valleys in the load curves of utility systems, as long as they persist they raise the problem of cost imputation as between on-peak and off-peak service. This problem is soluble under familiar principles of joint-cost imputation subject to a number of simplifying assumptions. The reason why it is soluble, despite the general principle that joint costs are unallocable costs from the standpoint of cost analysis, is that we now have that "limiting case" of joint production in which one of two products, the off-peak service, is a by-product in the strictest sense of that term, whereas the other product, the peak-time service, is the main product. Under this condition, no longer does the increase in capacity costs incurred in order to increase the output of the main product have to its credit any useful accomplishment in enhancing the further output of the by-product. For the plant is already redundant with respect to the by-product. Hence, at the margin, the by-product is costless save for its separable costs (energy costs and possible customer costs), and the main product is chargeable with the entire incremental capacity costs. Whether or not the by-product should nevertheless be sold at a profit over incremental cost, in order to help cover the company's total revenue requirements, is a problem of rate making, not a problem of cost analysis.

So far, then, the argument supports the system-peak responsibility formula of capacity-cost allocation. But the argument applies only to the allocation of *incremental* capacity cost—to the cost per kilowatt of enhancing the capacity rather than to the average cost per kilowatt of total capacity. To the extent to which this average cost either exceeds or falls short of incremental cost, it is unallocable on any principle of cost analysis. Unfortunately, this fact is ignored by fully distributed cost analysis of the public utility type.

Necessary use of stochastic methods by which to assign responsibility for capacity costs. The unqualified acceptance of a system-peak responsibility principle of capacity-cost apportionment (in so far as the average cost per kilowatt of capacity can be accepted as an approximation of incremental capacity cost) would imply that rates for off-peak service, if designed to cover only those costs allocable to this service, should include no allowance whatever for demand-related costs. And this statement would seem to apply to rates for services rendered at times that are only slightly off the system (or subsystem) peak, no less than to rates for services rendered at times when the system is operating at a mere fraction of its total capacity.

But the only costs that are directly relevant as measures of rates to be charged in the future are costs that will prevail in the future; and these costs must be estimated subject to the limitation that peaks and valleys in future loads are far from completely predictable. Hence utility rates, in so far as they are cost determined, must include charges for the *probability* that the service to which they apply will be taken at the time of the system peak—a probability which may be far lower if the service will be taken, say, at 3 A.M. daily than if it will be taken in a period which, during the past year or two, has been off system peak by only, say, 5 per cent. It follows that the sharp distinction between charges for completely and surely off-peak service and charges for completely and surely on-peak service will not serve all the purposes of rate making. Divisions of time into three or more periods, such as those observed in Europe, may be justified.[21]

[21] French rate experts have sanctioned a threefold division into "heures creuses," "heures pleines," and "l'époque de la pointe." Marcel Boiteux and Paul Stasi, "Sur la détermination des prix de revient de développement dans un système interconnecté de production-distribution." Report, Congrès de Rome, Sept., 1952,

So far as I am aware, the fully distributed cost apportionments of the type familiar to American gas and electric rate making do not attack directly this necessity of expressing costs, particularly capacity costs, in probabilistic terms. For they usually purport to reflect costs that have actually occurred in some past, test year, apportioned in the light of loads that have actually been experienced during that year. Hence, only to the extent to which history can be expected to repeat itself can these cost analyses be accepted as measures of those cost relationships which should receive weight in rate making to apply during future months or years.

In actual rate determinations, as distinct from special-purpose cost analyses which can make use of elaborate recording meters for testing purposes, the need to rely heavily on probability inferences, even of a pretty tenuous nature, in the design of the rate structure is greatly enhanced by the extreme crudeness of the accepted American devices for metering the individual consumers. Thus, most electric companies now use only watt-hour meters and no demand meters for residential and small commercial service; and even for large industrial power, the most frequently used demand meters are limited to the measurement of the customer's maximum demand during a given period of time (one month, one year, etc.), making no recorded distinction between those relatively harmless high individual loads which occur when the utility system has redundant capacity, and those very costly loads which coincide with system peak.

Handicapped by this absence of *recording* demand meters, and by the partial absence of *any* demand meters, a company must base its charges for different classes and different amounts of service on assumptions as to typical relationships between energy consumption and maximum loads, as well as between maximum loads and coincidence factors, that are sure to be wildly false in many cases. Thus, with the residential rate on a purely block-energy basis (subject only to a minimum monthly charge), any cost-based defense of

Union Internationale des Producteurs et Distributeurs d'Énergie Électrique. The British Technical Report (K/T 109), already cited in footnote 10, would meet this problem by an "improved method" for allocating demand-related costs to different classes of consumers under which (a) no account is taken of demand or consumption during "off-potential peak" periods; and (b) the allocation is to be based jointly on the consumption and the highest demand of the individual component loads within the potential-peak periods. This is a modified version of the "average-and-excess-demand method" (the "Greene Method"), already described.

lower charges for additional blocks of energy must rest on the assumption that the larger uses are increasingly off-peak uses—an assumption that was plausible in the earlier days when the main load was a lighting load but which is less tenable today.[22] The utility companies themselves are aware of this fact, and some of them have sought to minimize the deficiencies of noncompensatory terminal block-energy rates by the obviously crude device of a "stopper" provision, which prevents the consumer's *average* charge per kilowatt-hour from going below a certain minimum.

The great convenience and high popularity of residential rates based wholly on energy consumption (save for a modest minimum bill), combined with the economy of a simple and inexpensive watt-hour meter, probably justifies the continued use of this type of rate for traditional amounts of household consumption. But with the industry's drive toward the "fully electrified home," which uses electricity in large amounts for cooling in summer and for heating in winter, more elaborate metering devices will probably be called for. For this purpose the use of "time of day" and "time of season" differential energy rates, which has had more of a vogue in Europe than in America, may be well worth considering.[23] Here, use of service would still be metered in terms of kilowatt-hours rather than partly in terms of kilowatts of maximum demand. But the watt-hour meters may have two or three separate dials, by which to record the consumptions in different periods of time.

Special problems of capacity cost allocations to industrial power. Even if the allocation of capacity costs were not complicated by variations in the load curve of an electric power system during the day, week, and season of the year, the problem of rational allocation, and the related problem of rational rate making, would still be far from simple. For there would remain the problem of imputing to the services rendered during any given year proper shares

[22] A load-curve study of residential consumption by groups of customers using no electric ranges or water heaters is reported to have found that: "There is a slight general increase in annual load factor with increasing kwhr use, but this does not mean that increasing usage always improves annual load factor. Some added loads raise load factor, while others pull it down." *Electrical World*, Nov. 24, 1958, pp. 89–92, reporting on the findings of the Load Research Committee of the Association of Edison Illuminating Companies.

[23] In Chap. 10 of his book already cited in footnote 10, Davidson suggests this type of rate as preferable to the familiar Hopkinson-type rate. But among the objections to it is the danger that its sharp breaks will create surges in the loads imposed on a power station or on a distribution line.

of the capital costs incurred in the construction of plant and equipment designed to supply service for many years in advance. In order to be assignable to the services rendered during any one year, these construction costs are converted into terms of *annual* costs—costs composed of allowances for annual interest or "fair return" plus allowances for annual depreciation. But the conversion is necessarily largely arbitrary.[24] A cost analyst's statement, for example, that the total demand-related or capacity costs of an electric company have come to $12,500,000 (or $25 per kilowatt of capacity) during the past year does not report a cost that could have been saved if the plant had been shut down for this year; nor does it report a cost that was incurred merely and solely to supply the service required during this one year. Instead, it is an estimated or assumed *average* annual cost. And it would continue to be of an average character even if it were derived from an incremental cost of plant construction per kilowatt as against an average cost.

In order to illustrate the seriousness of this limitation of the typical cost imputation, let us assume that, because of a regular, five-year business cycle in the area served by the company, the plant capacity that is required in order to supply the demand for power during the two years of high prosperity will be unavoidably redundant during the remaining three years. What allocation or apportionment of annual capacity costs should be made under this assumption? The logic of the peak-responsibility method of capacity-cost imputation would require that all of this cost be imputed to the service rendered during the two prosperous years and that none of it should be assigned to the service of the other years—not even to the peak-time service of these years.

Indeed, just this kind of anticipatory cost imputation would be required if annual changes in system loads were regular and predictable. In fact, however, the changes are neither regular nor predictable, with the result that the cost analyst is pretty well limited to an attempt to report *annual* capacity costs based largely on such conventions as straight-line depreciation and on assumptions of uniformity in the annual allowances for "fair return."

In actual practice the serious consequences of these theoretical limitations imposed upon the cost analyst by the practical necessity

[24] On the arbitrary character of depreciation as an annual capital cost, see pp. 199–210, *supra*.

of converting capital costs of construction into annual capacity costs have been greatly reduced by the remarkable upward secular trend in the demand for many utility services including, notably, electric service. Hence even business depressions have not brought about many very long standing periods of plant redundancy—periods during which even the system peak falls for a long time far below plant capacity.

Yet the situation just mentioned—the arbitrary character of conversions of plant costs into annual costs or annual "rent values"—gives rise to serious problems on occasion, notably with respect to cost allocations or apportionments used in the support of rates for large-scale industrial power. Consider, for example, a large steel company with immediately prospective power requirements so large that the utility company, in order to supply these requirements, must add 50,000 kilowatts of otherwise unneeded capacity. Assume that the maximum load of this large customer coincides with system peak and that, in accord with the "peak-responsibility" principle, its annual charges for power during the following year will include 100 per cent of those interest charges, depreciation allowances, property taxes, etc., which constitute the annual costs of 50,000 kilowatts of capacity. Such charges might be supposed to constitute full indemnity for the cost to the utility of installing the additional capacity. And so they will if the steel company keeps on paying these charges, year after year, until the utility's investment in this capacity has been completely amortized.

But now let us suppose that, after the first year or two, the steel company cuts its demand for power to a maximum demand of 25,000 kilowatt-hours, or even shuts down its plant completely and moves to another location. In this event, the utility may find itself with a generating capacity in excess of its needs for several months or years to come, not to mention any investment in ancillary equipment that is no longer useful except for its possible salvage value. Should this possibility be actually experienced, the utility would not have been indemnified for the cost which it has incurred expressly in order to serve the particular customer and which it would not have incurred otherwise.

Public utility companies often seek to minimize these dangers of an unexpected drop in industrial-load requirements by the imposition of minimum, "billing demand" charges in excess of charges for the maximum demands imposed upon them by an industrial customer during any given twelve-month period. But the industrial consumers are not fond of these charges, since they quite naturally desire to maintain the maximum freedom to cut their power costs if their volume of business declines. Moreover, except under some special contracts between the industrial user and the utility company, the obligation of the consumer to continue the payment of *any* charges for service will cease, or will have only a limited duration, if it is ready to abandon the taking of service altogether. In consequence, the rates of charge for service rendered to a gigantic user of power, regardless of the formula used in the determination of his annual demand charges, may well result in less than barely compensatory rates when measured in the light of hindsight. If based on advance estimates of "cost of service" these rates should therefore incorporate an appropriate allowance for the risk factor —for a risk factor well in excess of that which would be appropriate in an estimate of the cost of supplying the more stable, residential service.

The unpredictability of future demands for certain types of utility services presents another problem alike for cost analysis and for rate determination: namely, the problem of an unpredictable change in the system load curve. It is this complication, more than any other, which may account for the reluctance of most American cost analysts to apply a system-peak responsibility formula of capacity cost allocation.

Assume, for example, that at the present time, the utility's system peak occurs on winter evenings, when the residential load is at its peak and when the industrial power load is negligible. On this assumption, one may argue that all incremental capacity costs should be charged against residential service as a matter of cost analysis, even if "rate-making policy" should dictate a rate structure which relieves residential users of some of these costs and throws the burden onto other users. But assume that, at some unpredictable time in the near future, demand for industrial power (a demand, by the way, which is said to be inelastic except for voracious power-using industries such as aluminum producers) is very likely to outpace the demand for residential service and that, in consequence, the system peak may quickly change from a winter residential peak to a summer power peak. Under this assumption, the analyst may hesitate to impute to the residential users full responsibility for being currently on a system peak that may shift over to the industrial-power load in another year.

To be sure, some writers who defend strict peak-responsibility methods of cost allocation, for purposes both of cost finding and of actual rate making, would deny that any such hesitation is justified. Let both the analyst and the rate maker, they argue, consider only those cost relationships that will prevail in the immediate future, without undertaking the hopeless and useless task of guessing what these relationships may be in the next year or in the next ten years. And this argument has much force. But it also has those limitations that were noted in the preceding chapter as basic to all short-run marginal-cost determination: namely, in its failure to recognize the practical desirability of "dampers" on excessive rate fluctuations and hence the practical desirability of types of cost allocation which measure cost in terms of longer-run rather than of very short-run variables. Indeed, if rapid and unpredictable shifts in system load curves were held to require equally rapid shifts in those cost allocations that are used as a basis of rate making, then the whole methodology, and not merely the detailed technique, of functional cost analysis would need to be changed. For in that event, the relevant methodology would be one of short-run cost analysis, which treats all capacity costs as constant costs and which therefore bases any distinction between on-peak and off-peak use entirely on differential energy costs of base-load versus peak-load generators.

Reduced to practical terms of rate-making policy, the question here at issue is whether services which are now supplied on system peak, and which will almost surely continue to be so supplied for the very near future (regardless of any proposed changes in rates), should be charged today as if they were expected to remain on peak indefinitely; or whether services that are presently off peak should nevertheless be made to pay some share of those capacity costs for which they are likely to become responsible in the next several years. To this question neither the general theory of cost analysis nor the general theory of rate making can supply a conclusive answer. At least at the present stage in the development of both theories, action must therefore be based largely on "judgment."

This chapter began by raising the question what, if any, significance should be attached to fully distributed cost apportionments as points of departure for public utility rate making. As a provisional answer, it suggested that the significance must lie in whatever claim can be made for the apportioned costs as indices, not of absolute costs but of relative differential or incremental or marginal costs. It then distinguished two types of full-cost apportionment, the double-step (or "railroad") type and the single-step (or "public utility") type, and proceeded to give major attention to the latter. A tentative opinion on the merits of this second type is now in order.

In my opinion, these merits are so dubious that they fully justify the skepticism with which utility cost analysis has been received by public utility companies and public service commissions. The basic deficiency of this analysis lies in its failure to distinguish between actual cost finding and mere cost apportionment—between those costs that can be imputed to specific classes or units of service by differential cost analysis and those other costs that should be deemed unallocable from the standpoint of cost determination even if they are somehow apportioned as a provisional step in rate determination. This failure seems to me critical.

Among the more specific deficiencies of the typical fully distributed cost analysis of the public utility type, three seem to me especially serious. In the first place, the capacity costs or demand-related costs are usually derived from book values of plant and equipment that reflect sunk costs in dollars of original investment, not costs that can be said to vary, except in a very indirect way, with present and future increases in plant capacity. In theory, this particular objection might be met by the use of appraised current values of the utility plant and equipment rather than by the use of book values. But if the appraised values were of no better quality than the "fair values" that are accepted as measure of the rate base in states applying a fair-value rule of rate making, their advantage over book values would be at least dubious.

In the second place, the cost analyst, faced with the necessity of apportioning all of his costs among three or four arbitrarily selected functional-cost categories, faces dilemmas such as that noted in the section of this chapter on customer costs. He is therefore bound to be impaled on one horn of a dilemma; and one may suspect that he chooses whatever impalement he believes to be less harmful in its consequences for sound rate making in view of noncost considerations.

And in the third place, most analysts, unwilling to follow the implications of joint-cost and by-product cost analysis in their treatment of demand-related costs, accept some compromise formula of apportionment, such as one which imputes capacity costs in proportion to noncoincidental maximum class demand. Here, too, one may suspect that the choice of the formula depends, not on principles of cost imputation but rather on types of apportionment which tend to justify whatever rate structure is advocated for noncost reasons.

What has just been said, however, is by no means meant to imply that cost analysis is useless for rate-making purposes. On the contrary, it is utterly essential.[25] But the really important analyses are not those which attempt to apportion total capital and operating costs among the different classes or units of service. Instead, they are the analyses designed to disclose differential, or incremental, or marginal, or escapable costs—costs which are not ordinarily derivable from total costs and which cannot be added together so as to equal this total.[26]

It is these costs which should be the primary object of study of the utility cost analyst. Whether or not, in addition, some kind of apportionment of unallocable cost residues is also worth making, along the lines followed by the staff of the Interstate Commerce Commission, is a secondary question, on which I venture no present opinion.

In short, then, a thorough re-examination of the whole philosophy of modern public utility cost analysis has long been overdue. If this country were Great Britain, the appointment of a governmental commission to make such a re-examination would be called for.

[25] In 1948, Major H. J. Flagg, Executive Officer of the Board of Public Utility Commissioners of the State of New Jersey, wrote: "One may venture a guess that scarcely a dozen electric utilities in the United States have any but the vaguest notion of the relative costs—or productivity in terms of net earnings—of the several classifications of their business. Many utilities really do not know whether their industrial load under present rates is a desirable form of business or not." "What of the Industrial Load?", 130 *Electrical World*, Aug. 28, 1948, p. 82. But the traditional fully distributed cost apportionments warrant little more than "the vaguest notion." As far back as 1921, Dr. G. P. Watkins wrote: "Were cost accountants willing to (or expected to) deal with less than total expenditures, so far as they claim to obtain the actual cost of a particular good or service, their true service might be more clearly perceived and therefore greater." *Electrical Rates*, p. 118.

[26] Compare a similar view expressed by some recent cost accountants with respect to the cost determination of manufactured products. R. Lee Brummet, *Overhead Costing* (Ann Arbor, 1957), esp. Chap. 5.

XIX

DISCRIMINATION, DUE AND UNDUE

THE LEGAL RULE AGAINST UNDUE DISCRIMINATION

One of the most nearly universal obligations imposed by Federal and state laws alike on railroads and on other public utilities is the obligation to furnish service and to charge rates that will avoid "undue" or "unjust" discrimination among shippers or customers, actual or potential.[1] Early writers often cited this obligation as a unique attribute of a business "affected with a public interest." In later years the legal right of even a "private" business to practice discrimination in the pricing of commodities has been materially restricted in this country, notably so by the Robinson-Patman Act provisions of the antitrust laws.[2] But the latter restrictions are much more limited in scope, although more severe in their application within this scope, than are the antidiscrimination standards of the public utility laws.

The legal obligation of public utilities to avoid unduly discriminatory rate relationships is distinguished from the equally general obligation to charge rates, each of which is "just and reasonable" in itself.[3] Needless to say, these two basic mandates are related; and a commission's finding of undue discrimination between rates is likely to go hand in hand with a finding that at

[1] "The principal evil at which the Interstate Commerce Act was aimed was discrimination in its various manifestations." New York v. United States, 331 U.S. 284, 296 (1947), citing Louisville & Nashville R. Co. v. United States, 282 U.S. 740, 749–750 (1931).

[2] Corwin D. Edwards, *The Price Discrimination Law: A Review of Experience* (Washington, D.C., 1959); Herbert F. Taggart, *Cost Justification* (Ann Arbor, Mich. 1959); Frederick M. Rowe, "Cost Justification of Price Differentials under the Robinson-Patman Act," 59 *Columbia Law Review* 584–616 (1959).

[3] Interstate Commerce Commission v. Alabama Midland Railway Co., 168 U.S. 144, 172–173 (1897).

least one of these rates is also "unreasonable." But not necessarily so; for undue discrimination may be held to exist between two rates, neither of which would be found "unreasonable per se" —possibly because the one rate may lie near the bottom of a "zone of reasonableness" whereas the other may lie nearer the top. In this event, the remedy may take the form of an increase in the former rate, of a decrease in the latter, or of a mutual adjustment.[4]

Readers of the treatises and the case law on public utility and railroad rates will often come across bald statements to the effect that, in these regulated industries, the practice of rate discrimination is unlawful. In fact, however, such statements are grossly inaccurate. What the law forbids is merely "undue" or "unjust" discrimination.[5] To be sure, the statutes also declare to be "unjust" without qualification a certain overt type of discrimination that had a disgraceful early American history, particularly in railroad rate making: namely, so-called "personal discrimination," whereby different customers are charged different rates for substantially the same service, rendered under similar conditions.[6] But beyond this point the legislatures have been content with general language forbidding "undue or unreasonable" preferences or prejudices, leaving the task of interpretation to the commissions, subject of course to review by the appellate courts.[7]

In view of this situation, one might suppose that the commissions and courts, in their efforts to interpret and apply the rules

[4] In reclamation cases under the Interstate Commerce Act, a shipper must prove injury in order to recover compensation under the sole count of unjust discrimination, but not in order to recover for the payment of unreasonable rates.

[5] New York v. United States, 331 U.S. 284, 305 (1947), quoting United States v. Illinois Central Railroad Co., 263 U.S. 515, 521 (1924). The statutes use other terms, such as "unjust" or "undue" or "unreasonable" "preferences" or "advantages" or "prejudices." But the tribunals bring them all under the heading of unjust discrimination.

[6] The recent relaxation of the British rules against unjust discrimination in railway transport rates does not even except personal discrimination. See Alastair M. Milne, *The Economics of Inland Transport* (London, 1955), p. 52, referring to Sec. 21 of the Transport Act, 1953. See also J. P. Carter, "Personal Discrimination in Transportation: A European Technique," 47 *American Economic Review* 372–381 (1957).

[7] "Whether a preference or advantage or discrimination is undue or unreasonable or unjust is one of those questions of fact that have been confided by Congress to the judgment and discretion of the Commission." Manufacturers Railway Co. v. United States, 246 U.S. 457, 481 (1918). To the same effect, Swayne & Hoyt, Ltd. v. United States, 300 U.S. 297, 304 (1937), quoted in New York v. United States, 331 U.S. 284, 347 (1947).

against undue discrimination, would have taken pains to develop two clear-cut distinctions: first, a precise distinction between discriminatory and nondiscriminatory relationships among rates for different classes and quantities of service; and secondly, a definite set of standards by which to draw the line between those practices of discrimination that are forbidden and those other practices that are approved or, at least, tolerated. Indeed, precisely this double-jump method of arriving at a decision on the merits of a complaint of unjust discrimination is revealed or suggested in some of the rate cases—notably, in cases upholding the challenged rates despite an overt or implied admission of their discriminatory character.[8]

By and large, however, the tribunals have shared a popular feeling that "discrimination," as a practice of rate making, is an odious term. They have therefore been reluctant to characterize as "discriminatory" any rates which they find lawful. Approval is more likely to be expressed by a finding of "no undue discrimination," without any commitment as to whether discrimination of even a due type was found to exist.

This tendency to treat "undue discrimination" as if it were one word has been encouraged by the failure of the rate regulators to observe any single and definitive distinction between discriminatory and nondiscriminatory rate differentials. At times, the cases suggest a distinction similar to that drawn by economists, in deeming "discriminatory" any rate differential not based on a cost differential. But at other times "discrimination" has been used as a mere synonym for *any* kind of rate differentiation;[9] whereas at still other times it has become a convenient, shorthand

[8] See Manufacturers Railway Co. v. United States, 246 U.S. 457, 481–482 (1918), letting stand a rate impliedly discriminatory. In the 1930s several electric utility companies sought commission approval of an "objective rate plan" initiated by the Commonwealth and Southern Corporation as a means of shifting to a lower level of rates without temporary loss of revenue pending the development of an enhanced demand for the service. Enjoyment of the lower rates was temporarily limited to consumers whose monthly bills under these rates would not be less than their bills for the previous year. Not all commissions approved. But even those that did admitted in words or by implication that the plan was discriminatory. William F. Kennedy, *The Objective Rate Plan* (New York, 1937).

[9] This appears to be its use under the Robinson-Patman Act despite some cryptic language suggesting a distinction by Representative Utterbach in the Congressional hearings on the bill. Thus, the Act treats a cost difference as justification for discrimination, not as proving the absence of discrimination. See the references cited in footnote 2.

term for *undue* discrimination—for what has been called "discrimination in a legal sense." The Interstate Commerce Commission, for example, has never to my knowledge declared that "value-of-the-service" rate making, which it has always sanctioned, constitutes discrimination.[10] Yet this practice is regarded as the very essence of discriminatory monopoly pricing by the academic economists, even by those economists who defend it as in the public interest.

In view of this confusion of language, something is to be said for the proposal to drop the word as a tool of public utility rate making or else to let it refer to *any* difference between rates of charge for physically similar types of service. But even if this proposal should win general acceptance, there would still be a need to understand the significance of the distinction that economists have attempted to draw, although with only partial success, between discriminatory and nondiscriminatory rate relationships. In the remainder of this chapter, we shall first discuss the nature of the distinction and shall then turn to the problem of setting practical limits between "due" and "undue" forms and degrees of discrimination.

ECONOMIC MEANING OF PRICE OR RATE DISCRIMINATION

In the literature of economics, one of the cardinal attributes of prices under assumed conditions of "perfect competition" is that of a uniform price for any one product at any given time and place. This uniformity precludes not only the price higgling that still characterizes many European markets but also the systematic practice of price differentiation designed to impose different charges on different groups of persons depending on differences in their capacity and willingness to pay. A purely competitive price is the same to the rich and the poor, to the powerful and the powerless, to the person who finds the product barely worth buying and to the person who would pay ten times the prevailing price rather than go without.

[10] For a possible exception, see the New York Harbor Case, 47 I.C.C. 643, 736 (1917). Utility representatives tend to deny that "value-of-service" pricing is "discriminatory." See, e.g., L. R. Nash, *The Economics of Public Utilities* (New York, 1931), p. 458. An opinion in a telephone rate case by the New Jersey Commission holds that *failure* to recognize relative "value-of-service" in exchange areas of different sizes would constitute (unjust) "discrimination." Re West Jersey Telephone Co., 91 PUR NS 33, 41 (1951).

This pressure, exercised on all sellers in a strictly competitive market, to sell any given product at a uniform price may be imposed more or less effectively even upon a monopolist unless he can divide the market for his product so as to prevent ready transfer among buyers. Otherwise, if the monopolist should attempt to discriminate, say, by selling lead pencils to general customers at 10¢ each while selling them to school children at only 5¢, there might arise a resale market beyond the seller's control, in which adults could buy pencils directly or indirectly from school children at less than 10¢.

But with many commodities and, particularly, with many personal services, transfer of the product among buyers is either impossible or inconvenient. Here the monopolist has the power, if he so chooses, to maintain substantial price differentials among different consumers or consumer groups. Not only does he possess this power; he will find it profitable to exercise it if, without too much expense and popular ill will, he can divide his customers into those who, at a uniform price, would have a relatively high elasticity of demand for the product, and into those who, at this same price, would have a relatively low elasticity. The profitability of the practice lies in the monopolist's ability to enjoy what would otherwise be the inconsistent advantage of doing a relatively large volume of business at a low unit rate of profits, and of doing a relatively small volume at a high rate. And public utilities, save when constrained by regulation or by public opinion, are in an especially favorable position to profit by this practice because of their monopoly status, their ability to prevent or restrict transfer of their services among consumers, and their tendency to operate at unit costs which decline as their rates of output increase.

If the term "price discrimination" were to be limited to the narrow meaning[11] given to it in the preceding paragraphs, it

[11] The outstanding modern analysis of price discrimination is by Mrs. Joan Robinson, *The Economics of Imperfect Competition* (London, 1948), Chaps. 15 and 16. She uses the term in the narrow sense, to refer to "the act of selling the same article, produced under a single control, at different prices to different buyers." P. 179. For the classical treatment of discrimination in its welfare aspects, see A. C. Pigou, *The Economics of Welfare*, 4th ed. (London, 1932), Chap. 17. Though supposedly impossible under *perfect* competition (by very definition of this latter term), price discrimination is a recognized practice under the competition of the actual market place. See, e.g., Eli W. Clemens, "Price Discrimination and the Multiple-Product

would apply only to the type of practice forbidden without qualification by the various public utility statutes and by Section 2 of the Interstate Commerce Act, namely, to the practice of charging different rates to different customers for substantially the same product. Hence, a broader definition is needed in order to cover "discriminatory" rate differentials for different types and quantities of service. But attempts by economists to supply this extended definition have run into difficulty and have never attained complete success.

Among economists there would probably be general agreement that the practice of exacting different charges for different classes of service rendered at the same marginal costs constitutes discrimination, and equally general agreement that failure to impose higher charges for services rendered at markedly higher marginal costs is also discriminatory. But if Service A (say, low-tension electric power) costs more to supply than Service B (say, high-tension power), just what price relationship between the two services can then qualify as nondiscriminatory? Here, two different answers have been suggested: the one that the rates must be proportional to marginal costs; the other that the differences in rates must be equal to the differences in these costs.

Of course, if all rates were to be set precisely at marginal costs, both of these suggested conditions of nondiscriminatory relationship would be met.[12] And this is the situation supposed to prevail under perfect competition, since prices in competitive equilibrium are equal both to average and to marginal costs. But particularly with public utilities, which must often charge rates

Firm," 19 *Review of Economic Studies* 1–11 (1950–1951), reprinted with alterations in Richard B. Heflebower and George Stocking, ed., *Readings in Industrial Organization and Public Policy* (New York, 1958). The history of railroad and public utility rate making would give strong support to the thesis that the condition most conducive to discrimination is not that of thoroughgoing monopoly but rather that of lopsided competition. See G. P. Watkins, *Electrical Rates* (New York, 1921), p. 199.

[12] However, there would remain the question whether the relevant marginal cost is of a short-run or a long-run variety. If one were to accept the price relationships of perfect competition as setting the norm of nondiscriminatory prices, the short-run cost would be the more nearly relevant, although this cost would itself tend to come into accord with long-run cost under conditions of competition. See pp. 331–332 of Chap. XVII. But in the regulation of public utility rates, the "practical" identification of nondiscriminatory rates with rates properly aligned with long-run marginal costs may be justified for reasons set forth in Chap. XVII and again to be discussed in Chap. XX.

substantially higher even than long-run marginal costs in order to cover their revenue requirements, the two answers just suggested are in conflict, and the choice between them presents a dilemma. Let us consider each choice in turn.

Discriminatory rates sometimes defined as rates not proportional to marginal cost. Under the proportionality definition, which has been favored by some economists,[13] the rates of a public utility company would be said to be nondiscriminatory if all of them were to exceed the marginal costs of supplying the respective services by some uniform percentage—say, by a 20 per cent minimum necessary to make total receipts cover the company's total revenue requirements. These hypothetical rates might then be used as points of departure by reference to which a rate maker might calculate whatever degrees and types of discrimination are deemed necessary or desirable on "value-of-the-service" principles of rate making.

As a practical, first-approximation standard of nondiscriminatory internal rate relationships, this standard is probably more widely applicable than any proposed alternative. Rates based thereon would go a considerable distance toward accomplishing the economic objectives of a "cost-price" system of rate making. For they would impose equal charges for services of equal costs, and higher charges for services of higher costs. And they would avoid the most significant *form* of discrimination, namely, value-of-service rate making.

Nevertheless, the proportionality definition has obvious deficiencies. For it would embrace some rate relationships that have the same distorting influences in affecting consumer choice among alternative services which economists associate with the practice of discrimination. The nature of these distortions will be noted in the following paragraphs.

[13] For example, by George J. Stigler, *The Theory of Price*, rev. ed. (New York, 1952), pp. 214–215; Tibor Scitovsky, *Welfare and Competition* (Chicago, 1951), p. 410; Ralph K. Davidson, *Price Discrimination in Selling Gas and Electricity* (Baltimore, 1955), p. 23 (but compare his different implied definition on p. 22); A. M. Henderson, "Prices and Profits in State Enterprise," 16 *Review of Economic Studies* 13–24, 23 (1948–1949). The proportionality definition is clearly the better choice with respect to the price relationship between services for which the demand is not interdependent. It is deficient when applied to complementary products (such as phonographs and phonograph records) and to substitute products (such as beef and mutton).

Alternative definition of discriminatory rates: rate differentials not equal to cost differentials. One of the major objectives of sound public utility and railroad rate-making policy is that of bringing rates for substitute services into proper relationship, so that consumers or shippers will not be led to make an economically "distortionate" choice between alternatives. This objective requires that equally costly services be sold at the same rates of charge. And it would also seem to require that when services are substituted for each other in fixed physical ratios (one coach trip versus one Pullman ride between New York and Chicago; one multi-party telephone versus one single-party telephone; one supply of off-peak electric power versus an equal supply of on-peak power; etc.), the differences in charges should be equal to the differences in costs. But as long as rates, while higher than marginal costs, are made proportional thereto, the price differentials will exceed the cost differences—an excess which may lead many consumers to make an "uneconomic" choice of the less costly alternative.

Recognition of this situation has led some writers to reject the proportionality definition of nondiscriminatory rates in favor of a cost-differential definition.[14] This point of view seems to have been accepted by the Cost Section of the staff of the Interstate Commerce Commission as a reason for its refusal to apportion railroad cost residues (excesses of total costs of railroad operation over estimated long-run out-of-pocket costs of specific classes of traffic) among the units of traffic (tons and ton-miles) in fixed percentages of the allocated out-of-pocket costs.[15]

[14] For example, Donald H. Wallace, "Joint and Overhead Cost and Railway Rate Policy," 48 *Quarterly Journal of Economics* 583–619 (1934); David G. Tyndall, "The Relative Merits of Average Cost Pricing, Marginal Cost Pricing, and Price Discrimination," 65 *Quarterly Journal of Economics* 342–372 at 343 (1951); Philip Locklin, *The Economics of Transportation*, 4th ed. (Homewood, Ill., 1954), p. 142.

[15] See the references to the Interstate Commerce Commission cost apportionments in footnote 4 of Chap. XVIII. In earlier days, before the *Class Rate Investigation*, Docket 28,300, the Cost Finding Section of the Commission's staff prorated constant costs in a fixed percentage of variable costs. But it now adds a uniform, pro rata allowance for "burden" (constant costs) to each ton (for terminal costs) and ton-mile (for line-haul costs). As a result, "it limits the differences in the fully distributed costs for any two shipments in the commodities hauled a given distance to the differences in the out-of-pocket costs, and it is the latter expenses which reflect those transportation conditions which have a bearing on costs." Samuel A. Towne, Chief, Cost Division, *I.C.C. Practitioners' Journal*, May, 1954, p. 702.

Unfortunately, however, the cost-difference standard of nondiscriminatory rate relationship can be given only a very limited application and is quite inapplicable to the general design of a rate structure. For the different services of any railroad or public utility company are not limited to substitutions in pairs or in any fixed physical ratios. Instead, they offer many different types of substitutions and at ratios that depend in part on relative rates of charge. This being the case, any attempt to make differences in rates for all classes and quantities of service equate with differences in costs between *some* alternatives (such as alternatives between more or less of the same kind of service) would frustrate attempts to accomplish the same objective with respect to *other* alternatives (such as alternatives between equal quantities of different grades of service). Moreover, such an attempt would throw an impossibly heavy burden of residue costs on the purchasers of the smaller quantities and on the less expensive grades of service.

Complete avoidance of discrimination is therefore impossible when rates in the aggregate are above marginal costs. Impressed with the difficulties just suggested, some of the more recent writers on price or rate theory have concluded that, as long as the prices charged by any given enterprise must exceed marginal costs, complete avoidance of discriminatory relationships among these prices is simply impossible.[16] And this is the view accepted in the present book. A certain amount of discrimination, of internal price distortion, would therefore be unavoidable even for an enterprise which would prefer to avoid it and which has neither the need nor the desire to practice value-of-service rate making as a means of maximizing profits.

In view of this somewhat frustrating situation, and not merely because some practices of discrimination are positively desirable, the law is indeed well advised to forbid only those discriminations which are "undue." This means that the attempt should be made to outlaw or minimize those price relationships which have a *serious* distortion effect on relative use of services, even at the expense of accepting other discriminatory price relationships which are less likely to have such an effect. In electric utility rate

[16] Joel Dean, *Managerial Economics* (New York, 1951), pp. 504–512; William Vickrey, "Some Implications of Marginal Cost Pricing for Public Utilities," 45 *American Economic Review, Proceedings* 605–620 (May, 1955).

making, for example, lower rates for industrial power than for residential service, even if not based on comparable cost differentials, will hardly lead consumers to reduce their residential consumption in favor of an increase in their industrial power load. The two services are noncompetitive, and there is no good reason to assume the existence of a high "cross-elasticity" of demand for them.[17] On the other hand, off-peak and on-peak electric services are, *to some extent,* alternative products. Hence, the relationship between on-peak and off-peak rates for similar classes of service may be a matter of special importance with respect to its substitution effects.

VALUE-OF-SERVICE RATE DISCRIMINATION

Most discussions of the merits and demerits of rate discrimination are not concerned primarily with those discriminatory relationships which are simply unavoidable under any full-cost system of rate making. Instead, they are concerned with that deliberate policy of rate differentiation known as "value-of-service" rate making or, in railroad parlance, as the practice of charging "what the traffic will bear." Here, the apportionment of the burden of covering any excess in total costs over marginal costs (or "out-of-pocket" costs) is deliberately "biased" so as to impose on the services in relatively inelastic demand (services of high "value" in popular speech) higher surcharges, over and above marginal costs, than would be imposed by *any* rule of apportionment based exclusively on cost relationships.

The most familiar use of the value-of-service principle is found in railroad rate making, where higher rates are charged for the shipment of goods of higher value, the rate differentials being far greater than those which could be justified by any cost differentials.[18] Among the nontransport utilities, this principle is perhaps

[17] "Cross-elasticity of demand" refers to the tendency of an increase (or decrease) in the price of any given product to result in an increase (or decrease) in the demand for a substitute product. Thus, an increase in the price of gas may lead to an increase in the demand for electricity, absent any change in the price of the latter product.

[18] See the study on *Value of Service* dated Nov., 1959, prepared by the staff of the Interstate Commerce Commission. The readiness of the Commission to accept the value of the shipped commodity as an index of the value of the transportation service has been a subject of much comment and of some speculation as to *rationale*. Certainly the correlation between the two "values" must be poor. It seems likely that the Commission has been aware of this poor correlation but has tended to support a commodity-value standard of rate classification on its own

most openly avowed by the telephone companies, which adduce it as the major defense of their practice of charging for local service higher rates in larger cities than in smaller communities.[19] But electric and gas rate making is also admittedly under the influence of the principle, although the lack of any really satisfactory differential-cost analysis in these industries precludes quantitative statements as to the force of this influence in the determination of class rates and block rates.

From the standpoint of a corporate monopoly with the assumed objective of profit maximizing, the case for this type of discrimination is obvious, although the most profitable extent of the practice is by no means easily determined. Even from the standpoint of the public interest, the argument in its favor is clear enough under the assumption that some discrimination is utterly essential in order to meet revenue requirements.

But the more interesting and important question concerns the desirability of value-of-service rate making even for an enterprise which could cover its total costs of production, including a capital-attracting rate of return, by basing its rates entirely on cost differentials. Here the advantage must lie in the resort to discrimination as a means of encouraging the maximum economic use of a company's services consistent with the so-called "full-cost" requirement. For this purpose, the rate structure imposes the higher surcharges, over and above marginal costs, on classes and quantities of service for which demand is relatively unresponsive to price changes.

Ideally, this type of discrimination might permit a company to cover its total costs without repressing the demand for any service for which consumers would be ready to pay at least marginal costs—to pay what rate makers have called "compensatory" rates. In practice, this ideal is unattainable; for the only feasible types of discrimination are crude types, which must necessarily

---

merits and not merely as a presumptive measure of service value. But the recent competition of motor, water, and air transport has hit this standard hard, and its days as a major basis of railroad rate structure may be numbered. In an unpublished paper, my colleague Professor Joel Dean suggests that the imposition of higher charges for more valuable shipments (over and above insurance-cost differentials) is particularly bad for the railroads since the trucks, being faster on delivery, have a special quality-of-service advantage in the transportation of high-valued commodities.

[19] See pp. 83–84, *supra*.

impose relatively high charges for classes and quantities of service, for *portions* of which the demand is highly elastic.[20] But discrimination need not be "ideal" in this respect in order to be less repressive of economical demand than would a rate structure based entirely on cost relationships.

Value-of-service rate discrimination versus joint-product pricing. Before turning to the question what restrictions should be imposed upon a utility company's freedom to practice rate discrimination of a value-of-service character, we must first emphasize the distinction between this practice and the practice of joint-product pricing. This distinction was drawn in the previous chapter with special reference to railroad rate making. But it is so often ignored, despite its high importance for rate theory, that a further reference to it is in order.

Unlike price discrimination, joint-product price differentiation is a phenomenon characteristic of the strictest forms of competition. In one respect the two practices are akin, in that the price relationships of the joint products—cotton fiber and cotton seed, hides and meat, superior cuts of meat and inferior cuts, etc. —depend on demand relationships and are not determined solely by cost relationships.[21] But they differ in that the joint-product relationships are designed to "clear the markets" for the joint products in view of largely fixed ratios of feasible output. Thus, the different charges for the different cuts of meat will adjust relative demands so that no meat is wasted, not even the inferior cuts.

On the other hand, value-of-service rate discrimination, instead of adapting relative rates to unavoidable relative scarcities of the different types of service, curtails the output of the service in less elastic demand without bringing this output into an optimum balance with the available output of the service in more elastic demand. This procedure may be justified for reasons already suggested. But the justification must lie in the choice of the lesser evil—in the view that the resulting repression in the demand for

[20] In Pigou's terms, this amounts to the statement that the only feasible discrimination is discrimination of the "third degree." *The Economics of Welfare*, Chap. 17. Perhaps the closest practical approach to completely nonrepressive forms of discrimination in public utility rates may be found in skillfully designed block rate making, which allow lower rates to a given consumer for successive blocks of service. See Davidson, *Price Discrimination in Selling Gas and Electricity*, Chap. 2.

[21] For a qualification, see footnote 18 of Chap. XVIII.

the service discriminated against, while unfortunate, will be less unfortunate than the repression in the demands for one or both services that would result if both were priced above marginal cost but without reference to "what the traffic will bear."

Thus, if railroad traffic is not evenly balanced between forward and backward hauls, the principle of joint-product pricing requires the imposition of differential charges in favor of the lighter traffic. But this differential is not discriminatory at all; on the contrary, uniform pricing would be discriminatory even though commissions and courts have not often so regarded it.[22] Or again, if an electric plant is operating near its full capacity, the imposition of higher charges for on-peak than for off-peak service would actually be required in order to avoid discrimination.

I stress this point in the present chapter because many of the practices of rate differentiation which are called discriminatory in popular parlance are nondiscriminatory in an economic sense. By the same token, many of the rate uniformities widely accepted in current public utility regulation, such as the use of "blanket" electric rates throughout large areas within which costs of service differ markedly, are discriminatory from the viewpoint of an economist. Again we come back to the point that the law, so far from completely forbidding, actually approves or even requires discrimination of many types and degrees.[23]

THE PROBLEM OF DISTINGUISHING BETWEEN DESIRABLE AND UNDESIRABLE RATE DISCRIMINATION

Even if the sole object of public utility rate making were assumed to be that of maximizing the profits of a given company,

[22] On the tendency of the Interstate Commerce Commission to forbid or restrict the quotation of low, return-haul rates, see Marvin L. Fair and Ernest W. Williams, *The Economics of Transportation*, rev. ed. (New York, 1959), pp. 600–601. At times the Commission has even required that the rates be lower in the direction of the greater volume of traffic. I. L. Sharfman, *The Interstate Commerce Commission*, Vol. III B (New York, 1936), pp. 570–571.

[23] Compare the statement by M. A. Adelman that the Robinson-Patman Act, although called an antidiscrimination law, consistently enforces discrimination in that it declines to go beyond severe limits in validating price differences justified by cost differences. For it (a) expressly forbids discounts to buyers who avoid brokerage costs, (b) permits the Federal Trade Commission to set maximum quantities for discounts regardless of savings in cost, and (c) imposes what Adelman regards as a hopeless burden of proof of cost differentials. "The Consistency of the Robinson-Patman Act," 6 *Stanford Law Review* 1–22 (1953).

the question how far it would pay this company to "bias" its relative rates of charge by rate differentiation of a value-of-service character would be a question of great practical difficulty. To be sure writers on the theory of monopoly pricing have offered formal answers based on simplified assumptions. But their direct application would require a knowledge of demand and cost functions which rate makers simply do not and cannot possess.[24]

Even more formidable are the problems raised by the question to what extent, and subject to what restrictions, the practice of this kind of discrimination is in the public interest. Here too, theoretical answers have been offered by recent economists.[25] They take the form of a statement of conditions for "optimum rates subject to a budgetary constraint"—subject, for example, to the requirement that the rates of any one company must yield total revenues equal to total operating expenses plus a "fair return" on capital investment. But just as with the problem of price determination for the purpose of securing maximum profits, so with this distinct though related problem, their application would call for unattainable information about elasticities of demand and supply. Moreover, they depend for their validity on the acceptance of highly artificial standards of social welfare. At the present stage of their formulation, they can hardly serve as practical guides to the rate maker or rate regulator in his attempt to draw

[24] The problem of estimating price elasticity of demand is especially intractable. Professor Ernest Williams writes: "Unfortunately, little is known in a quantitative way about the elasticity of demand for the railroad freight service. Since it is a derived demand its analysis is fraught with the utmost difficulty." "Railroad Rate Levels and Earning Power in an Era of Competitive Transport," 25 *Land Economics* 405–413 at 409 (1949). On the elasticity of demand for electricity, see D. J. Bolton, *Electrical Engineering Economics*, Vol. 2: *Costs and Tariffs in Electricity Supply*, 2d ed. (London, 1951), Chap. 1; Hubert F. Havlik, *Service Charges in Gas and Electric Rates* (New York, 1938), Appendix C, "The Relationship between Rates and Consumption"; Samuel Ferguson, "Do Rate Reductions 'Per Se' Bring Increased Use?", *Public Utility Papers* (Hartford, Conn., 1947), II, 757–778; H. S. Houthakker, "Electricity Tariffs in Theory and Practice," 61 *Economic Journal* 1–25 (1951).

[25] Marcus Fleming, "Optimal Production with Fixed Profits," 20 N.S. *Economica* 215–236 (Aug., 1953); Marcel Boiteux, "Sur la gestion des mônopoles publics astreints à l'équilibre budgétaire," 24 *Econometrica* 22–40 (1956). For a closely related problem in taxation, see F. P. A. Ramsey, "A Contribution to the Theory of Taxation," 37 *Economic Journal* 47–61 (1927). The general conclusion, based on simplified assumptions including assumed independent demands for the various services, is that rates should be proportional to the demand elasticities of the various services at these rates. Applied to utility rates, Ramsey's article implies that the surcharges above marginal costs should diminish in the same proportion as the demand for each service. I take it that these two propositions are essentially the same.

lines between "due" and "undue" forms and degrees of price discrimination.

How, then, should these lines be drawn in practice? Much of the entire literature on railroad rates might be viewed as a discussion of this extremely complex question.[26] But all that will be attempted here is a suggestion of a number of partial, tentative answers applicable even to the nontransport utilities.

First, the legitimate role of the value-of-the-service principle is a role limited in its application to a public utility company which must charge rates in excess of marginal costs in order to meet revenue requirements. Hence, this principle has its most clear-cut defense for companies with chronically redundant plants, as is the case with most American railroads.[27]

Secondly, as a wise, practical rule, value-of-service rate differentials should not often be permitted unless they can be expected to result in lower rates even for those consumers who are discriminated against. The possibility of such a happy result was illustrated in the concluding section of Chapter XVI. It arises when special, low rates are granted for types or quantities of service which could not otherwise be attracted but which will make some contribution to total revenue requirements over and above mere incremental costs. True, utility rate theory offers no positive proof that discrimination should stop at the point just suggested. Conceivably, the public interest might be served by a decrease in the rates for services in highly elastic demand even at the expense of necessary increases in the rates for services in inelastic demand. But considerations of income-distributive fairness would be likely to weigh against any such arrangement.

Thirdly, permission to discriminate in favor of some uses of service that could not otherwise be attracted should seldom be granted in the absence of good evidence that the favored rates will

[26] For a review of the case law on unjust discrimination, see Isaac B. Lake, *Discrimination by Railroads and Other Public Utilities* (Raleigh, N.C., 1947).

[27] The principle also applies to companies in a position to enjoy long-run economies of scale. Telephone companies are not supposed to be in this position, at least not if the unit of output is taken to be a telephone station rather than a telephone message. This fact has led some writers to question the validity of the value-of-service principle as applied to telephony. But the telephone industry rests its defense on what amounts to a "social" theory of reasonable rates—a theory based on the importance to the community of individual access to the telephone. See pp. 114–115, *supra*.

cover, not just those short-run incremental costs due to a temporary redundancy of plant capacity, but rather those "long-run" incremental costs (including incremental capital costs) which can be expected to persist for the indefinite future. Otherwise, there arises the danger, noted in Chapter XVII, that the favored consumers will secure a kind of vested interest in the maintenance of their preferential rate relationships even after the economic excuse for this preference has ceased to be valid.[28]

Fourthly, even if a utility company, by the skillful practice of discrimination, could and would thereby reduce *all* of its rates, including the rates which it charges to those consumers who are discriminated against, the practice should nevertheless be forbidden if it would seriously prejudice the competitive business relationships between these consumers and those other consumers who would receive a preference. This maxim, which is based on a recognition of the high importance of rules of "fair competition," applies with special force to transportation rates. But it also applies to nontransport utility rate making with respect to relative charges to competitive businesses for larger and smaller amounts of electric, gas, telephone, and water-supply services.

At least in rate regulation under the Interstate Commerce Act, a finding of unlawful discrimination among rates is said to be contingent on a finding of competitive injury, actual or potential, except for the absolute statutory taboo of Section 2 against charges of different rates to different persons for a "like and contemporaneous service in the transportation of a like kind of traffic under substantially similar circumstances and conditions." [29] I am not sure whether a similar statement would apply to the laws restricting the practice of discrimination by the utilities. But no

[28] On vested-interest claims against rate increases, see the Interstate Commerce Commission staff report on *Value of Service*, Nov., 1959; Antioch Milling Co. v. Public Service Company of Northern Illinois, 4 Ill., 2d 200, 123 N.E. 2d 301 (1954), citing other cases; Southern Pacific Co. v. Interstate Commerce Commission, 219 U.S. 433 (1911). It is my distinct impression that vested-interest arguments have carried much more weight with commissions than with the appellate courts.

[29] And except for long- and short-haul discrimination. See I. L. Sharfman, *The Interstate Commerce Commission*, Vol. III B (New York, 1939), p. 529. But Sharfman notes that on occasion, the Commission has suggested that a rate may be "unduly preferential" merely because it is so low as to throw an unjust burden on the entire complex of other traffic. And see New York v. United States, 331 U.S. 284, 308 (1947) for an indication that the assertion cited in the text should at least be qualified.

one can doubt that the competitive-injury injustice of "unjust discrimination" is the injury that receives major emphasis by courts and commissions.

In the railroad rate cases, which offer the bulk of the cases on undue discrimination under public utility law, the decisions concerned with fair versus unfair competitive relationships among rates for different types of traffic and for different points of origin and destination seem confusing and inconsistent. A discussion of these decisions is a major task of the treatises on transportation rates. Here let me note merely that the Interstate Commerce Commission has only occasionally permitted railroads to quote lower rates for freight shipments in trainload lots than for single carload shipments despite substantial economies in the former type of traffic.[30] This means that the Commission, probably desiring to make some limited use of railroad rates as a positive device by which to discourage the growth of dangerously large-scale business, has actually made mandatory a certain degree of discrimination in favor of the smaller shipper. The wisdom or lack of wisdom of this rule, which is similar to a well-known provision of the Robinson-Patman Act,[31] is a question beyond the realm of utility rate theory.

Enough has been said in this very elementary chapter on the character, the advantages, and the abuses of rate discrimination to suggest reasons why the law of rate making, no less than the literature on utility rate theory, recognizes alike the justification for the practice and the need for its strict limitation. To the economist, this practice is viewed as a means of minimizing the harmful effect that would otherwise result from the acceptance of a so-called full-cost standard of rate making. But the harm can only be reduced thereby, not completely avoided. In the next and final chapter, we shall therefore turn to a proposed alternative system of rate making designed in principle to outlaw, and not merely to restrict, discrimination.

[30] See American Bound Bale Press Co. v. A.T. & S.F. Ry. Co., 32 I.C.C. 458, 463–464 (1914); Sharfman, *The Interstate Commerce Commission*, Vol. III B, p. 514.

[31] The provision authorizing the Federal Trade Commission to place limits on quantity sales entitled to further discounts by virtue of cost savings.

XX

THE PHILOSOPHY OF MARGINAL-COST PRICING

With only minor exaggeration, this entire book may be viewed as an attempt to play variations on a main theme first expressly set forth in Chapter III, "The Role of Public Utility Rates." This theme runs to the effect that utility rates, like other prices, are designed to perform multiple functions as instruments of economic control. To a high degree, these functions can be performed in harmony; necessarily so, indeed, since they are partly complementary. But the harmony is far from complete, for the most efficient performance of any one function would require the acceptance of a system of rates not also best designed to perform any of the others. In consequence, one of the most frustrating problems of rate theory and of practical rate making is that of suggesting and applying principles of workable compromise.

Among these conflicts of rate-making objectives, one of the most serious is that between the usually accepted principle that the rates of any public utility, in the aggregate, should cover its total costs of service, including fixed charges or a "fair return," and the also widely approved principle that specific rates should be based on the costs of specific amounts and types of service. For reasons stated in Chapter XVI, these two goals of rate-making policy are incompatible except under a somewhat rare coincidence in corporate operation and finance. In the absence of such a coincidence, any attempt to attain them both completely would be as hopeless as would be an attempt to draw a square circle.

How has rate-making practice undertaken to face this dilemma? In the main, except when circumstances have made the gap between the two objectives too wide to jump, it has done so by a somewhat qualified grant of priority to the first goal—that of a level of rates adequate for financial self-sufficiency. Even here, cost-of-service criteria of specific rates and rate relationships have received "due consideration." But these criteria have been relegated to a subordinate position, yielding precedence to the total-cost standard partly through overt deviations of a value-of-service nature, partly through loose "interpretations" of cost of service in terms of ill-defined and arbitrary "fully distributed" costs rather than in terms of differential or incremental or marginal costs.

Accepting, at least for the purpose of discussion, the traditional priority of the full-cost-coverage goal of rate-making policy, the four preceding chapters have considered methods of rate differentiation designed to permit the attainment of this goal with the least harmful departures from the principle of specific cost pricing. These methods include the limited but substantial resort to price discrimination. But while, under most conditions, their skillful application may bring about a reasonable degree of harmony between the two above-noted criteria of a sound rate structure, the results are bound to be far from ideal. Indeed, in some situations, including those now faced by the American railroads and, possibly, by the natural-gas industry, the results may become almost intolerable.

Impressed with both the theoretical and practical difficulties of any attempt to secure a sound structure of individual rates subject to the constraint that rates as a whole must cover costs as a whole, one important group of modern economists has proposed to seek riddance from this constraint. Let all rates be set at marginal costs. But if the resulting revenues should fall short of meeting total financial requirements, let the deficiency be made good by a tax-financed subsidy. On the other hand, if the total revenues should prove excessive, as they may well do if current unit costs of additional plant are far in excess of the unit costs of the existing plant, let at least a large part of the excess be recaptured by the community or by the nation through special taxation or else through outright public ownership.

What these economists here propose is a narrowing of the role of public utility rates as instruments of economic control. Relieved of any obligation to supply whatever funds may be required in order to maintain the credit of the enterprise as a going concern, rates can now be designed as single-purpose, precision instruments by which to control the demands of consumers for services of different kinds and in different amounts.[1] Even here, to be sure, rates will play an ancillary role in helping to finance the enterprise. But they need no longer be tailored with this subsidiary function in mind, since their deficiencies in this respect can be made good by another powerful instrument of economic control—that of taxation.

We may restate this proposed limitation of the role of public utility rates by saying that these rates should be called upon to perform only those functions which they will perform when designed as if their sole purpose were to control the effective demand for the services. But an even more restricted role is contemplated by those more thoroughgoing marginalists whose view will be set forth in the next section. For, in their concern to use utility rates as devices by which to secure the optimum utilization of an *existing* utility plant, however excessive or deficient may be its present capacity in relation to the potential demand for its services, they would apply a *short-run* measure of marginal costs—a measure which may seriously detract from the usefulness of rates in the determination of the demand for and supply of utility services over extended periods of time.

Before turning to the basic philosophy of marginal-cost pricing, we may comment briefly on its recent history.[2] The underlying idea is by no means new and is derived from the well-known principles of competitive-price determination developed in England by Alfred Marshall and other "neoclassical" economists of the late nineteenth century.[3] Under perfect competition as de-

[1] A French authority on marginal-cost pricing writes: "Vu sous ce jour, le prix n'a pas pour objet de rémunérer les dépenses du producteur au nom d'un quelconque principe d'équité, mais de motiver les décisions à venir des consommateurs de telle manière que celles-ci ne causent pas dans l'économie de dépenses irrationelles." M. Boiteux, "La Tarification des demandes en pointe: Application de la théorie de la vente au coût marginal," 58 *Revue générale d'électricité* 321–339 at 323 (1949).

[2] This history down to 1950 is well presented by Nancy Ruggles, "The Welfare Basis of the Marginal Cost Pricing Principle," 17 *Review of Economic Studies* 29–46 (1949–1950); "Recent Developments in the Theory of Marginal Cost Pricing," *ibid* 107–126.

[3] But Hotelling, in an article cited in footnote 6 below, gives to a French engineer, Jules Dupuis, the primary credit for developing the principle in its application to

fined by these economists, the prices of all products "tend" to come into equality with their average unit costs of production, in that such an equality is one of the conditions of static equilibrium. But these prices also "tend" to equal marginal costs, both of short-run and of long-run varieties. There is no inconsistency among these multiple conditions of equilibrium as long as the products in question are produced at unit costs that either stay constant despite a change in the rates of output, or else increase with increases in output.[4] But if any product should be produced under a condition of *decreasing* unit costs, the maintenance of perfect competition is impossible.

Public utilities belong to a group of industries which are supposed to operate under the latter condition, at least in the typical case—a fact often cited as accounting for the need for their regulation as a substitute for competition. But the very condition which rules out actual competition also rules out any attempt to secure by regulation *all* of the good attributes of competitive prices including those of an equilibrium position in which prices are simultaneously equal to a whole variety of costs including average total costs, short-run marginal costs, and long-run marginal costs.

As already indicated, public opinion, insofar as it has been made aware of the very existence of the dilemma between a full-cost principle of competitive pricing and a marginal-cost principle, has tended to accept the former principle and to reject the latter. Indeed, until well into the twentieth century, the wisdom of this choice was not seriously challenged, head on, by the professional economists, most of whom accepted for public utility services, as for commodities in general, the traditional precept,

---

public works: "De la mesure de l'utilité des travaux publics," *Annales des ponts et chaussées*, 2nd Series, Vol. 8, 1884. An English translation appears in No. 2 *International Economic Papers* 83–110 (1952).

[4] One might suppose that, if a firm is producing under conditions of increasing unit cost, marginal costs will *exceed* average costs. And so they may if the average costs are an average of total costs defined and measured as a regulating commission would do for purposes of determining corporate revenue requirements under a fair-return standard of rate making. But under the theory of a competitive price, the total costs of production are defined as the costs to an enterpriser who must buy or rent all of his factors of production, including land sites, water rights, rights to use the public streets, etc., in a current, competitive market. In consequence, any excess earnings which might otherwise go to the enterpriser because of an ability to sell his product at a marginal cost in excess of average cost are obligingly gobbled up by landlords and by other recipients of rents or quasi rents.

"Let the beneficiaries bear the burden." Objections to this standard of financial self-sufficiency, while not completely ignored, were met with proposals for "practical" solutions, such as that for a type of rate base which hopefully would bring the average total costs of public utility services within a tolerable range of long-run marginal costs; [5] or such as that for the adroit use of rate discrimination designed to bring down toward marginal costs the charges for those classes and quantities of service for which the demand is *unusually* responsive to changes in price.

In 1937, however, this cautious—some writers would even say timid—attitude of the academic economists was sharply challenged by Professor Harold Hotelling, then at Columbia University, in one of the most brilliantly written articles in the history of utility rate theory.[6] In a sense, the economic philosophy of this article was orthodox in that it accepted a competitive-price rather than a "social-value" standard of rate making. But in another sense it was heretical, in that it regarded the marginal-cost attribute of a competitive price as more important for the purpose at hand than the average-cost attribute. This unconventional preference was based largely on the belief that the price system is almost uniquely qualified to perform the function of resource allocation or consumer rationing with maximum efficiency, whereas, in a regulated monopoly operating under conditions of decreasing cost with increasing output, it is not well qualified to perform the other functions of a competitive price (those other functions set forth in Chapter III, on the role of public utility rates) except in an auxiliary manner.

Both in this country and in Europe, Hotelling's thesis soon attained fame among the academic economists and led to a substantial flow of monographs and articles, some supporting his position and some defending the traditional principles of "full-cost" pricing.[7] The number of his *unqualified* supporters has been

[5] See p. 230, n. 7, *supra*, referring to an early debate between Professor Harry Gunnison Brown and myself on the relative merits of an original-cost or a reproduction-cost rate base.

[6] "The General Welfare in Relation to Problems of Taxation and of Railway and Utility Rates," 6 *Econometrica* 242–269 (1938). See also his later note, replying to Professor Ragnar Frisch, on "The Relation of Prices to Marginal Costs in an Optimum System," 7 *ibid.* 151–160 (1939). Hotelling had presented his main thesis at a meeting of the Econometric Society in December, 1933, though I believe not with his newer mathematical proof of its validity.

[7] See the articles and books noted by Nancy Ruggles in her previously cited re-

very small. But this fact by no means belies the significance of his contribution to the theory of public utility rates. For he has substantially influenced the development of this theory, even on the part of those writers who still insist that "rates as a whole should cover costs as a whole."

It is a most disturbing commentary on the lack of communication in America between writers on the economic theory of public utility rates and persons engaged in the actual practice of rate making or rate regulation that few of these latter persons are familiar with, or interested in, the philosophy of marginal-cost pricing. Partly in the hope of bringing this philosophy to the attention of the practitioners, I close this book with the following elementary exposition and appraisal.

SHORT-RUN MARGINAL-COST RATE MAKING

For reasons already stated in Chapter XVII, the significance of the distinction between marginal-cost and average-cost pricing is far greater when the former alternative is taken to mean pricing based on *short-run* marginal cost than when it is taken to mean pricing based on a persistent or chronic or *long-run* marginal cost. While the broad distinction is significant even under the latter interpretation, it is revolutionary only under the former. We shall therefore first consider the case for marginal-cost pricing in its former, uncompromising sense—the sense accepted by Hotelling. A technical treatment of its rationale would be an elaborate procedure, involving mathematical analysis based on a host of simplifying assumptions. But a general idea of the argument in its favor is easily presented—all the more so here since it is implicit in the more familiar acceptance of "out-of-pocket cost," the popular version of marginal cost, as a measure of *minimum* rates.

By way of illustration, let us borrow Hotelling's example of a very simple type of public utility plant, that of a toll bridge for motor vehicles. The bridge, let us say, is owned and operated by a public authority, which has financed its construction by the sale

view (footnote 2) of the literature. Compare, e.g., the defense of full-cost pricing by Professor R. H. Coase with the qualified support of marginalism by Professor William S. Vickrey: Coase, "The Marginal Cost Controversy," 13 N.S. *Economica* 169 (1946); Vickrey, "Some Objections to Marginal-Cost Pricing," 56 *Journal of Political Economy* 218 (1948). For a significant book on the subject by a socialist economist, see Burnham P. Beckwith, *Marginal-Cost, Price-Output Control* (New York, 1955).

of revenue bonds. In line with the traditional principle that public utility rates must cover total costs, the toll must be set at the minimum deemed necessary to pay the annual maintenance costs of the bridge plus fixed charges, which are here composed of interest charges plus amortization. Ignoring for simplicity the limited possibilities of differential tolls, let us assume that the required toll is $1 per crossing. This toll reflects the average total cost of the service.

Now let us note the economic harm done by this orthodox attempt to base utility rates on the financial principle that "every tub should stand on its own bottom." Assume, first, that the bridge has a capacity at least sufficient to accommodate all traffic forthcoming on a toll-free basis. On this assumption, the levy of the $1 toll is economically unsound, since its deterrent effect will prevent the bridge from being put to its fullest feasible use. Indeed, on the only slightly inaccurate assumption that even the maintenance costs of the bridge will not be affected by the amount of use, the imposition of any toll whatever is unsound.[8] All of the costs being "sunk costs," they should have been paid for, directly or indirectly, entirely by taxation, just as are city streets and sidewalks.

But what about the situation that will probably arise, sooner or later, when the bridge traffic, especially if stimulated by freedom from tolls, threatens to become too congested for safe handling by the present bridge? When this time arrives, a toll should be im-

[8] This statement requires qualification since it makes the simplifying assumption that, up to some definite point of "serious traffic congestion," the quality of the bridge-crossing service remains unaffected by the traffic. In fact, however, this quality, as measured by speed, comfort, and safety, will deteriorate gradually with increases in traffic, since each additional driver will get in the way of other drivers. As long as the bridge remains toll free, every user therefore imposes a cost upon other users which he is not required to defray. This means that the marginal *social* cost of bridge service then exceeds the cost imposed upon any one user. In theory, there should be a toll sufficient to deter use of the bridge by anyone to whom the value of the service is not sufficient to warrant the extra traffic congestion which he imposes on other users. See a much cited discussion of this point by Professor Frank W. Knight, taking issue with a position previously held by Professor Pigou: "Some Fallacies in the Interpretation of Social Cost," 38 *Quarterly Journal of Economics* 582–606 (1924), reprinted in Knight, *The Ethics of Competition* (New York, 1936). The subject is very important in the field of highway-transportation economics and, as will be noted later, has an important bearing on rate-making policies for local transit. See Martin Beckmann, C. B. McGuire, and Christopher B. Winsten, *Studies in the Economics of Transportation,* a series of studies for the Cowles Commission (New Haven, Conn., 1956). Telephony presents a situation reverse from that here mentioned in that the addition of another telephone subscriber may *add* to the "value" of the service enjoyed by the other users.

posed. But the amount of the toll should be in no way determined or limited by the authority's financial needs. Instead, it should be made high enough to preclude serious traffic congestion by limiting the use of the bridge to those persons who are ready to pay the potential "market-clearing" price for this use. In an extreme situation in which an unexpected or unprovided-for growth in potential demand for river crossings makes the present capacity of the bridge grossly inadequate, a toll as high as, say, $10 might well be required. This toll should be removed completely if and when a later enlargement or duplication of the bridge again makes the capacity redundant.

But would not the imposition of this market-clearing toll violate the very principle of marginal-cost pricing which Hotelling has advanced as a general substitute for pricing based on average total cost? The answer is no, although the definition of "marginal cost" must here be extended so as to give it a relevant meaning without sacrifice of its economic significance. For, if the bridge traffic has now reached its uppermost limit, the marginal social cost of permitting any one vehicle to use the bridge is measured by the value of the opportunity of use that must be denied to the highest excluded bidder. This bidder is the potential user who is barely deterred by the obstacle of the $10 toll. In other words, the relevant marginal cost, instead of being a production cost, is here an exclusion cost as measured by the marginal value of the service.

Later writers have elaborated on Hotelling's bridge illustration by noting the "economic distortion" that may result from the imposition of different tolls on different bridges or highways designed to meet separate tests of financial self-sufficiency, project by project.[9] Where optional routes are available for trucks and passenger cars, the resulting mixture of high-toll, low-toll, and no-toll routes is almost sure to lead to serious economic wastes, because it motivates the road users to base their choices on relative money costs that do not reflect relative social costs.

But the toll-bridge illustration is merely a simple example of the asserted advantages of marginal-cost pricing over full-cost pricing applicable to all public utilities—applicable, in short, to a

[9] Particularly William S. Vickrey, "Some Implications of Marginal Cost Pricing for Public Utilities," 45 *American Economic Review, Proceedings* 605–620 (1955). Professor Vickrey has become the leading American authority on marginal-cost pricing in its application to public utilities.

vitally important group of noncompetitive industries with respect to which the gap between the two types of pricing is especially wide and especially menacing. To be sure, marginal costs even of a short-run variety are less likely to be merely trivial for these other utilities than for toll bridges. Moreover, opportunities for rate discrimination as a means of full-cost recovery are likely to be much better. But the general principle still applies. And, as to the use of discrimination as a device by which to jump the gap between average-cost and marginal-cost standards, Hotelling cites some unhappy consequences of the attempts by railroads to make these jumps as failing to justify any complacency toward this device for the attainment of essentially inconsistent advantages.

In recent years, many railroad properties, including rights of way, tracks, and passenger terminals, have become redundant as a result of the growth of competing forms of transportation, combined with technological progress in signalling, etc., with respect to main line hauls. This redundancy, moreover, may last for a long time—some of it as long as the structures remain standing. Meanwhile, attempts to put the existing properties to their best available use are seriously handicapped by the largely hopeless attempts of the railroads to secure a "fair return" thereon and by the pressure upon the Interstate Commerce Commission to sanction rates designed to yield such a return.[10] Professor Hotelling's proposal would meet this situation by a general reduction of railroad rates to short-run marginal costs. This reduction would apply to entire railroad rate levels instead of being confined to a discriminatory and distortionate reduction of those particular rates that must be reduced in order to meet the immediate and direct competition of the road and water carriers. Any resulting fair claims by railroad investors for restitution for this retroactive change in rate-making policy would be cared for, presumably, by a government-financed indemnity.

[10] The seriousness of the handicap can be appreciated when one recognizes that, if a railroad or utility company were entitled to annual changes in rate levels designed to yield a stable "fair rate of return" on the net investment in, or so-called value of, its property, year after year, it would need to raise rates during years of depression and lower them during years of prosperity—a viciously cyclical procedure. Indeed, the Interstate Commerce Commission actually gave the railroads permission to make emergency rate-level increases in the depression period of the 1930s. Marvin L. Fair and Ernest W. Williams, Jr., *Economics of Transportation*, rev. ed. (New York, 1959), p. 576.

CRITIQUE OF PROPOSAL TO FIX RATES AT SHORT-RUN MARGINAL COSTS

Reserving for a later section a discussion of the much milder proposal to base rates on marginal costs of a long-run character, let us now consider critically the merits of the far more drastic proposal to base rates on short-run marginal costs.[11] Already some of the more serious objections have been noted in Chapter XVII, which discusses the relative merits of the two major types of marginal costs as measures of *minimum* rates.

First, let us recall that, with most public utilities, the really significant choice is not a simple choice between *marginal* cost and *average* cost as the basis of rate making. To be sure, the assumption that the rate maker faces this dire dilemma is not too far from reality in the toll-bridge example, since here the practical opportunities for rate differentiation are severely limited. Hence the bridge example presents an unusually forcible case for the adoption of marginal-cost pricing or, at least, for the abandonment of any attempt to make each particular bridge rest on its own financial foundations. But with railroads and most other utilities, there exists a wide variety of plausible rate structures, including those which resort to multi-part rate making, block rate making, and various forms of discriminatory pricing. Most of the rate structures now in effect are subject to material improvement with advances in the technique of rate design but without abandoning the total-cost principle. While none of them can be expected to have *all* of the consumer-rationing advantages of unqualified marginal-cost pricing, neither can they be assumed to result in economic losses of the order of magnitude of those suggested by an attempt to make a particular toll bridge financially self-sufficient through a uniform charge of so many cents or dollars per vehicle per crossing. Unfortunately, however, the measures of the relative gains and losses of marginal-cost pricing versus any given type of discriminatory, full-

[11] For one of the most well-balanced critical appraisals of marginal-cost pricing, both of the short-run and the long-run varieties, see Robert W. Harbeson, "A Critique of Marginal Cost Pricing," 31 *Land Economics* 54–74 (1955). Harbeson comments on one criticism not discussed in this brief chapter: that the supporters of marginal-cost pricing for regulated monopolies ignore the supposed failure of unregulated prices to come into accord with marginal costs under the most widely prevailing types of competition, namely, "imperfect" or "monopolistic" competition.

cost pricing that are suggested by economic theory are impossible to apply in terms of present factual knowledge.

Secondly, we must consider whether or not the almost undeniably superior efficiency of short-run marginal-cost pricing as a means of securing the optimum utilization of a plant of temporarily redundant capacity [12] warrants the surrender or impairment of all of the other important functions of utility rates, even the function of aiding in the control of the demand for and supply of utility services in the longer run. By and large, the major influence exercised on consumer demand for utility services by any current rates of charge for these services is an influence based on the expectation that these rates indicate, at least in a general way, the rates that will remain in effect over a considerable period of time. For it is the anticipated, fairly long-run costs of service which a potential consumer wisely takes into account when he faces a decision whether to commute from New Jersey to New York despite the daily payment of tolls on the George Washington Bridge; or whether to equip his home with an electric range or with electric space heating; or whether to locate his aluminum plant on the St. Lawrence River rather than in the state of Washington. Once having become dependent on the services required for the operation of expensive complementary equipment, the consumer's responsiveness to temporary changes in rates of charge will probably be very limited. In short, the price elasticity of demand for utility services can be expected to be much greater in the fairly long run than in any very short period of time. But if utility rates were to be made as volatile as would be required by the mandate of conformity to short-run marginal costs, they would deprive consumers of those expectations of "reasonable continuity" of rates and of rate relationships on

[12] Even this claim of superiority must be conceded only on the assumption that the better-than-nothing use of temporarily excess capacity will not materially interfere with possible emergency use. Instant readiness to serve may well be the best use of idle capacity. Professor Eli W. Clemens had this point in mind in doubting the wisdom of proposed attempts by electric utilities to encourage three-shift factory loads by the concession of very low rates for off-peak industrial service. 23 *Southern Economic Journal* 92–93 (1956). Resort to three shifts, he recalled, was one of the major ways by which the country avoided a menacing power shortage during the Second World War. "One day's loss of lives," he added, "constitutes quite a lot of marginal disutility." To the same effect, see Emery Troxel, "Reserve Plant Capacity of Public Utilities," 26 *Land Economics* 145–161 (1950), quoting at p. 150 from a significant book by the South African economist, Professor W. H. Hutt, entitled *The Theory of Idle Resources* (London, 1939).

which they must rely in order to make rational advance preparations for the use of service.

Thirdly, and closely related to the objection just mentioned, there is the probability that short-run marginal-cost rate making would deprive utility managements of an almost essential guide to intelligent decisions as to the needs for plant expansion. Under prevailing systems of rate making, managements base their estimates of future service requirements on a projection of past growth trends in the consumption of different classes of service. Their provisional assumptions that these same trends will persist in the future are premised on the expectation, not necessarily that rates will stay fixed, but at least that they will remain in fairly stable relationship to the prices of other products. But if current rates were to rise and fall with changes in current marginal costs, the resulting unpredictability of future demands might seriously handicap managements in timing their programs of construction.[18]

In his defense of short-run marginal-cost pricing, Professor Hotelling argued that the construction and expansion of public utility plants and of other public works are seriously retarded by the traditional and fallacious assumption that the economic justification of a proposed project depends on a finding that it can be made to yield revenues at least sufficient to cover its total operating and capital costs. This assumption, he declared, ignores the failure of the anticipated revenues to reflect those benefits that will accrue to consumers in the form of "consumers' surplus"—benefits measured by the excess in the highest prices that consumers would be willing to pay for different amounts of service rather than go without, over the prices that they will be required to pay under any feasible system of more or less uniform prices. While this point certainly suggests a complicating factor in a cost-benefit analysis of a large public-works project, it is not a convincing argument for the gen-

[18] But my colleague Professor Vickrey, who has kindly read this chapter in manuscript, disagrees sharply. "Far from making the timing of construction more difficult," he writes, "fluctuating rates would provide an instrument whereby demand could be adjusted to supply, so that the consequences of an unexpectedly large surge in demand, or failure of supply due, say, to drought, would be much less disastrous. Therefore a much smaller margin of excess demand to protect against these contingencies would be necessary. At the other end of the scale, excess capacity provided by overoptimistic planning would be utilized more fully so that mistakes in this direction would be less wasteful also. Compare Ontario Hydro in the depression." I have cited the Ontario Hydro experience in an earlier chapter, p. 335, *supra*.

eral abandonment of a financial self-sufficiency test of utility plant expansion. For, if consumers' surplus is to be included in a calculation of the benefits derivable from any proposed expansion, then it must also be included on the cost side of the cost-benefit balance sheet, in the estimate of the "opportunity costs" of public utility plant construction—costs which may include the withholding of economic resources from the construction of plants designed to produce other products, the sale of which would also yield a consumers' surplus.[14]

In the fourth place, we must consider the question whether the claimed advantages of short-run marginal-cost pricing as a means of improving utility rate structures, even if substantial, would still be great enough to warrant the required resort to tax-financed subsidies. The most popular objection to subsidies is based on the declared unfairness of a system of rates which requires nonconsuming taxpayers to subsidize the beneficiaries. But quite aside from any such considerations of income-distributive justice, there are serious political objections to wholesale extensions of public subsidies: namely, that legislative grants of subsidies come so largely through the efforts of pressure groups and of regional interests which are under little impulse to weigh the benefits to themselves against the costs to other people. Even so, of course, many subsidized public projects will be justified, since many of them will have great social value even though their benefits are not of the type that can be made vendible except, perhaps, to a limited degree. But a general proposal to bring the country's railroad and other public utility systems into this category for the sake of the possible benefits of short-run marginal-cost pricing seems to me too portentous politically to deserve favorable consideration.

Finally, and apart from any general objections to a scheme of rate making which taxes nonbeneficiaries for the benefit of the beneficiaries, there is another reason for a refusal to impose upon

[14] The traditional cost-benefit analyses of public-works projects by the U.S. Corps of Army Engineers and the other government agencies include no allowances for consumers' and producers' surpluses as such, although they take account of the same phenomena in their estimates of indirect benefits. A pioneering study of public-project evaluation by Otto Eckstein seems to support the conclusion that, on the whole, the traditional analyses follow techniques tending to result in overoptimistic conclusions as to the ratios of benefits to costs. *Water-Resource Development* (Cambridge, Mass., 1958). See also John V. Krutilla and Otto Eckstein, *Multiple Purpose River Development* (Baltimore, 1958).

the country's existing tax system a part of the burden now borne with only moderate efficiency by its price system. This reason is that the only forms of taxation still available to carry the extra load are of a highly imperfect nature, with the result that the additional taxes might have more serious repressive or distorting effects on the nation's economy than full-cost-recovery rate making has on the outputs and prices of utility services. In his article in 1938, Hotelling recognized this danger and proposed to meet it by the use of certain taxes arguably not subject to shifting, such as taxes on land-site values, inheritance taxes, and income taxes. But in view of the critical subsequent rise in the nation's tax burdens, any hope of finding additional sources of public subsidies in the form of yet unused taxes that are "neutral" in their effect on the allocation of the country's economic resources seems extremely remote. Certainly neither the corporate nor the individual income tax could any longer be regarded as such a source. The most promising alternative might well be the resort to a widely spread excise tax, such as a sales tax, the use of which would permit the reduction of the wide gaps between the rates and the marginal costs of utility services in exchange for a slight increase in the narrower gaps between the prices and the marginal-production costs of commodities and services throughout a much larger sector of the economy.[15]

THE ALTERNATIVE STANDARD OF LONG-RUN MARGINAL COSTS

One may summarize the foregoing criticisms of the proposal to base utility rates on short-run marginal costs by saying that, in giving sole consideration to one very limited though important function of prices, that of securing the optimum utilization of whatever plant capacity exists at a particular time, it would sur-

[15] The sales tax or some similar, widely applied, excise tax, once so largely opposed by economists in view of its regressive character, has regained status among many economists, who recognize the serious distortion effects of the current high income taxes and of other taxes of a designedly progressive nature. But in view of the nation's need to devote far more of its resources to nonvendible products of high social value, such as in public education, health service, and recreation—a need so persuasively argued by Professor John Kenneth Galbraith in his *The Affluent Society* (Boston, 1958), one may doubt the wisdom of resort to excise taxes in order to subsidize the production of vendible public utility services. Indeed, the mere cost of administering such taxes might offset whatever value they might otherwise have in securing a better allocation of resources.

render other functions of even greater importance including, particularly, that of the long-run control of the demand for and supply of utility services. Mainly for this reason, I take it, the proposal has won little support even from those economists who are most impressed with the shortcomings of full-cost standards of rate making.

But another version of the marginal-cost principle has enlisted more sympathetic interest among academic economists: the principle of rate making based on long-run marginal costs. As yet, at least, even this far milder version has made no appeal to the legislatures or rate-making practitioners of this country or of England. But on the European continent it has made some headway among the administrators of the nationalized railroads and utilities.[16] And in France the principle has already been put into partial application, particularly by the electric power system (Électricité de France).[17]

The basic distinction between short-run and long-run marginal cost has been discussed at length in Chapter XVII and need not be restated here. Advocates of either type of cost as a measure of reasonable rates accept, as the primary objective of rate-making policy, an optimum-allocation or consumer-rationing objective even if its realization calls for the surrender of the traditional principle of financial self-sufficiency. But the long-run marginalists emphasize the need for a relatively stable and continuous level or trend of rates, in the belief that the rates which have the most important effects on the demand for and provisions for utility services are rates that may be expected to persist over a considerable period of

[16] See *The Theory of Marginal Cost and Electricity Rates*, published by the Organization for European Economic Co-operation (Paris, 1958).

[17] See Thomas Marschak, "Capital Budgeting and Pricing in the French Nationalized Industries," 33 *Journal of Business of the University of Chicago* 133–156 (1960). The engineer–economists of these industries—notably Mr. Marcel Boiteux, Vice President in charge of economic studies, Électricité de France—have been in advance of any other writers in their contributions to the principles of rate making based on marginal cost. In addition to the articles cited in footnotes 10 and 21 of Chapter XVIII, footnote 25 of Chapter XIX, and footnote 1 of the present chapter, see Boiteux, "La Vente au coût marginal," *Revue française de l'énergie*, Dec., 1956; "Le Tarif vert d'Électricité de France," *ibid.*, Jan., 1957; "Les Tarifs de la concession d'alimentation générale d'E.D.F.," in a later issue of the *Revue* for 1957. See also Gabriel Dessus, "The General Principles of Rate-Making in Public Utilities," translated, *International Economic Papers*, No. 1, 5–22 (1951); Dessus and Fleurquin, "Les Tarifs de gaz et d'électricité et l'orientation du consommateur," 58 *Revue d'économie politique* 513–546 (1948).

time. Hence, the most important marginal costs for purposes of rate control are the normal or persistent marginal costs rather than the very short-lived marginal costs that may fall almost to zero in some brief period of time, only to rise to several times average total costs soon thereafter. For this purpose, however, "long-run" marginal costs must be given a flexible and frankly indefinite interpretation, since any attempt to fix rates today by reference to cost functions that may not materialize, say, for twenty-five years or more would be utterly foolish. In short, the costs that should be covered by the rates are the marginal costs that are "permanent" in the sense used by a dentist when he refers, optimistically, to a permanent rather than a temporary filling.

Although long-run marginal-cost rate making must rest its claim for acceptance primarily on its asserted superiority from the standpoint of optimum resource allocation, it has an important secondary advantage over the short-run alternative in requiring far less drastic departures from the orthodox requirement of full-cost pricing. As noted in Chapter XVII, this conclusion follows from the very definition of long-run marginal cost—a definition which treats all costs as varying with rates of output, even the so-called fixed costs. Indeed, with utility enterprises that have already attained the major economies of large scale, rates set at long-run marginal costs may be sufficient, or more than sufficient, to yield total revenue requirements. Under these conditions, the proposal to adopt marginal-cost pricing of the type now under review would not conflict with the maintenance of the traditional principle that rates in the aggregate must cover total costs. But down to the present time, both the theory and the practice of utility cost analysis are too primitive and too meager to supply an answer to the question how prevalent such conditions may be.[18]

The very reasons, however, which make the proposal to set utility rates at long-run marginal costs more "practical" than the proposal to adjust rates to short-run changes in marginal costs also constitute reasons for doubting its net advantages over full-cost

[18] In an unpublished study based on Interstate Commerce Commission data for 1936, the results of which he presented at the meeting of the Econometric Society in Montreal, Sept., 1954, Professor William S. Vickrey has estimated that, for American railroads as a whole, long-run marginal costs of freight service may well constitute from 75 to 80 per cent of average total costs. He notes that, as his study was based on cross-section comparisons at different actual densities of traffic, "it represents the extreme long-run end of the spectrum of marginal costs."

pricing as a generally applicable basis of rate control. Its best claim for serious consideration can be made in those situations in which the traditional attempt to make "rates as a whole cover costs as a whole" must be judged hopeless or intolerably wasteful in the light of experience. A verdict of this nature would almost surely apply to the city rapid-transit systems and may well apply to the entire American railroad industry, taken as a whole. The intercity railroads are vainly struggling to earn capital-attracting rates of return against the competition of heavily subsidized alternative forms of transport. Their chances of survival might be better if all of their rates, and not merely the rates for commodities or routes that face direct competition, could be brought down to levels set by standards of marginal costs over an extended period of time.

But one must not assume that marginal-cost rate making, whether short-run or long-run, offers a general solution of all of those rate-making problems raised by the inability of a utility to cover its full costs by the sale of its services. For there are many possible explanations of this inability other than the presence of a gap between marginal costs and average total costs. The local transit systems, for example, face competition from drivers of private cars, each of whom gets in the way of the others by adding to the traffic congestion. Under these conditions, subway fares should probably be set at less than marginal costs if the lower fare will serve to diminish congestion on the surface. A somewhat similar situation prevails on the railroads, which must compete with road and water users whose marginal private costs of using public highways and waterways understate the marginal social costs of maintenance and expansion. For these reasons, as well as for the reason that national-defense considerations may require the maintenance of railway facilities that are excessive for peacetime use, the marginal-cost-pricing philosophy of rate making is subject to important deviations in the direction of "social" principles of rate making.

MARGINAL-COST PRICING AND PUBLIC UTILITY TAXATION

Even though marginal-cost pricing may never win, or deserve to win, such widespread and unqualified acceptance as to supersede the traditional principle of full-cost coverage for entire rate levels, it may nevertheless have increasing influence of an indirect nature as it gains in familiarity, in theoretical development, and hopefully in the prestige of successful partial application in France and elsewhere in Europe. One possibility—remote, I fear, for the immediate future—is that it may influence utility and railroad taxation in a downward direction. This possibility calls for a comment. But the comment will be brief, since the subject of taxation is far beyond the scope of the present book.

In a study made in the early 1920s,[19] Dr. Herbert D. Simpson of Northwestern University concluded that the history of American railroad and utility taxation could be divided into four stages. The first stage, which was in the development period of the canals, turnpikes, and railroads, was one of "reverse taxation"—of outright governmental subsidy. The second stage was that of complete or partial tax exemption. In this period, which ended around the middle of the nineteenth century, proposals to tax the railroads were regarded as "radical." The third stage was that of attempted uniformity of taxation—a uniformity that proved impossible to achieve because of the difficult problems of utility property-tax assessment. The fourth and final stage was that of special types of utility taxation, many of which discriminated harshly against the utility companies as compared to other taxpayers. Thus, said Dr. Simpson, "the pendulum in this field of taxation has swung from outright subsidy and exemption at one extreme, through the uniformity period, and over to high and discriminatory taxes at the other extreme."

When Dr. Simpson wrote, he believed that the time was ripe for a return to a policy of uniformity, although this uniformity might best be approached by modification of the special utility taxes rather than by the inclusion of utility properties under general-property taxation. And his view may well have reflected the general

[19] Manuscript of lecture on "Taxation of Public Service Companies" published in Herbert B. Dorau, ed., *Materials for the Study of Public Utility Economics* (New York, 1930), pp. 471–490. Professor Harold M. Groves divides utility-tax history into three periods: those of (a) subsidy, ending at the end of the nineteenth century, (b) neutrality, ending with the depression of the 1930s, and (c) "special burdens" from 1930 to date. *Financing Government*, 5th ed. (New York, 1958), p. 336. One tax expert whose name escapes me writes that public utility companies "are taxed, for the most part, as though the regulating commissions do not exist; tax authorities and regulators scarcely seem aware of their conflicting purposes."

trend of American tax philosophy at that time. But there soon took place a tendency to revert to discriminatory taxation. This tendency may have been stimulated by a growing belief that utility rate regulation, handicapped by the Supreme Court's current rulings on a "fair-value" rate base and by the growth of uncontrollable holding companies, was proving impotent to prevent utility companies from earning excessive profits. Taxation was therefore looked upon as a device by which to offset the shortcomings of regulation.

Since the Second World War, this excuse for high utility taxation has lost most of its force. But its place has been taken by the legislative recognition of the utility companies as unusually convenient and stable sources of revenue for hard-pressed government treasuries. Hence, utility services have been subject to special taxes, including excise taxes some of which are charged directly to the consumer, the others being charged to them indirectly through their inclusion as legitimate operating deductions whenever rates are readjusted by public service commissions.

At the present time, the sum total of local, state, and Federal taxes paid by the utility companies amounts to a substantial fraction of their gross revenues. For the private electric companies in the aggregate, this fraction was reported as 21 per cent in 1957 but came to 25 per cent or more for specific companies. For telephone companies the ratio of taxes to operating revenues has been even higher, on the order of 30 per cent. Typically, somewhat more than half of a utility's tax bill is for Federal corporate income taxes. But, under a rule laid down by the Supreme Court in the 1920s,[20] allowance for these taxes is also included in a commission's calculation of a company's total revenue requirements.

Assuming, as it seems plausible to assume, that this complex of utility taxes imposes a proportionately higher tax burden on the prices of utility services than nonutility taxes impose on nonutility commodities and services, there tends to result a misallocation of economic resources—an allocation of inadequate resources to the production of electricity, railroad transportation, etc. And, to the extent to which utility services are produced under conditions of declining unit costs, this discrimination compounds the shortcomings of a rule of rate making which requires a utility company to recover its total costs by charges for its various services. Indeed,

[20] Galveston Electric Co. v. Galveston, 258 U.S. 388, 399 (1922).

the most practical way to avoid or minimize these shortcomings may well be through tax discrimination in the opposite direction.

But important as is the likelihood that, on the whole, utility taxes are higher than they should be, relative to the taxes embodied indirectly in the prices of nonutility commodities and services, even more important is the high probability that these taxes discriminate against the supply of *particular types* of utility services. The most outstanding example of this situation is that in the field of transportation, where the heavily taxed railroads must compete with other, subsidized forms of transportation. As long as this situation of subsidy exists for road, water, and air carriers, sound economic policy dictates a closer approach to symmetry by the grant of complete or partial tax exemption to the railroads.

I close this book by raising one other serious question about utility taxation in its relation to rate-making policy. It concerns the marked differences between the high taxes imposed upon the private electric utilities of this country and the relative freedom from taxes or tax equivalents enjoyed by the electric plants operating under Federal, state, or local ownership. Needless to say, opponents of public ownership bitterly resent this tax differential. But even those who, like the present writer, favor a substantial but limited number of publicly owned electric systems, such as that of the Tennessee Valley Authority, cannot fail to be concerned with a tax situation that makes so difficult any fair comparison between private-plant and public-plant performance.

By all means the most desirable way by which to secure greater comparability of performance would be substantially to reduce, if not to eliminate, private utility taxation. But if the only politically feasible way to secure equalization is to raise the tax burden imposed on the services of the public plants, there then arises a dilemma, and the choice of the less seriously penetrating horn is not easily made.

In earlier years, when the country's great Federal hydroelectric projects were in their development stages, tax exemption of these projects seems to me clearly to have been the wiser choice. Today, when these public projects are now full-fledged going concerns, the issue is far less clear. There still remains one strong argument in favor of continued tax exemption: namely, that the country as a whole may benefit from the presence of a limited number of tax-free plants which can serve as experiment stations in developing and supplying the demand for power by means of promotional rates lower than those with which a tax-burdened enterprise can afford to experiment. The Tennessee Valley Authority's experience in the widespread promotion of a house-heating load may be of this nature. But when public plants receive a high tax advantage, whether for this purpose or any other, they lose much of their value as yardstick plants against which to measure private-plant performance. Moreover, freedom of public plants from the Federal taxation imposed on private utilities gives to public ownership a factitious advantage over private ownership, viewed from the self-interested standpoint of a local community, that must interfere with an unbiased appraisal of the relative merits of the two forms of organization from the standpoint of the nation. My present, tentative, opinion is that the weight of the argument favors attempts to put publicly and privately owned electric power systems more nearly on a par, tax-wise, by a combination of a heavier public tax load and a lighter private tax load. But this is not a book on public utility taxation or on public ownership, and the questions raised here demand thoroughgoing separate studies.

PUBLICATIONS CITED IN FOOTNOTES

Adams, Walter, and Horace M. Gray. Monopoly in America. New York: Macmillan, 1955.

Adelman, M. A., "The Consistency of the Robinson-Patman Act," 6 *Stanford Law Review* 1–22 (1953).

Aitchison, Clyde B., "Fair-Return-on-Value-Theory in Rate Making Loses Force," 25 *I.C.C. Practitioners' Journal* 11–15 (Oct., 1957).

American Institute of Accountants. *Accounting Research Bulletin*, No. 22, reprinted in The Institute's *Accounting Terminology Bulletin*, No. 1. New York, 1953.

American Telephone & Telegraph Co. Profit, Performance and Progress—A Study of Regulated and Non-Regulated Industry for Bell System Use. May, 1959.

Arthur Andersen & Co. Principles Underlying Interest during Construction. Chicago, March 1, 1953.

—— Return Allowed in Public Utility Rate Cases; Supplement No. 2, 1953–1957; Supplement No. 3, 1958–1959.

Barriger, John W., Jr. Super-Railroads for a Dynamic American Economy. New York: Simmons-Boardman, 1956.

Bary, Constantine, "Coincidence-Factor Relationships of Electric-Service-Load Characteristics," in American Institute of Electrical Engineers, *Transactions*, Vol. 64, 1945.

—— "Dollars and Sense of Electric House Heating," in Association of Edison Illuminating Companies, *75th Annual Meeting*, Nov., 1959.

—— "Economic Significance of Load Characteristics as Applied to Modern Electric Service," in Association of Edison Illuminating Companies, *53d Meeting*, Jan., 1938.

—— "Load Characteristics of 'Small Light and Power' Customers—Their Economic Significance," *Electric Light and Power*, Oct. 15, 1955.

—— "Load Structure of a Modern Electric Utility System," *Electric Light and Power*, May, 1954.

Bauer, John, "Rate Base for Effective and Non-Speculative Railroad and Utility Regulation," 34 *Journal of Political Economy* 479–513 (1926).

—— Transforming Public Utility Regulation. New York: Harper, 1950.

Bauer, John, and Nathaniel Gold. Public Utility Valuation for Purposes of Rate Control. New York: Macmillan, 1934.

Baumol, William J. Welfare Economics and the Theory of the State. Cambridge: Harvard University Press, 1952.

Beatty, F. M. "Current Regulatory Status of Deferred Tax Accounting for Utilities," Address before the New York Society of Security Analysts, Oct. 16, 1957.

Becker, Arthur P., "Fixed Dividends for all Public Utility Stock," 21 *Journal of Land and Public Utility Economics* 243–249 (1945).

Beckmann, Martin, C. B. McGuire, and Christopher B. Winsten. Studies in the Economics of Transportation. New Haven: Yale University Press, for the Cowles Commission, 1956.

Beckwith, Burnham Putnam. Marginal-Cost, Price-Output Control. New York: Columbia University Press, 1955.

Bernstein, E. M. Public Utility Rate Making and the Price Level. Chapel Hill: University of North Carolina Press, 1937.

"Bills Are Lower with 'Fair Value.'" *See* Electrical World.

Bliss, John Alden, "Maybe the Utilities Have Got Something in the Way They Make Rates," *Railway Age*, Aug., 1952.

—— "A Model of Make-or-Buy Competition." Contributed paper, Third Annual Meeting of the Operations Research Society of America, June 3 and 4, 1955.

—— "The Picture Is So Different with a Two-Dimensional Price," *The Analysts Journal*, Aug., 1954, pp. 89–95.

—— "A Short Survey of Non-Linear Pricing Forms." Contributed paper, Eighth National Meeting of the Operations Research Society of America, Jan. 9 and 10, 1956.

—— "Some Comments on Tobacco—North Carolina Points to Southern Points (Rail)," *I.C.C. Practitioners' Journal*, Dec., 1951, pp. 281–301.

Boiteux, Marcel, "Sur la gestion des monopoles publics astreints à l'équilibre budgétaire," 24 *Econometrica* 22–40 (1956).

—— "La Tarification des demandes en pointe: Application de la théorie de la vente au coût marginal," 58 *Revue générale de l'électricité* 321–339 (1949).

—— "Les Tarifs de la concession d'alimentation générale d'E D F," *Revue française de l'énergie* (1957).

—— "Le Tarif vert d'Électricité de France," *Revue française de l'énergie*, Jan., 1957.

—— "La Vente au coût marginal," *Revue française de l'énergie*, Dec., 1956.

Boiteux, Marcel, and Paul Stasi, "Sur la détermination des prix de revient de développement dans un système interconnecté de production-distribution." Union Internationale des Producteurs et Distributeurs d'Énergies Électriques. Congrès de Rome, Sept., 1952.

Bolton, D. J. Electrical Engineering Economics. 1st ed. London: Chapman & Hall, 1928.

—— Electrical Engineering Economics, Vol. 2: Costs and Tariffs in Electricity Supply. 2d ed. London: Chapman & Hall, 1951.

Bonbright, James C., "Contributions of the Federal Power Commission to the Establishment of the Prudent Investment Doctrine of Rate-Making," 14 *George Washington Law Review* 136–151 (1945).

—— "Depreciation and Valuation for Rate Control," published simultaneously in 4 *Quarterly Journal of Economics* 185–211 (1927), and 27 *Columbia Law Review* 113–131 (1927).

—— "The Economic Merits of Original Cost and Reproduction Cost," 41 *Harvard Law Review* 593–622 (1928).

—— "Major Controversies as to the Criteria of Reasonable Public Utility Rates," 31 *American Economic Review, Proceedings* 379–389 (May, 1941).

—— "Original Cost as a Rate Base," 20 *The Accounting Review* 441–447 (1945).
—— "Public Utility Rate Control in a Period of Price Inflation," 27 *Land Economics* 16–23 (Feb., 1951).
—— "Railroad Valuation with Special Reference to the O'Fallon Decision," 18 *American Economic Review, Supplement* 182–205 (1928).
—— "Utility Rate Control Reconsidered in the Light of the Hope Natural Gas Case," 38 *American Economic Review, Proceedings* 465–482 (May, 1948).
—— Valuation of Property. 2 vols. New York: McGraw-Hill, 1937.
Boulding, Kenneth E. Economic Analysis. 3d ed. New York: Harper, 1955.
—— "Welfare Economics," in Bernard F. Haley, ed., A Survey of Contemporary Economics, Vol. II. Homewood, Ill.: Irwin, 1952.
The British Electrical and Allied Industries Research Association. An Improved Method for Allocating to Classes of Consumers the Demand-Related Portion of the Standing Cost of Electricity Supply. Technical Report K/T 109. London, 1945.
—— Methods of Allocating to Classes of Consumers or Load the Demand-Related Portion of the Standing Costs of Electricity Supply. Technical Report K/T 106. London, 1943.
British Electricity Authority. Third Report and Accounts, 1950–1951. London, 1951.
Brooks, F. Warren, "Needed Reform for Utility Tax Depreciation," 52 *Public Utilities Fortnightly* 407–415 (1953).
Brown, Harry Gunnison, "Economic Basis and Limits of Public Utility Regulation," 53 *American Bar Association Reports* 717–737 (1928).
—— "Railroad Valuation and Rate Regulation," 33 *Journal of Political Economy* 505–530 (1935); 34 *ibid.* 500–508 (1936).
—— Transportation Rates and Their Regulation. New York: Macmillan, 1916.
Brown, Hillyer, "Value of the Service as a Ceiling on Public Utility Rates," 33 *California Law Review* 283–297 (1945).
Brummet, R. Lee. Overhead Costing. Ann Arbor: University of Michigan, Michigan Business Series, 1957.
Bryant, J. M., and R. R. Herrmann. Elements of Utility Rate Determination. New York: McGraw-Hill, 1940.
Burns, Arthur R. Comparative Economic Organization. New York: Prentice-Hall, 1955.
Bussing, Irvin. Public Utility Regulation and the So-Called Sliding Scale. New York: Columbia University Press, 1936.
Cabot, Philip, "The Dangers of Rigid Rate Structures," 12 *Public Utilities Fortnightly* 183–191 (1933).
—— "Ethics and Politics," *Atlantic Monthly*, Nov., 1929, pp. 686–694.
—— "Four Fallacious Dogmas of Utility Regulation," 7 *Public Utilities Fortnightly* 719–729 (1931).
—— "Public Utility Rate Regulation," 7 *Harvard Business Review* 257–266, 413–422 (1929).
—— "Rate Making and Rate Regulation," *American Bar Association Journal*, Oct. 11, 1932.
Cahn, Edmond N. The Sense of Injustice; an Anthropocentric View of the Law. New York: New York University Press, 1949.

Carter, J. P., "Personal Discrimination in Transportation: A European Technique," 47 *American Economic Review* 372–381 (1957).
Caywood, Russell E. Electric Utility Rate Economics. New York: McGraw-Hill, 1956.
Chamberlin, Edward H. The Theory of Monopolistic Competition. 6th ed. Cambridge: Harvard University Press, 1950.
Clark, John Maurice, "Competition: Static Models and Dynamic Aspects," 45 *American Economic Review, Proceedings* 450–462 (May, 1955).
—— Preface to Social Economics. New York: Farrar & Rinehart, 1936.
—— "Rates for Public Utilities," 1 *American Economic Review* 473–487 (1911).
—— Studies in the Economics of Overhead Costs. Chicago: University of Chicago Press, 1923.
—— "Toward a Concept of Workable Competition," 30 *American Economic Review* 241–256 (1940).
Clemens, Eli W. Economics and Public Utilities. New York: Appleton-Century-Crofts, 1950.
—— "Price Discrimination and the Multiple-Product Firm," 19 *Review of Economic Studies* 1–11 (1950–1951).
—— Review of Ralph Kirby Davidson's Price Discrimination in Selling Gas and Electricity (Baltimore: Johns Hopkins Press, 1955), in 23 *Southern Economic Journal* 92–93 (1956).
—— "Some Aspects of the Rate-of-Return Problem," 30 *Land Economics* 32–43 (1954).
Coase, R. H., "The Economics of Uniform Pricing Systems," 15 *The Manchester School*, May, 1947, pp. 139–156.
—— "The Marginal Cost Controversy," 13 N.S. *Economica* 169–182 (1946).
Davidson, Ralph Kirby. Price Discrimination in Selling Gas and Electricity. Baltimore: Johns Hopkins Press, 1955.
Davidson, Sidney, "Accelerated Depreciation and the Allocation of Income Taxes," 33 *The Accounting Review* 173–180 (1958).
Dean, Arthur H., "Provision for Capital Exhaustion under Changing Price Levels," 65 *Harvard Law Review* 1339–1360 (1954).
Dean, Joel. Managerial Economics. New York: Prentice-Hall, 1951.
De Melverda, H. A. A. "The Illusion of Fixed Costs," in International Economic Papers, No. 2, pp. 155–177. London and New York: Macmillan, 1952.
Dessus, Gabriel. "The General Principles of Rate-Making in Public Utilities," in International Economic Papers, No. 1, pp. 5–22. London and New York: Macmillan, 1951.
Dessus, Gabriel, and —— Fleurquin, "Les Tarifs de gaz et d'électricité et l'orientation du consommateur," 58 *Revue d'économie politique* 513–546 (1948).
Devantéry, Pierre. Les Prix de l'électricité. Lausanne: Held, 1950.
Dirlam, Joel B., "Natural Gas: Cost, Conservation, and Pricing," 48 *American Economic Review, Proceedings* 491–501 (May, 1958).
Dohr, James L., "Limitations on the Usefulness of Price Level Adjustments," 30 *The Accounting Review* 198–205 (1955).
—— "The Next Step in Depreciation Accounting," 89 *Journal of Accountancy* 114–119 (1950).
—— "Power Price Fixing," 79 *Journal of Accountancy* 432–439 (1945); 80 *ibid.* 15–22, 111–117 (1945).

Dorety, Frederic G., "The Function of Reproduction Cost in Public Utility Valuation and Rate Making," 37 *Harvard Law Review* 173–200 (1923).
Dorfman, Joseph, ed. Relations of the State to Industrial Action. New York: Columbia University Press, 1954.
Dow, Alexander. Some Public Service Papers. Detroit: Detroit Edison Co. (?), 1927.
Dupuis, Jules, "De la mesure de l'utilité des travaux publics," *Annales des ponts et chaussées,* 2d Series, Vol. 8, 1884. English translation in International Economic Papers, No. 2, 83–110. London and New York: Macmillan, 1952.
Eckstein, Otto. "Natural Gas and Patterns of Regulation," 36 *Harvard Business Review* 126–136 (1958).
—— Water Resource Development. Cambridge: Harvard University Press, 1958.
Edgerton, Henry H., "Value of the Service as a Factor in Rate Making," 32 *Harvard Law Review* 516–556 (1919).
Edison Illuminating Company of Detroit. The Development of Scientific Rates for Electricity Supply. Detroit, 1915.
Edwards, Corwin D. The Price Discrimination Law: A Review of Experience. Washington, D.C.: Brookings Institution, 1959.
Edwards, Ford K. "Cost Analysis in Transportation," 37 *American Economic Review, Proceedings* 441–461 (May, 1947).
—— "Transportation Costs, Value-of-Service and Freight Rates," 21 *I.C.C. Practitioners' Journal* 494–510 (1954).
Eisner, Robert, "Accelerated Amortization, Growth, and Net Profits," 60 *Quarterly Journal of Economics* 533–544 (1952).
Electrical World, "Bills Are Lower with 'Fair Value,'" Vol. 149, June 16, 1958, pp. 71–74.
—— "'Fair Value' Yields Top Returns," Vol. 149, May 5, 1958, pp. 97–100.
Ely, Owen, "Deferred Taxes—How the Issue Is Working Out in Rate Cases," 62 *Public Utilities Fortnightly* 104–109 (1958).
Epstein, Ralph C., and John D. Sumner, "Effect of the Depression upon Earnings and Prices of Regulated and Nonregulated Industries," 26 *American Economic Review, Supplement* 36–45 (March, 1936).
Fair, M. L., and Ernest W. Williams. Economics of Transportation. Revised ed. New York: Harper, 1959.
"'Fair Value' Yields Top Returns." *See Electrical World.*
Federal Communications Commission. Final Report of the Telephone Rate and Research Department. Washington, D.C., June 15, 1938.
Federal Coordinator of Transportation. Report on Cost Finding in Railway Freight Service for Regulatory Purposes. Washington, D.C., June, 1936.
Federal Power Commission. Report on the Reclassification and Original Cost of Electric Plant of Public Utilities and Licensees. Washington, D.C., 1950.
—— State Commission Jurisdiction and Regulation of Electric and Gas Companies. Washington, D.C., 1954.
Ferguson, Samuel, "Further Comments on the NARUC Report," 21 *Journal of Land and Public Utility Economics* 181–184 (May, 1945).
—— Public Utility Papers. 3 vols. Hartford, Conn.: Hartford Electric Light Co., 1947.
Fisher, Clyde O., "Value of the Service and Public Utility Rates," 12 *Journal of Land and Public Utility Economics* 76–83 (1935).

Fisher, Joseph I., "Natural Resources and Technological Change," 29 *Land Economics* 57–71 (1953).
Flagg, H. J., "What of the Industrial Load?" 130 *Electrical World*, Aug. 28, 1948, p. 82.
Fleming, Marcus, "Optimal Production with Fixed Profits," 20 N.S. *Economica* 215–236 (Aug., 1953).
Frankfurter, Felix, and Henry M. Hart, Jr., "Rate Regulation," in Encyclopaedia of the Social Sciences, Vol. XIII. New York, 1934.
Friedman, Milton. Essays in Positive Economics. Chicago: University of Chicago Press, 1953.
Galbraith, John Kenneth. The Affluent Society. Boston: Houghton Mifflin, 1958.
—— "Economic Preconceptions and the Farm Policy," 44 *American Economic Review* 40–52 (1954).
—— "Farm Policy: The Current Position," 37 *Journal of Farm Economics* 292–304 (1955).
—— "Monopoly Power and Price Rigidities," 50 *Quarterly Journal of Economics* 456–475 (1936).
Glaeser, Martin G. Public Utilities in American Capitalism. New York: Macmillan, 1957.
Goldberg, Louis, "Concepts of Depreciation," 30 *The Accounting Review* 468–484 (1955).
Gort, Michael, "The Planning of Investment: A Study of Capital Budgeting in the Electric Power Industry," 24 *Journal of Business of the University of Chicago* 79–95, 181–202 (1951).
Graaff, J. de V. Theoretical Welfare Economics. Cambridge: Cambridge University Press, 1957.
Graham, Benjamin, and David L. Dodd. Security Analysis. 3d ed. New York: McGraw-Hill, 1951.
Grant, E. L., and Paul T. Norton, Jr. Depreciation. New York: Ronald Press, 1949.
Gray, Horace M., "The Passing of the Public Utility Concept," 16 *Journal of Land and Public Utility Economics* 8–20 (1940).
Great Britain. Committee of Inquiry into the Electricity Supply Industry (The "Herbert Committee"). Report, Jan., 1956. Cmd. 9672.
—— Committee on National Policy for the Use of Fuel and Power Resources (The "Ridley Committee"). Report, Sept., 1952. Cmd. 8647.
Groves, Harold M. Financing Government. 5th ed. New York: Holt, 1958.
Guercken, C. P., "An Economic Appraisal of the Private Electric Utility Industry," 55 *Public Utilities Fortnightly* 739–754 (1955).
Hale, Robert L., "Commissions, Rates, and Policies," 53 *Harvard Law Review* 1103–1144 (1940).
—— Freedom through Law. New York: Columbia University Press, 1952.
—— "Non-Cost Standards in Rate Making," 36 *Yale Law Journal* 56–67 (1926).
—— "Utility Regulation in the Light of the Hope Natural Gas Case," 44 *Columbia Law Review* 488–530 (1944).
Hamilton, Walton H., "Affectation with Public Interest," 39 *Yale Law Journal* 1089–1112 (1929–1930).
—— Price and Price Policies. New York: McGraw-Hill, 1938.

Harbeson, Robert W., "A Critique of Marginal Cost Pricing," 31 *Land Economics* 54–74 (1955).

Harriss, C. Lowell. The American Economy. Homewood, Ill.: Irwin, 1953.

—— "Sources of Injustice in Death Taxation," 7 *National Tax Journal* 289–308 (1954).

Havlik, Hubert. Service Charges in Gas and Electric Rates. New York: Columbia University Press, 1938.

Heflebower, Richard B., and George Stocking, eds. Readings in Industrial Organization and Public Policy. Homewood, Ill.: Irwin, 1958.

Henderson, A. M., "Prices and Profits in State Enterprise," 16 *Review of Economic Studies* 13–24 (1948–1949).

Henry, Donald M., "What Is the Cost of Service: What Is Equitable Allocation?" *Gas Age*, May 4, 1944, pp. 58–60, 160.

Heyman, Eleanor, "Value of the Service; Its Various Meanings and Uses," 9 *Journal of Land and Public Utility Economics* 252–265 (1933).

Hirshleifer, Jack, H. S. Houthakker, and Peter Steiner, "Peak Loads and Efficient Pricing," 71 *Quarterly Journal of Economics* 585–610 (1957); 72 *ibid.* 451–468 (1958).

Hotelling, Harold, "The General Welfare in Relation to Problems of Taxation and of Railway and Utility Rates," 6 *Econometrica* 242–269 (1938).

—— "The Relation of Prices to Marginal Costs in an Optimum System," 7 *Econometrica* 151–160 (1939).

Houthakker, H. S., "Electricity Tariffs in Theory and Practice," 61 *Economic Journal* 1–25 (1951).

—— Review of Ralph Kirby Davidson's Price Discrimination in Selling Gas and Electricity (Baltimore: Johns Hopkins Press, 1955), in 46 *American Economic Review* 734–735 (1956).

Howell, Paul L., "The Rate of Return in Air Transport," 24 *Law and Contemporary Problems* 677–701 (Autumn, 1959).

Hutt, W. H. The Theory of Idle Resources. London: Jonathan Cape, 1939.

Interstate Commerce Commission. "Depreciation Charges of Telephone Companies" (No. 14700), and "Depreciation Charges of Steam Railroad Companies" (No. 15100), 118 I.C.C. 295–411 (1926); 177 *ibid.* 351–499 (1931).

—— Bureau of Accounts, Cost Finding and Valuation. Explanation of Rail Cost Finding Procedures and Principles Relating to the Use of Costs. Washington, D.C., Nov., 1954.

—— Bureau of Transport Economics and Statistics. Value of Service in Rate-Making. Washington, D.C., Nov., 1959.

Jeming, Joseph, "An Actual Application of the Reproduction-Cost-of-Service Principle in Rate Making," 18 *Journal of Land and Public Utility Economics* 188–203 (1942).

—— "Public Utility Rates on a Reproduction-Cost-of-Service Principle," 17 *Journal of Land and Public Utility Economics* 138–150 (1941).

Jevons, W. Stanley. The Theory of Political Economy. 4th ed. London: Macmillan, 1911.

Johnston, J. Statistical Cost Analysis. New York: McGraw-Hill, 1960.

Kahn-Freund, Otto. The Law of Carriage by Inland Transport. 3d ed. London: Stevens, 1956.

Kapp, K. William. The Social Costs of Private Enterprise. Cambridge: Harvard University Press, 1950.

Kennedy, William F. The Objective Rate Plan for Reducing the Price of Residential Electricity. New York: Columbia University Press, 1937.

Keynes, John Maynard. The General Theory of Employment, Interest, and Money. New York: Harcourt, Brace, 1936.

Knapp, Charles W., "Economics of Transit Operating Ratio," 56 *Public Utilities Fortnightly* 467–479 (1955).

Knappen, Laurence S., "Transit Operating Ratio—Another View," 51 *Public Utilities Fortnightly* 485–497 (1953).

Knight, Bruce W., "Why Not Regulate Investment Instead of Return?" 6 *Public Utilities Fortnightly* 406–419 (1930).

Knight, Frank W., "Some Fallacies in the Interpretation of Social Cost," 38 *Quarterly Journal of Economics* 582–606 (1924).

Kripke, Homer, "A Case Study in the Relationship of Law and Accounting; Uniform Accounts 100.5 and 107," 57 *Harvard Law Review* 433–478, 693–727 (1944).

Krutilla, John V., and Otto Eckstein. Multiple Purpose River Development. Baltimore: Johns Hopkins Press, 1958.

Lake, Isaac B. Discrimination by Railroads and Other Public Utilities. Raleigh, N.C.: Edwards & Broughton, 1947.

Langdon, Jervis, Jr., "Regulation of Competitive Business Forces: The Obstacle in Transportation," 41 *Cornell Law Quarterly* 57–92 (1955).

Leake, P. D. Depreciation and Wasting Assets. 3d ed. London: Pitman, 1920.

Lerner, Abba P. The Economics of Control. New York: Macmillan, 1947.

Lerner, Joseph, "Capital Costs of Private vs. Public Power for AEC—A Review," 33 *Land Economics* 113–126 (1957).

Lewis, Ben W., "The Duty of a Public Utility to Reduce Its Income Tax Liability by Using Accelerated Depreciation," 35 *Land Economics* 104–114 (1959).

—— "State Regulation in Depression and War," 36 *American Economic Review, Proceedings* 384–404 (May, 1946).

Lewis, W. Arthur. Overhead Costs—Some Essays in Economic Analysis. London: Allen & Unwin, 1949.

Lilienthal, David E., "Regulation of Public Utilities during the Depression," 66 *Harvard Law Review* 745–775 (1933).

Lippit, Henry F. II., "Net Investment Rate Making—The Deduction for Depreciation," 62 *Harvard Law Review* 1155–1179 (1949).

Little, I. M. D. A Critique of Welfare Economics. Oxford: Clarendon Press, 1950.

—— The Price of Fuel. Oxford: Clarendon Press, 1953.

Locklin, D. Philip. The Economics of Transportation. 5th ed. Homewood, Ill.: Irwin, 1960.

Loeb, Benjamin S. Electric Power Supply and National Security: a Study in Public Policy. Ph.D. dissertation, Columbia University, 1959. Reproduced. Office of Technical Services, Dept. of Commerce, Washington, D.C.

Machlup, Fritz. Economics of Sellers' Competition. Baltimore: Johns Hopkins Press, 1952.

Marschak, Thomas, "Capital Budgeting and Pricing in the French Nationalized Industries," 33 *Journal of Business of the University of Chicago* 133–156 (1960).

Marshall, A. C., and H. A. Snow, "Distribution Costs—Residential Service," in Samuel Ferguson, Public Utility Papers, Vol. I, 354–381. Hartford, Conn.: Hartford Electric Light Co., 1947.

Marshall, Alfred. Principles of Economics. 8th ed. London: Macmillan, 1930.
Mason, Perry. Principles of Public-Utility Depreciation. Chicago: American Accounting Association, 1937.
May, George O. Financial Accounting. New York: Macmillan, 1943.
Milne, Alastair M. The Economics of Inland Transport. London: Pitman, 1955.
Modigliani, Franco, and M. H. Miller, "The Cost of Capital, Corporation Finance and the Theory of Investment," 48 *American Economic Review* 261–297 (1958). Critical comments and reply, 49 *ibid.* 638–669 (1959).
Morgan, Charles S. Regulation and the Management of Public Utilities. Boston: Houghton Mifflin, 1923.
Morrissey, Fred P., "Current Aspects of the Cost of Capital to Utilities," 62 *Public Utilities Fortnightly* 217–227 (1958).
—— "Dividend Policy and Reduction of Tax Liability," 59 *Public Utilities Fortnightly* 15–22 (1958).
—— "Inflation and Public Utility Regulation," *California Management Review*, Vol. I, No. 4, 1959, pp. 74–88. Summarized in 65 *Public Utilities Fortnightly* 625–627 (1960).
—— "A Reconsideration of Cost of Capital and a Reasonable Rate of Return," 31 *Land Economics* 229–244 (1955).
—— "Relation of Growth and Rate of Return for Utilities," 60 *Public Utilities Fortnightly* 361–373 (1957).
Morton, Walter A., "Rate of Return and the Value of Money in Public Utilities," 28 *Land Economics* 91–131 (1952).
Munby, D. L., "The Price of Fuel," 6 N.S. *Oxford Economic Papers* 226–242 (1954).
Nash, L. R. Economics of Public Utilities. 2d ed. New York: McGraw-Hill, 1931.
—— Public Utility Rate Structures. New York: McGraw-Hill, 1933.
National Association of Railroad and Utilities Commissioners. Report of Committee on Depreciation, 1943. Washington, D.C., 1943.
—— Report of the Special Committee to Study Principles of Rate Regulation in the Motor Bus Industry. Proceedings, 64th Annual Convention, 461–470. Washington, D.C., 1952.
National Bureau of Economic Research. Policies to Combat Depression. Princeton: Princeton University Press, 1956.
—— Regularization of Business Investment. Princeton: Princeton University Press, 1954.
—— Committee on Price Determination. Cost Behavior and Price Policy. New York: National Bureau of Economic Research, 1943.
New York State, Commission on Revision of the Public Service Commissions Law. Reports and Hearings. 3 vols. Albany, N.Y., 1930.
—— Power Authority of. Fourth Annual Report (1934), Appendix 8: Report on Cost of Distribution of Electricity to the President of the United States, the Governor of New York, and the Chairman of the Federal Power Commission. Albany, N.Y., 1934.
Nichols, Ellsworth. Ruling Principles of Utility Regulation: Rate of Return. Washington, D.C.: Public Utility Reports, Inc., 1955.
Nissel, Hans, "Allocation of Costs in Gas Systems," *American Gas Journal*, Oct. 15, 1953.
—— "The Impact of Cost Allocations upon Future of the Natural Gas Industry," 66 *Public Utilities Fortnightly* 512–524 (1960).

O'Connor, Leonard A., "Some Critical Thoughts on Cost of Capital," 62 *Public Utilities Fortnightly* 93–97 (1958).
Olds, Leland, "The Economic Planning Function under Public Regulation," 48 *American Economic Review, Proceedings* 553–561 (May, 1958).
Organization for European Economic Cooperation. The Theory of Marginal Cost and Electricity Rates. Paris, 1958.
Orgel, Lewis. Valuation under the Law of Eminent Domain. 2d ed. 2 vols. Charlottesville, Va.: The Michie Co., 1953.
Owen, Wilfred. The Metropolitan Transportation System. Washington, D.C.: Brookings Institution, 1956.
Paton, William A., "Accounting Policies of the Federal Power Commission—A Critique," 77 *Journal of Accountancy* 432–460 (1944).
Paton, William A., and Howard C. Greer, "Utility Rates Must Recognize Dollar Depreciation," 51 *Public Utilities Fortnightly* 333–356 (1953).
Paton, William A., and William A. Paton, Jr. Asset Accounting. New York: Macmillan, 1952.
Pegrum, Dudley F., "The Economic Basis of Public Policy for Motor Transport," 28 *Land Economics* 244–263 (1952).
Pigou, A. C. The Economics of Welfare. 4th ed. London: Macmillan, 1932.
Power Authority of the State of New York. *See* New York State, Power Authority of.
Presidential Advisory Committee on Transport Policy and Organization. Revision of Transportation Policy: A Report to the President (The "Weeks Report"). Washington, D.C., 1955.
Ramsey, F. P., "A Contribution to the Theory of Taxation," 37 *Economic Journal* 47–61 (1927).
Raymond, William G. What Is Fair?—A Study of Some Problems of Public Utility Regulation. New York: Wiley, 1918.
Reder, Melvin W. Studies in the Theory of Welfare Economics. New York: Columbia University Press, 1947.
Roberts, Merrill J., "Maximum Freight Rate Regulation and Railroad Earnings Control," 35 *Land Economics* 125–138 (1959).
Robinson, Joan. The Economics of Imperfect Competition. London: Macmillan, 1948.
Robson, William A., ed. Problems of Nationalized Industry. New York: Oxford University Press, 1952.
Ross, J. R., and David Durand, "The Cost of Capital and the Theory of Investment," 49 *American Economic Review* 638–669 (1959).
Rowe, Frederick M., "Cost Justification of Price Differentials under the Robinson-Patman Act," 59 *Columbia Law Review* 584–617 (1959).
Ruggles, Nancy, "The Welfare Basis of the Marginal Cost Pricing Principle," 17 *Review of Economic Studies* 29–46 (1949–1950); "Recent Developments in the Theory of Marginal Cost Pricing," *ibid.* 107–126.
Salin, Edgar, "Just Price," in Encyclopaedia of the Social Sciences, Vol. VIII. New York, 1932.
Samuelson, Paul A. Economics, an Introductory Analysis. 3d ed. New York: McGraw-Hill, 1955.
Scharff, Maurice R., "Valuation and Depreciation of Public Utility Property," in American Society of Civil Engineers, 114 *Transactions* 907–925 (1949).
Scitovsky, Tibor. Welfare and Competition; The Economics of a Fully Employed Economy. Chicago: Irwin, 1951.

Sharfman, I. Leo. The Interstate Commerce Commission. 5 vols. New York: The Commonwealth Fund, 1931–1937.
Shomacher, Larry, "Is FPC Gas Cost Allocation Equitable?" 50 *Public Utilities Fortnightly* 680–688 (1952).
Simons, Henry C. Economic Policy for a Free Society. Chicago: University of Chicago Press, 1948.
Simpson, Herbert D., "Taxation of Public Service Companies," in Herbert B. Dorau, ed., Materials for the Study of Public Utility Economics 471–490. New York, 1930.
Sleeman, J. F. British Public Utilities. London: Pitman, 1953.
Smith, Nelson Lee. The Fair Rate of Return in Public Utility Regulation. Boston: Houghton Mifflin, 1932.
Sporn, Philip, "Observations on Private Versus Public Power," 53 *Public Utilities Fortnightly* 717–733 (1954).
Staudinger, Hans, "Planning in Electricity," *Social Research*, Nov., 1937, pp. 417–439.
—— "Social Rates in Electricity," *Social Research*, Aug., 1936, pp. 259–281.
Stigler, George J. The Theory of Price. 1st ed., New York: Macmillan, 1946; revised ed., New York: Macmillan, 1952.
Study Group on Business Income. Changing Concepts of Business Income, Report. New York: Macmillan, 1952.
Sumner, J. D., "Public Utility Prices and the Business Cycle: A Study in the Theory of Price Rigidity," 21 *Review of Economic Statistics* 97–109 (1939).
—— "Public Utility Rate Making in Depression: A Comment," 52 *Quarterly Journal of Economics* 713–717 (1937).
Sutton, Francis X., et al. The American Business Creed. Cambridge: Harvard University Press, 1956.
Taggart, Herbert F. Cost Justification. Michigan Business Studies, Vol. 14, No. 3. Ann Arbor: University of Michigan, Bureau of Business Research, 1959.
Tatham, Charles, "The Growth Factor in Electric Utility Earnings," *The Analysts Journal*, March, 1952.
Taussig, Frank W., and A. C. Pigou, "Railway Rates and Joint Costs," 27 *Quarterly Journal of Economics* 378–384; *ibid.* 535–538; *ibid.* 687–694 (1912–1913).
Taylor, F. M., "The Guidance of Production in a Socialist State," 19 *American Economic Review* 1–8 (1929).
Temporary National Economic Committee (TNEC). Investigation of Concentration of Economic Power, 76th Congress, 3d Session. Monograph No. 32, Economic Standards of Government Price Control. Washington, D.C., 1941.
Terborgh, George. The Bogey of Economic Maturity. Chicago: Machinery and Allied Products Institute, 1945.
—— Realistic Depreciation Policy. Chicago: Machinery and Allied Products Institute, 1954.
Thatcher, Lionel W., "Cost-of-Capital Techniques Employed in Determining the Rate of Return for Public Utilities," 30 *Land Economics* 85–111 (1954).
Towne, Samuel A., "Cost Level Guides to Rate-Making," 21 *I.C.C. Practitioners' Journal* 697–707 (1954).
Troxel, Emery. Economics of Public Utilities. New York: Rinehart, 1947.

Troxel, Emery (*cont.*) "Reserve Plant Capacity of Public Utilities," 26 *Land Economics* 145–161 (1950).

—— "Valuation of Public Utility Property—A Problem in Efficient Resource Use and Efficient Regulation," 23 *Journal of Business of the University of Chicago* 1–21 (1950).

Twentieth Century Fund. Electric Power and Government Policy. New York, 1948.

Tyndall, David G., "The Relative Merits of Average Cost Pricing, Marginal Cost Pricing, and Price Discrimination," 65 *Quarterly Journal of Economics* 342–372 (1951).

United States. Commission on Organization of the Executive Branch of the Government (The "Second Hoover Commission"). Task Force Report on Water Resources and Power. Washington, D.C., 1955.

—— Congress. House of Representatives. Committee on Interstate and Foreign Commerce. Report of the Subcommittee on Surface Transportation, April 30, 1958. Washington, D.C., 1958.

—— *See also* entries under Federal; Interstate Commerce Commission; Presidential Advisory Committee; Temporary National Economic Committee.

Vickrey, William S. The Revision of the Rapid Transit Fare Structure of the City of New York. Technical Monograph No. Three, Finance Project, Mayor's Committee on Management Survey of the City of New York. New York, Feb., 1952.

—— "Some Implications of Marginal Cost Pricing for Public Utilities," 45 *American Economic Review, Proceedings* 605–620 (May, 1955).

—— "Some Objections to Marginal-Cost Pricing," 56 Journal of Political Economy 218–238 (1948).

Viner, Jacob, "Cost," in Encyclopaedia of the Social Sciences, Vol. IV. New York, 1931.

—— "Cost Curves and Supply Curves," 3 *Zeitschrift für National-Ökonomie* 23–46 (1931). Reprinted in George J. Stigler and Kenneth E. Boulding, eds., Readings in Price Theory. Chicago: Irwin, 1952.

Von Neumann, John, and Oskar Morgenstern. The Theory of Games and Economic Behavior. Princeton: Princeton University Press, 1944.

Wallace, Donald H., "Joint and Overhead Cost and Railway Rate Policy," 48 *Quarterly Journal of Economics* 583–619 (1934).

Watkins, G. P. Electrical Rates. New York: Van Nostrand, 1921.

Waugh, Frederick V., "Does the Consumer Benefit from Price Stability?" 58 *Quarterly Journal of Economics* 602–614 (1944).

Weintraub, Sidney, "Revised Doctrines of Competition," 45 *American Economic Review, Proceedings* 463–479 (May, 1955).

Whitten, Robert H., "Regulation of Public Service Companies in Great Britain," Appendix F, Vol. 1, Annual Report of Public Service Commission of New York, First District, 1913.

—— Valuation of Public Service Corporations. Revised ed. by Delos F. Wilcox. 2 vols. New York: Banks Law Publishing Co., 1928.

Wilcox, Clair. Public Policies toward Business. Revised ed. Homewood, Ill.: Irwin, 1960.

Williams, Ernest W. "Railroad Rate Levels and Earning Power in an Era of Competitive Transport," 25 *Land Economics* 405–413 (1949).

Williams, John Davis. The Compleat Strategyst; Being a Primer on the Theory of Games of Strategy. New York: McGraw-Hill, 1954.

Wilson, G. Lloyd, and J. R. Rose, "'Out-of-Pocket' Cost in Railroad Freight Rates," 60 *Quarterly Journal of Economics* 546–560 (1946).

Wisconsin Public Service Commission. Depreciation—A Review of Legal and Accounting Problems. Madison, 1933.

Wiseman, J., "The Theory of Public Utility Rates—An Empty Box," 9 *N.S. Oxford Economic Papers* 56–74 (1957).

Wright, Alan, "Operating Ratio—A Regulatory Tool," 51 *Public Utilities Fortnightly* 24–29 (1953).

TABLE OF CITED CASES

The references to pages and footnotes in this book are at the end of each case listed, *in italics.*

INDEX

Bei Fragen zur Produktsicherheit wenden Sie sich bitte an:
If you have any questions regarding product safety,
please contact:

Walter de Gruyter GmbH
Genthiner Straße 13
10785 Berlin
productsafety@degruyterbrill.com